T0205853

The Concise Industrial Flow Measurement Handbook

The Concise Industrial Flow Measurement Handbook

A Definitive Practical Guide

Michael A. Crabtree

CRC Press is an imprint of the
Taylor & Francis Group, an **informa** business

CRC Press
Taylor & Francis Group
6000 Broken Sound Parkway NW, Suite 300
Boca Raton, FL 33487-2742

First issued in paperback 2021

Printed on acid-free paper

ISBN-13: 978-0-367-36655-1 (hbk)

ISBN-13: 978-1-03-208498-5 (pbk)

Visit the Taylor & Francis Web site at
http://www.taylorandfrancis.com

and the CRC Press Web site at
http://www.crcpress.com

Contents

Preface

First and foremost, this book, *The Concise Industrial Flow Measurement Handbook: A Definitive Practical Guide* is a 'Primer' – a book providing a basic introduction to the subject of flow measurement. Second, it's complete in that it looks at all the current technologies of flow measurement. Finally, it is a 'definitive guide' because it represents a wealth of experiential knowledge gleaned from the author's experience working within a systems integration company and also feedback from more than 4,000 technicians and engineers who have attended the author's workshops.

No single attribute of the book is unique. However, because it incorporates several distinctive features, at a number of different levels, these agents combine to make it one-of-a-kind.

In this book, I use a building block approach in such a manner that it is presented in a form suitable for two distinct classes of readers: the beginner, with no prior knowledge of the subject; and the more advanced technician.

The complete text is suitable for the advanced reader. However, those parts of the text, which involve a mathematical treatment that is not required by the beginner, are indicated by a mark (►) at the beginning and (◄) at the end. Consequently, for the beginner the text may be read, with full understanding, by ignoring the marked sections.

I offer no apologies for my preference for metric-based measurement – the SI system. Apart from the United States, only two other countries in the world still adhere to the fps (foot–pound–second) system – the so-called Imperial system first defined in the British Weights and Measures Act of 1824 – Myanmar and Liberia.

I've tried to mix it up as far as possible and I've included a unit conversion table in the front of the book. But for the moment just try for the following:

1 bar = 100 kPa ≈ 1 atmosphere ≈ 14.7 psi
1 inch = 25.4 mm
20°C = 68°F
100°C = 212°F

Finally, I reserve my right to spell according to the British system:

English	USA
Metre	Meter
Litre	Liter
Fibre	Fiber
Colour	Color

DISCLAIMER

Whilst all reasonable care has been taken to ensure that the description, opinions, programs, listings, software, and diagrams are accurate and workable, neither the author nor the publisher accept any legal responsibility or liability to any person, organisation or other entity for any direct loss, consequential loss or damage, however caused, that may be suffered as a result of the use of this publication or any associated workshop or software.

Neither the author nor the publisher endorse any referenced commercial products nor make any representation regarding the availability of any referenced commercial product at any time. The manufacturer's instructions on the use of any commercial product must be followed at all times, even if in conflict with the information in this publication.

Author

Michael (Mick) A. Crabtree, joining the Royal Air Force as an apprentice, trained in aircraft instrumentation and guided missiles. Completing his service career seconded to the Ministry of Defence as a technical writer, he emigrated to South Africa in 1966 where he worked, for many years, for a local manufacturing and systems integration company involved in industrial process control, SCADA (Supervisory Control and Data Acquisition) and PLC (Programmable Logic Controller)-based systems.

Later, as editor and managing editor of a leading monthly engineering journal, Mick wrote and published hundreds of articles as well as eight technical resource handbooks on industrial process control: *Flow Measurement, Temperature Measurement, Analytical On-line Measurement, Pressure and Level Measurement, Valves, Industrial Communications*, and *The Complete Profibus Handbook*.

He subsequently founded his own PR and advertising company and was retained by a number of leading companies involved in the process control industry including: Honeywell, Fisher-Rosemount, Krohne, Milltronics, and AEG. Apart from producing all their press releases and articles, he also undertook the conceptualisation and production of a wide range of advertisements and data sheets, as well as newsletters.

For the past 12 years Mick has been involved in technical training and consultancy and has run workshops on industrial instrumentation and networking throughout the world (USA, Canada, UK, France, Africa, Trinidad, Middle East, Far East, Australia, and New Zealand).

During this period, he has led more than 5,000 engineers, technicians and scientists on a variety of practical training workshops covering the fields of Process Control (loop tuning), Process Instrumentation, Data Communications, Fieldbus, Safety Instrumentation Systems, Project Management, On-Line Liquid Analysis, and Technical Writing and Communications.

Mick's skill sets include:

- Technical and non-technical authoring
- Course development
- Face-to-face training facilitation
- One-on-one mentoring
- Development of mentoring programmes
- Distance and e-learning

Having relocated to Britain several years ago, he now lives in Wales, just outside Cardiff.

Mick earned a Master's degree from the University of Huddersfield. His hobbies and pastimes include: cycling, rambling, history, and reading.

Units Conversion

Quantity	SI	US Customary
Distance	25.5 mm	1 in
	1 millimetre (mm)	0.03937 in
	1 m	39.37 in
	1 m	3.281 ft
	0.9144 m	1 yd
Area	1 square metre (m^2)	1,550 in^2
	1 square metre (m^2)	10.76 ft^2
	1 square millimetre (mm^2)	0.00155 in^2
Volume	1 cubic metre (m^3)	61.02 in^3
	1 cubic metre (m^3)	35.31 ft^3
	0.02832 m^3	1 ft^3
	1 litre (L)	61.02 in^3
	1 litre (L)	0.03531 ft^3
	1 litre (L)	0.2642 gal
	3.785 litres	1 gal
Mass	1 kg	2.205 lb
	454 g	1 lb
Force	1 N	0.2248 lbf
	4.448 N	1 lbf
Pressure	1 bar	14.504 lbf/in^2 (psi)
	1 kPa (kN/m^2)	0.145 lbf/m^2 (psi)
	6.895 kPa	1 psi
Temperature	K	1.800°R
	°C	1.8°C + 32 = °F
Flow rate	1 m^3/h	4.403 gal/min (gpm)
	1 kg/h	2.205 lb/h

CRUDE OIL MEASUREMENT

Oil barrel (bbl): 42 US gallons = 158.9873 litres

1 Background and History

1.1 INTRODUCTION

Over the past 60 years, the importance of flow measurement has grown, not only because of its widespread use for accounting purposes, such as the custody transfer of fluid from the supplier to consumer, but also because of its application in manufacturing processes. Throughout this period, performance requirements have become more stringent – with unrelenting pressure for improved reliability, accuracy,* linearity, repeatability, and rangeability.

These pressures have been caused by major changes in manufacturing processes and because of several dramatic circumstantial changes such as the increase in the cost of fuel and raw materials, the need to minimise pollution, and the increasing pressures being brought to bear in order to adhere to the requirements for health and safety.

Industries involved in flow measurement and control include:

- Food and beverage
- Medical
- Mining and metallurgical
- Oil and gas transport
- Petrochemical
- Pneumatic and hydraulic transport of solids
- Power generation
- Pulp and paper
- Distribution

Fluid properties can vary enormously from industry to industry. The fluid may be toxic, flammable, abrasive, radio-active, explosive, or corrosive; it may be single-phase (clean gas, water, or oil) or multi-phase (e.g. slurries, wet steam, well-head petroleum, or dust-laden gases).

The pipe carrying the fluid may vary from less than 1 mm to many metres in diameter. Fluid temperatures may vary from close to absolute zero to several hundred degrees Celsius and the pressure may vary from high vacuum to many thousands of bar.

Because of these variations in fluid properties and flow applications, a wide range of flow metering techniques have been developed with each suited to a particular area. However, of the numerous flow metering techniques that have been proposed in the past, only a few have found widespread application and no one single flowmeter can be used for all applications.

---

* In the field of process instrumentation the term 'accuracy' is generally defined as the ratio of the error to the full-scale or actual output, expressed as a percentage. Strictly speaking, the term should be confined to generalised descriptions and not to specifications – where the term 'error' is preferred. However, the vast majority of instrumentation manufacturers continue to use the term 'accuracy'.

1.2 WHY MEASURE FLOW?

There is of course no single answer. Flow measurement is normally concerned with the question of 'how much' – how much is produced or how much is used. For small quantities this can normally be achieved by volumetric measurement (e.g. pulling a pint of beer). But as the amount grows larger this becomes impractical and, for example, it becomes necessary to measure the volumetric flow (e.g. dispensing fuel from a garage petrol pump).

However, most petrol pump calibration is carried out using a test measuring can, which is a purely volumetric measurement. On this basis, during hot weather, it would be prudent to purchase petrol early in the morning when the temperature is low so that you would end up with more mass per litre. Alternatively, a more practical and consistent approach would be to make use of a mass flow metering system.

A further use of a flow measuring system is to control a process. In closed-loop regulatory control, there are several instances where the prime consideration is repeatability rather than accuracy. This is particularly true in a cascaded system where the prime objective is not to control the inner loop to an absolute value but rather to increment up or down according to the demands of the master. Another instance of where absolute accuracy is relatively unimportant is in controlling the level of a surge tank. Here, the requirement is to allow the level to vary between an upper and lower value and absorb the upstream surges – thus preventing them from being passed downstream.

Accuracy is of prime importance in automatic blending control, batching and, of course, custody transfer and fiscal metering. In fluid measurement, custody transfer metering involves the sale, or change of ownership, of a liquid or gas from one party to another. On the other hand, fiscal measurement involves the levering of taxes – again relating to the production or sale of a liquid or gas.

1.3 BACKGROUND

The book, *Principles and Practice of Flow Meter Engineering* by L.K. Spink, first published in 1930, is generally recognised as the first, and for many years the only, definitive collected 'body of knowledge' appertaining to industrial flow measurement. Undergoing nine revisions, the last addition was printed in 1978 – 21 years after Spink's death.

In the flyleaf of this last addition the publishers lay claim to the book covering '... the latest developments in flow measurement'. A weighty tome, by anyone's standards, the book comes in at 575 pages. However, in essence *Principles and Practice of Flow Meter Engineering* is a eulogy '... devoted primarily to the characteristics of flow rate measurement based on a differential pressure generated by the flow of liquid through a restriction (such as an orifice) inserted in a line'.

Only a single page is devoted to the magnetic flow meter. A single page is likewise devoted to the turbine meter. And barely a single paragraph is used to allude to ultrasonic, thermal, and vortex-shedding meters – already key players in the field of flow measurement. And barely a single page is dedicated to the electrical pressure transmitter (already in common use in 1978) – contrasting noticeably with long descriptions covering a variety of mechanical–pneumatic-type transmitters.

Readers, lured into purchase of the ninth edition of his book by the flyleaf promise of 'new data on the target meter', might be disappointed to discover less than two pages devoted to the subject. A similar enticement extends to the promise of new data on the Lo-Loss tube, which is similarly dismissed in approximately a page and a half.

Of course, it could be said of Spink's work that he spent most of his life in the oil and gas industries and was instrumental in the early work of the American Gas Association. In the oil and gas industry, in particular, conservatism is rife. A case in point is that in the United States, most graduate facility engineers are taught in, and make use of, the metric SI system (abbreviated from the French Le Système International d'Unités). And yet, because of their mentoring programme, they will have a reverted back to fps (the Imperial 'foot–pound–second' system used extensively in the United States) within a few years.

Generally regarded as the heir apparent to Spink, R.W. Miller's *Flow Measurement Engineering Handbook* weighs in at more than 1,000 pages. Although still referenced as a standard for orifice plate sizing, the second edition, published in 1989 still devoted less than 15 pages in total to the combined technologies of magnetic, ultrasonic, and Coriolis metering.

Too many process engineers, having had extensive experience with measuring instruments and systems that have stood the test of time, see no reason to change. Consequently, they will cling to the familiar, despite numerous shortcomings when compared with the benefits offered by newer systems. And so, more than 50 years after Spink's death, the orifice plate still reigns supreme – not because of its technological superiority but because of the industry's unwillingness to accept and implement new ideas and new technologies.

1.4 HISTORY OF FLOW MEASUREMENT

Early flow measurement was centred round the question of disputation: 'how much has he got' versus 'how much have I got'. As early as 5000 BC flow measurement was used to control the distribution of water through the ancient aqueducts of the early Sumerian civilisations from the Tigris and Euphrates rivers. Such systems were very crude, based on volume per time: for example diverting flow in one direction from dawn to noon, and diverting it in another direction from noon to dusk. And although not fully comprehending the principles, the Romans devised a method of charging for water supplied to baths and residences, based on the cross-sectional areas of pipes.

The first major milestone in the field of flow technology occurred in 1738 when the Swiss physicist Daniel Bernoulli published his *Hydrodynamica* (Bernoulli 1738) in which he outlined the principles of the conservation of energy for flow. In his book, he produced an equation showing that an increase in the velocity of a flowing fluid increases its kinetic energy while decreasing its static energy. In this manner, a flow restriction causes an increase in the flowing velocity and a fall in the static pressure – the basis of today's differential pressure flow measurement.

The word 'turbine' is derived from the Latin *spinning top* and although the ancient Greeks ground flour using horizontal turbine wheels, the idea of using a spinning rotor or turbine to measure the flow did not come about until 1790 when the German engineer, Reinhard Woltman, developed the first vane-type turbine meter for measuring flow velocities in rivers and canals.

Other types of turbine meters followed. In the late 1800s, Lester Pelton built the first Pelton water wheel that turned as a result of water jets impinging on buckets attached around the outside of the wheel. And in 1916 Forrest Nagler designed the first fixed-blade propeller turbine.

A third milestone occurred in 1832 when Michael Faraday attempted an experiment to use his laws of electromagnetic induction to measure the flow. With the aim of measuring the water flow of the River Thames, Faraday lowered two metal electrodes, connected to a galvanometer, into the river from Waterloo Bridge. The intent was to measure the induced voltage produced by the flow of water through the Earth's magnetic field. The failure of Faraday to obtain any meaningful results was probably due to electrochemical interference and polarisation of the electrodes.

It was left to a Swiss Benedictine monk, Father Bonaventura Thürlemann, working in a monastery in Engelberg, to lay the foundation of this technology, with the publication of his scientific work, *Methode zur elektrischen Geschwindigkeitsmessung in Flüssigkeiten* (Method of Electrical Velocity Measurement in Liquids) (Thürlemann 1941).

Unfortunately, although his work was sound, the technology of the time was insufficient to develop a practical system. Consequently, it was not until the mid-1950s that sufficient progress had been made in electronics to make it possible to produce a low voltage, interference free, measuring amplifier that was sufficiently sensitive and drift free. Despite the many advantages of this technology, initial conservatism slowed down its acceptance for use in industrial applications. The impetus required to initiate further research and general acceptance, came in 1962 when J. A. Shercliff published his decisive book *The Theory of Electromagnetic Flow-Measurement* (Shercliff 1962), setting down a firm theoretical foundation on the principles of magnetic flowmeters.

The last milestone occurred only 3 years after Faraday conducted his original experiment when, in 1835, Gaspar Gustav de Coriolis made the discovery of what is now termed the Coriolis effect, which led, nearly a century and a half later, to the development of the highly accurate direct measurement mass flow Coriolis meter.

1.5 CATEGORISATION OF FLOW METERING TECHNOLOGIES

Many attempts have been made to categorise flow metering technologies. Several have merely split the technologies into two divisions: head-loss metering and non-head-loss metering – a simple enough categorisation but one that is not only far too simplistic but one that is also somewhat dismissive of many of the modern additions to the technology as a whole.

Another approach is to classify them into new- and traditional-technology flow meters. In its way, this approach is equally dismissive of the traditional technologies that include head-loss meters. In some instances considerable work has been carried out to dramatically improve the shortcomings of these previously restricted technologies. In this book, the author has used neither of these two approaches – seeking rather to look at each technology in turn and to weigh them according to their importance in industry as a whole, rather than any specific sector. To this effect the technologies are discussed in the following categories:

- Positive displacement
- Head loss
- Indirect volumetric
- Oscillatory
- Electromagnetic
- Ultrasonic
- Coriolis mass flow
- Thermal mass flow
- Open channel

When examined in detail these nine divisions encompass a total of nearly 50 different technologies.

2 Fluid Mechanics

2.1 INTRODUCTION

Fluid mechanics is simply the study of forces and flow within fluids. The fluid properties can vary enormously from industry to industry. The fluid may be toxic, flammable, abrasive, radio-active, explosive or corrosive; it may be single-phase (clean gas, water, or oil) or multi-phase (e.g. slurries, wet steam, unrefined petroleum, or dust-laden gases). The pipe carrying the fluid may vary from less than 1 mm to many metres in diameter. Fluid temperatures may vary from close to absolute zero to several hundred degrees Celsius, and the pressure may vary from high vacuum to hundreds or even thousands of atmospheres.

2.2 MASS VERSUS WEIGHT

One of the most basic properties of any liquid or gas is its mass. And this in turn raises the question: mass or weight – what's the difference? Both terms are frequently misunderstood and, consequently, frequently misused.

Let's start with 'mass'. Mass is a fundamental measure of the amount of material in an object and is directly related to the number and types of atoms present (Figure 2.1). It would seem self-evident that in (a) there are more atoms enclosed by the volume than there are in (b) and thus it would contain more material and would, consequently, have a higher mass.

'Weight', on the other hand, is a measure of the gravitational force acting on the mass of the object and may be given by:

$$w = m \cdot g \qquad (2.1)$$

where

w = weight
m = mass
g = acceleration due to gravity

From this we should be able to deduce that the mass of an object remains constant wherever it is measured. Its weight on the moon would only be a sixth of that measured on the Earth since the Moon's gravitational force is six times weaker.

In essence, there are two systems of measurement used to describe mass and weight:

- The SI or metric system
- The FPS system – alternatively known as the 'US customary system', The 'Standard' system, or even, erroneously (especially in the United States), the 'Imperial' or 'English' system

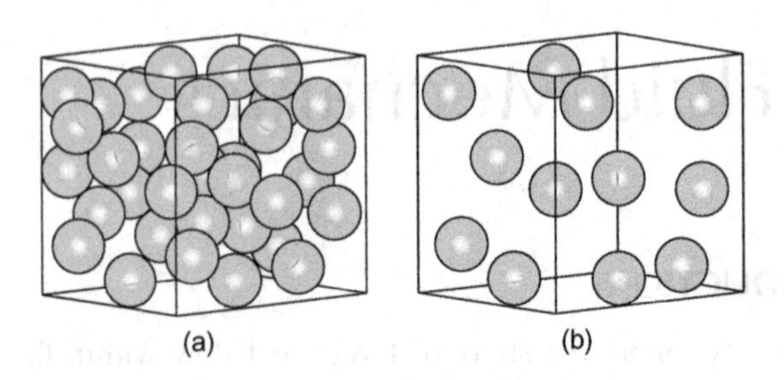

FIGURE 2.1 There are more atoms enclosed by the volume in (a) than there are in (b) and thus it would contain more material.

2.2.1 SI Units

In the SI system the unit of mass is *kilogram* (*kg*) and, since weight is a force, the unit of weight is *Newton* (*N*).

► According to Newton's Second Law the fundamental relationship between mass and weight is defined as:

$$F = m \cdot a \tag{2.2}$$

where
F = force (N)
m = mass (kg)
a = acceleration (m/s^2)

Consequently, for a body having a mass of 1 kg:

$$F = (1\ kg) \cdot (9.807\ m/s^2) = 9.807\ N$$

where
$9.807\ m/s^2$ = standard gravity close to Earth in the SI system. ◄

2.2.2 FPS Units

If the correct units were adhered to, the FPS system would be equally straightforward in which the unit of mass is *slug* and the unit of force/weight is *pound* or *pound-force* (*lb*).

The slug is derived from pound-force by defining it as the mass that will accelerate at 1 foot per second per second when a 1 pound-force acts upon it:

$$1\ lb = (1\ slug) \cdot (1\ ft/s^2)$$

In other words, 1 lb (pound) force acting on 1 slug mass will give the mass an acceleration of 1 ft/s^2.

Since standard gravity (g) in FPS = 32.17405 ft/s^2 a mass of 1 slug weighs 32.17405 lb (pound-force).

Unfortunately, lack of standardisation has also led to the basic unit of mass being defined as *pound-mass* (lb_m) or just (*lbm*) and the unit of force as *pound-force* (lb_f) or (*lbf*).

2.3 DENSITY

The density of an object is simply its *mass per unit volume* and the symbol most frequently used is the Greek symbol rho (ρ).

Mathematically, density is defined as the mass divided by volume:

$$\rho = \frac{m}{V} \tag{2.3}$$

where

ρ = density
m = mass
V = volume

Again, when comparing Figures 2.2a and b, for the given volume enclosing the two substances there are more atoms enclosed by the same volume in (a) than in (b) and thus the mass would be much higher and, consequently, its density.

In SI units, density is usually described in *kg/m³* – with several SI-derived units that include: *g/cm³* or *kg/L*.

In the FPS system there are again a number of units used to describe density including: *lb/ft³*, *lb/inch³* and *lb/gal.*

2.4 SPECIFIC GRAVITY

The *specific gravity* (often referred to as the *relative density*) is a dimensionless figure in which the density of the material is referenced to water – simply, is it heavier or lighter than water.

Since the densities of the sample and of water vary with temperature and pressure, both the temperatures and pressures at which the densities were determined must be specified. Measurements are normally made at 1 nominal atmosphere (1,013.25 mbar). However, since specific gravity is usually related to highly incompressible fluids, the variations in density caused by pressure are usually neglected.

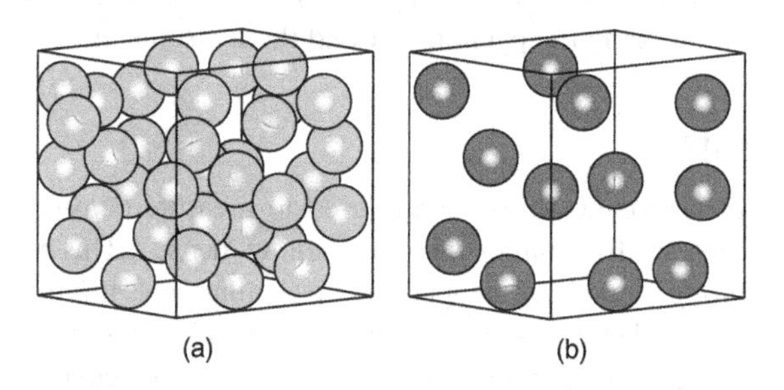

FIGURE 2.2 Since there are more atoms enclosed by the same volume in (a) than in (b) the mass would be much higher as would its density.

TABLE 2.1
Some Typical Values of Specific Gravity for a Variety of Incompressible Liquids

Fluid	Temperature (°C)	Specific Gravity (SG)
Alcohol, ethyl (ethanol)	25	0.787
Alcohol, methyl (methanol)	25	0.791
Ammonia	25	0.826
Benzene	25	0.876
Butane, liquid	25	0.601
Crude oil, California	15.55	0.918
Crude oil, Texas	15.55	0.876
Ethane	−89	0.572
Ether	25	0.716
Ethylene glycol	25	1.100
Hexane	25	0.657
Kerosene	15.55	0.820
Methane	−164	0.466
Octane	25	0.701
Pentane	25	0.755
Propane	25	0.495
Sea water	25	1.028
Toluene	25	0.865
Turpentine	25	0.871
Water, pure	4	1.000

The density of water is normally taken at its densest ($3.98 \approx 4$°C). Table 2.1 shows some typical values of specific gravity for a variety of incompressible liquids.

2.5 MEASUREMENT OF DENSITY/SPECIFIC GRAVITY

Whilst there are a wide range of instruments used to measure density/specific gravity online, they would need to be calibrated against some form of laboratory standard – the two most common being the pycnometer and the hydrometer.

2.5.1 Pycnometer

The standard laboratory calibration method used to measure density and/or specific gravity is the *pycnometer* (from Greek *puknos* meaning 'dense'). This comprises a glass flask with a close-fitting ground glass stopper with a capillary hole through it (Figure 2.3).

The capillary hole releases any spare liquid or air after closing the stopper and allows a given volume of the fluid to be obtained with a high accuracy. This enables liquid density or SG to be measured accurately with reference to for example water, using an analytical balance.

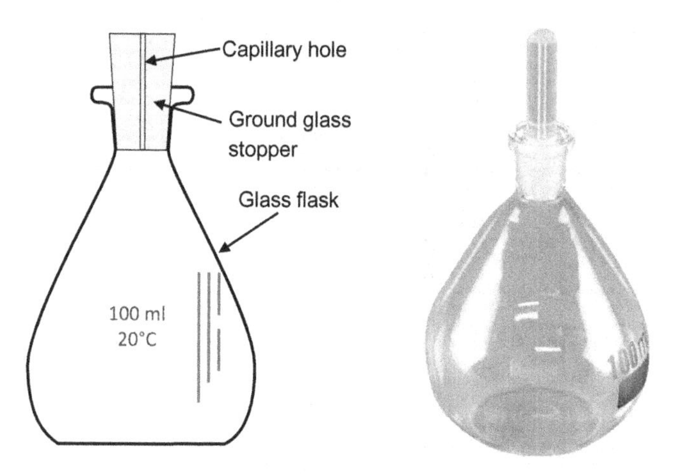

FIGURE 2.3 Pycnometer is used to measure the density and/or specific gravity and comprises a glass flask with a close-fitting ground glass stopper with a capillary hole through it.

2.5.2 Hydrometer

Although the hydrometer is an indirect measuring system, and requires to be calibrated, it is also the most widely used. It comprises a sealed, long-necked glass bulb that is weighted with mercury or lead, to make it float upright, and which is then immersed in the liquid to be measured (Figure 2.4).

The depth of flotation gives an indication of liquid density, and the neck can be calibrated to read density, specific gravity, or some other related characteristics.

2.6 DENSITY MEASUREMENT SCALES

2.6.1 Baumé Scale

One of the earliest hydrometer scales was the *Baumé* scale (generally notated as °B or °Bé) developed by the French pharmacist Antoine Baumé in 1768.

Traditionally, the Baumé scale has been used in industries where hydrometer readings have long been used to indirectly determine the concentration of a solution. Examples include brewing, wine-making, honey production, acid production and, in early days, the petrochemical industry. Its use in the wine industry is still extensive since the °Bé of settled grape juice closely correlates with the potential alcohol, when the juice is fermented to dryness.

In reality the Baumé scale comprises two independent and mutually exclusive (non-overlapping) hydrometer scales that cover liquids with a specific gravity greater than 1.0 and liquids having a specific gravity less than 1.0. The Baumé of distilled water is 0.

For liquids *heavier* than water:

0°Bé = distance the hydrometer sinks in pure water
15°Bé = distance the hydrometer sinks in a solution that is 15% sodium chloride (salt, NaCl) by mass

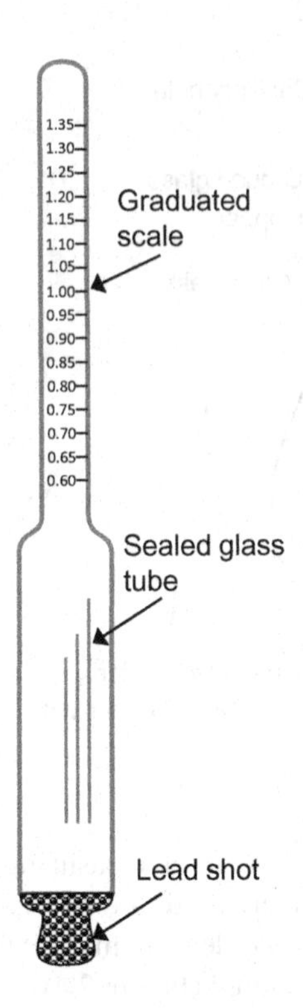

FIGURE 2.4 A hydrometer typically comprises a sealed, long-necked glass bulb that is weighted with mercury or lead, to make it float upright, and which is then immersed in the liquid to be measured.

► To convert from °Bé to specific gravity at 60°F:

$$SG = \frac{145}{145 - {}^{\circ}Bé} \tag{2.4}$$

For liquids *lighter* than water:

0°Bé = distance the hydrometer sinks in a solution that is 10% sodium chloride (salt, NaCl) by mass
10°Bé = distance the hydrometer sinks in pure water

To convert from °Bé to specific gravity at 60°F:

$$SG = \frac{140}{130 - {}^{\circ}Bé} \tag{2.5}$$

◄

It's important to remember to specify whether you are using the 'heavy' or 'light' scale when stating a Baumé value since the two scales cannot be overlapped: 30°Bé (heavy) and 30°Bé (light) are *not* the same.

2.6.2 API Gravity Scale

As mentioned previously, the Baumé scale had been traditionally used in the petrochemical industry for the measurement of crude oils. Indeed, in 1916 the US National Bureau of Standards (later to become American National Standards Institute (ANSI)) established the Baumé scale as the standard for measuring the specific gravity of liquids less dense than water.

Unfortunately, later investigation revealed major errors in salinity and temperature that had caused serious variations. However, hydrometers in the United States had been manufactured and distributed widely with a modulus of 141.5 instead of the Baumé scale modulus of 140. Because, by 1921, this scale was so firmly established, the remedy implemented by the API was to create the *API gravity scale*.

In essence API gravity indicates how heavy or light a petroleum liquid is compared with water. If its API gravity is greater than 10, it is lighter and floats on water; if less than 10, it is heavier and sinks.

More specifically API gravity measures the density of crude oil at a specific temperature compared with the density of water at a standard temperature of 60°F.

The relationship between specific gravity (SG) and API gravity is:

$$\text{API Gravity} = \frac{141.5}{\text{SG at } 60^\circ\text{F}} - 131.5 \tag{2.6}$$

API gravity is graduated in degrees on a hydrometer and was designed so that most values fall between 10° and 70°API gravity. Thus, a heavy oil with a specific gravity of 1.0 (i.e. with the same density as pure water at 60°F) would have an API gravity of:

$$\text{API Gravity} = \frac{141.5}{1} - 131.5 = 10^\circ\text{API} \tag{2.7}$$

Crude oil is classified as:

Light crude oil: Higher than 31.1°API
Medium oil: Between 22.3°API and 31.1°API
Heavy oil: API gravity below 22.3°API
Extra heavy oil: API gravity below 10.0°API

Generally speaking, the higher the API gravity the greater the commercial value. This general rule only holds up to 45°API gravity – beyond this value the molecular chains become shorter and less valuable to a refinery.

Oil that will not flow at normal temperatures, or without dilution, is called bitumen. Bitumen derived from the oil sands deposits in Alberta, Canada, has an API gravity of around 8°API and is 'upgraded' to an API gravity of 31–33°API and is known as synthetic oil.

2.7 PRESSURE

As distinct from solids, which have the intrinsic ability to maintain a fixed shape, liquids, and gases, collectively called fluids, have the ability to flow and tend to fill whatever solid containers they are held in. When a gas is in a closed container the molecules are in constant, random motion whose average speed increases with increasing temperature. In the course of their movements they collide elastically – not only with each other but also with the container walls (Figure 2.5).

All the collisions that occur over a given area combine to result in a force that is distributed over the complete internal area of the container. Now consider what a fluid would do when subjected to an external compressional force as illustrated in Figure 2.6.

Given the freedom of a fluid's molecules to move about, any external compressional force will be directed everywhere against the inside surface of the cylinder. Since a liquid is virtually incompressible the piston will remain in its resting position. The gas, on the other hand, will compress and the piston will move down as the external force increases.

In any event, the force applied to the fluid is evenly dispersed in all directions to the containing surface and how much force (F) is distributed across how much area (A) is defined as *pressure* (P):

$$P = \frac{F}{A} \tag{2.8}$$

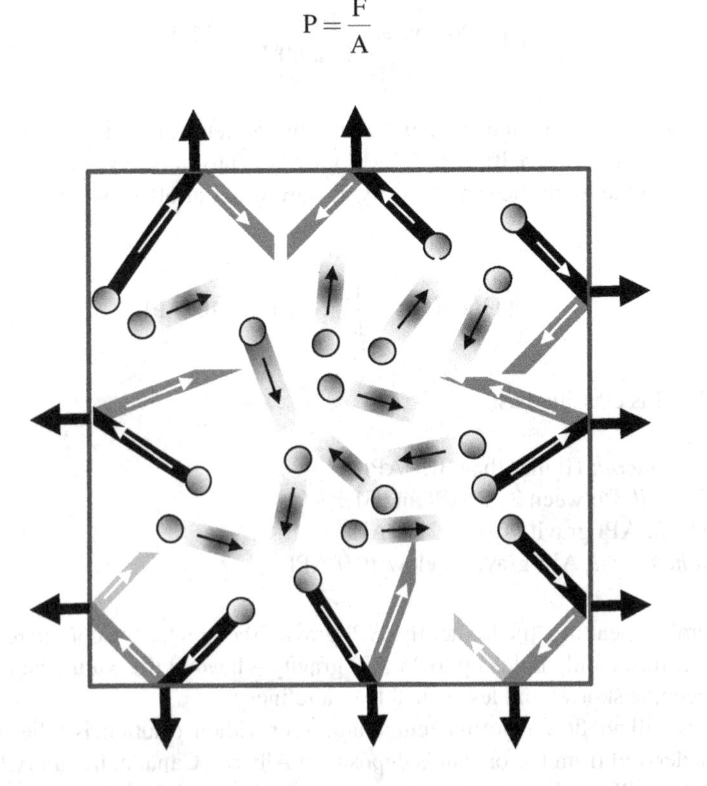

FIGURE 2.5 Pressure is exerted by atoms or molecules colliding with the inner walls of the container and is the sum of the collisional forces.

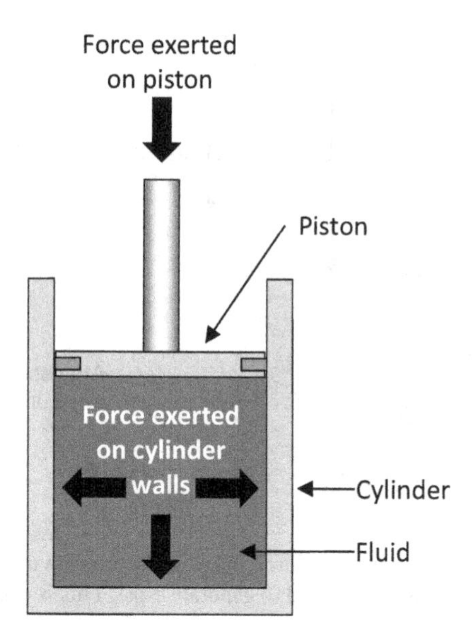

FIGURE 2.6 When the fluid is subjected to an external compressional force, it will be directed everywhere against the inside walls of the cylinder.

Since pressure is defined as the force per unit area, for many the older definition of *pounds per square inch* (*psi*) seems very much more descriptive of pressure than, the SI unit for pressure: the *Pascal.*

At first glance it might be thought that we could use something a little more description in SI units: for example kilograms per square metre. But as we've already seen, the kilogram is a unit of mass not force. And as we've already seen the SI unit of force is the Newton. Pressure, therefore, is measured in Newton per square metre (N/m^2) and the unit is called the Pascal.

Arguably, a more practical unit of measurement is the bar, named after the French physicist Bar, since it approximates to atmospheric pressure (1.013 bar = 1 atmosphere and 1 bar = 100 kPa). For this reason, the German Deutsches Institut für Normung (DIN) standards use the bar almost exclusively and its application is also spreading internationally, particularly in European standardisation.

Pressure can also be stated in terms of the height of a liquid column. If one slug of water was poured into a glass tube having a cross-sectional area of 1 square in, the weight of the water on that area at the bottom of the glass tube is one pound and the pressure is therefore 1 psi. At 39°F, the height of the water column would be 27.68 inches which is annotated as 27.68″ WC (for water column) or, more simply, 27.68″ H_2O. With a liquid that is heavier than water the pressure increases and for example only 2.036″ of mercury is required to generate 1 psi (Figure 2.7). This is usually just annotated as 2.036″ Hg.

Where the metric system is prevalent, inches are replaced with millimetres and thus 1 psi is the equivalent of 703.07 mm H_2O or 51.7149 mm Hg. 1 mm Hg ($\approx$1/760 atm) is also known as the Torr in honour of Evangelista Torricelli who did a lot of early work in pressure measurement and invented the barometer. Table 2.2 provides a guide to various pressure conversions.

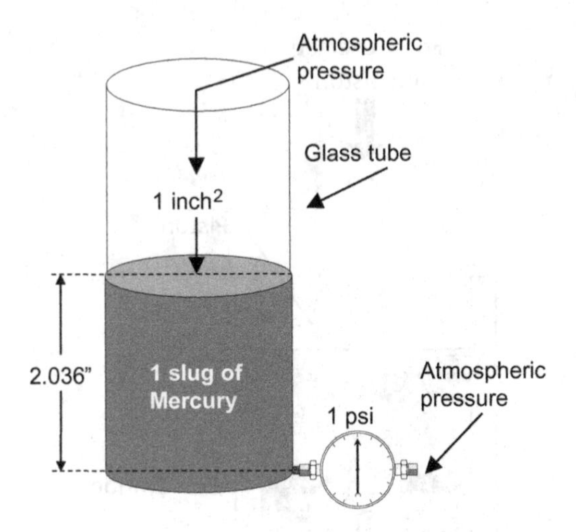

FIGURE 2.7 If 1 slug of mercury is poured into a glass tube having a cross-sectional area of 1 in² a height of only 2.036″ is required to generate 1 psi. This is usually just annotated as 2.036″ Hg.

► We have already seen that since pressure is defined as the force per unit area, a liquid within a container produces a pressure determined by the cross-sectional area (A) and the weight (w) of the fluid:

$$P = \frac{w}{A} \tag{2.9}$$

We have also already seen that weight is given by mass times the acceleration (Equation 2.1). And thus with gravitational force (g) acting on the object:

$$w = m \cdot g \tag{2.10}$$

and therefore:

$$P = \frac{m \cdot g}{A} \tag{2.11}$$

As we have seen, the density (ρ) of an object is simply its mass per unit volume, that is its mass divided by its volume:

$$\rho = \frac{m}{V} \tag{2.12}$$

Consequently, we can also say that:

$$m = \rho \cdot V \tag{2.13}$$

TABLE 2.2
Pressure Conversions between Various Standards

	psi	kPa	Inch H_2O	mm H_2O	Inch Hg	mm Hg	bar	mbar	kg/cm²	g/cm²
psi	1	6.8948	27.7296	704.332	2.0360	51.7149	0.0689	68.9476	0.0703	70.3070
kPa	0.1450	1	4.0218	102.155	0.2953	7.5006	0.0100	10.0000	0.0102	10.1972
Inch H_2O	0.0361	0.2486	1	25.4000	0.0734	1.8650	0.0025	2.4864	0.0025	2.5355
mm H_2O	0.0014	0.0098	0.0394	1	0.0029	0.0734	0.0001	0.0979	0.00001	0.0998
Inch Hg	0.4912	3.3864	13.6195	345.936	1	25.400	0.0339	33.8639	0.0345	34.532
mm Hg	0.0193	0.1333	0.5362	13.6195	0.0394	1	0.0013	1.3332	0.0014	1.3595
bar	14.5030	100.00	402.184	10,215.5	29.5300	750.062	1	1,000	1.0197	1,019.72
mbar	0.0145	0.1000	0.4022	10.2155	0.0295	0.7501	0.001	1	0.0010	1.0197
kg/cm²	14.2233	98.0665	394.408	100,018	28.9590	735.559	0.9607	980.665	1	1,000
g/cm²	0.0142	0.0981	0.3944	10.0180	0.0290	0.7356	0.0010	0.9807	0.001	1

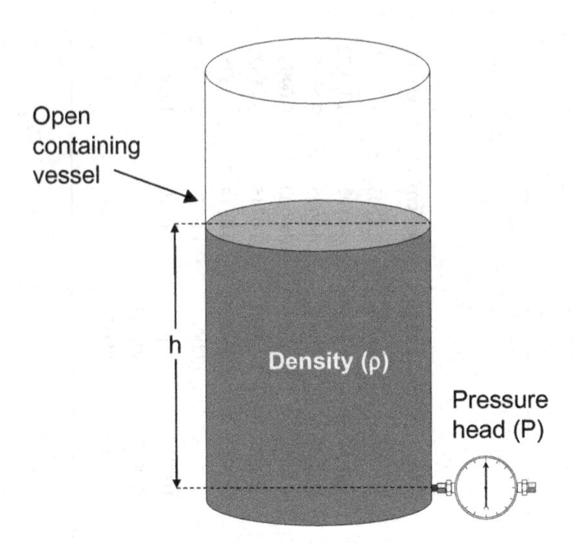

FIGURE 2.8 Pressure head (P) is determined by the liquid height (h), its density (ρ), and the acceleration due to the force of gravity (g).

and that the volume (V) is equal to the height (h) of the liquid times the cross-sectional area (A):

$$m = \rho \cdot h \cdot A \tag{2.14}$$

therefore:

$$P = \frac{\rho \cdot h \cdot A \cdot g}{A} \tag{2.15}$$

or:

$$P = \rho \cdot h \cdot g \tag{2.16}$$

◄

This means that, as shown in Figure 2.8, the pressure exerted as a result of a head of liquid, known as the pressure head (P) or *hydrostatic pressure* is simply determined by the liquid height (h), its density (ρ), and the acceleration due to the force of gravity (g).

Thus for example if the fluid was water, then a depth (h) of 1 ft would be equivalent to 0.4333 psi – based on the density of water and the force that the water at a depth of 1 ft exerts at the bottom of the tank.

2.8 PRESSURE REFERENCES

In the measurement of pressure, there are three commonly used basic pressure references, which, in essence, describe the zero point on the scale.

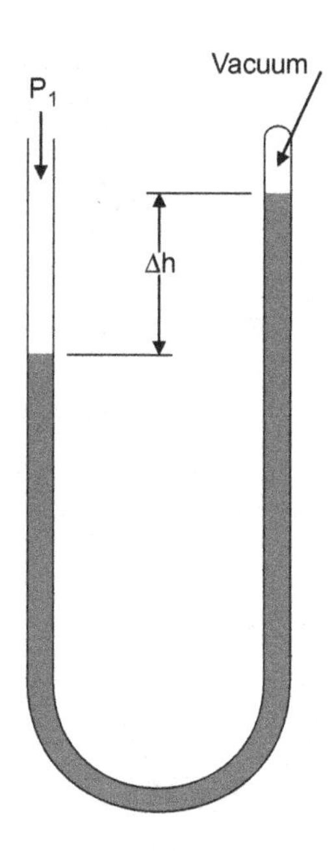

FIGURE 2.9 Δh is determined by the difference in zero pressure (the reference) and the applied pressure P_1.

2.8.1 Absolute Pressure

Absolute pressure is referenced to absolute zero, that is a vacuum and all absolute pressure measurements are therefore positive. Formerly absolute pressure measurements were abbreviated with an 'a' for example *10 psia* or *10 bara.* In the SI system the abbreviation should not be used and would thus be expressed as for example 10 kPa (absolute).

The U-tube manometer is used to illustrate this principle in Figure 2.9 where Δh is the difference in height due to the difference in zero pressure (the reference) and the applied pressure P_1.

2.8.2 Gauge Pressure

Gauge pressure references the measured pressure to the ambient atmospheric pressure – thus ignoring the effects of changing weather, altitude, or depth. Again, gauge pressure measurements were formerly abbreviated with a 'g' for example *10 psig* or *10 barg.* In the SI system it should be expressed as for example *10 kPa (gauge).*

Gauge-referenced transducers are usually constructed by opening a hole in the pressure sensor so that the ambient atmospheric pressure can enter the unit and oppose the pressure being measured. This reference or 'breather' hole is usually specified as 'dry' and only a clean, dry gas should be allowed to enter it.

In Figure 2.10 Δh is the difference in height due to the difference in atmospheric pressure (the reference) and the applied pressure P_1.

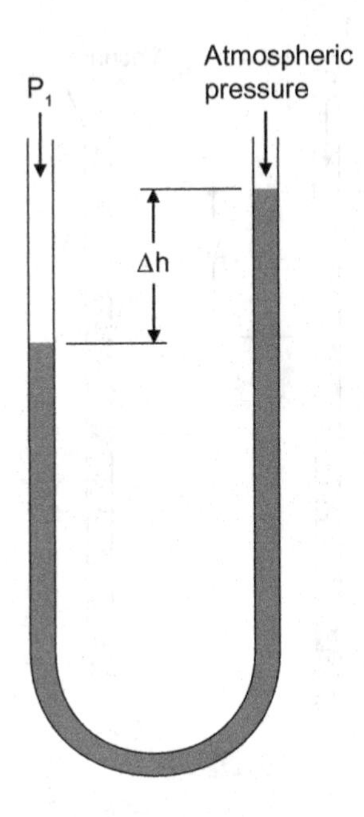

FIGURE 2.10 Δh is determined by the difference in atmospheric pressure (the reference) and the applied pressure P_1.

2.8.3 Differential Pressure

Differential pressure measurement is the difference between two unknown pressures and thus, the output is zero when the two pressures are the same – regardless of magnitude.

Differential pressure is usually abbreviated in a number of ways: *DP*, *dP*, or ΔP. Again, the SI system prefers to spell it out in the form for example *10 kPa (differential)*.

A comparison of the three commonly used pressure references is shown in Figure 2.11.

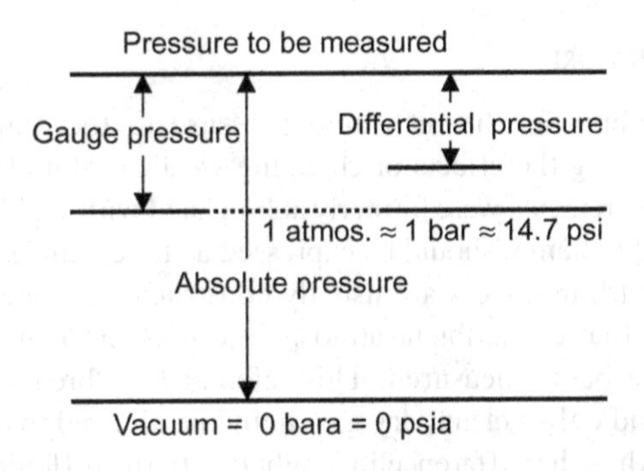

FIGURE 2.11 Comparison of the three commonly used pressure references.

2.9 PRESSURE IN GASES

We saw earlier that the tendency of a gas to fill the entire volume of space available (referred to as gas expansion) is due to what is termed molecular mobility. It also means that a pressure exerted on a point of the container is equally distributed to all sides.

When a gas is heated, its average molecular velocity increases and the gas pressure rises.

► This relationship between volume (V) and pressure (P) is described in Boyle's Law, which states that at a constant temperature, the volume of a gas is inversely proportional to its pressure:

$$V \propto \frac{1}{P} \tag{2.17}$$

Charles' Law states that if the pressure is not too high and is kept constant, the volume of a gas varies linearly with temperature:

$$V \propto T \tag{2.18}$$

And Gay Lussac's Law states:

$$P \propto T \tag{2.19}$$

When combined these give rise to the ideal gas law:

$$PV = nRT \tag{2.20}$$

where

n = number of moles (abbreviation 'mol')
R = universal gas constant (8.315 J/mol K)

Note: 1 mol = amount of substance that contains as many atoms or molecules as there are in 12 g of carbon-12.

◄

2.10 PRESSURE IN LIQUIDS

For most applications, liquids can be assumed to be incompressible (i.e. volume does not change with pressure) and when pressurised in a closed container, the pressure is distributed equally to all sides – as with gases.

However, in what is normally regarded as a non-compressible fluid such as oil, we would expect that the speed of response would be virtually instantaneous. In reality, all fluids have some degree of compressibility that can lead to a delayed response resulting for example in an actuator failing to move until the upstream fluid has been compressed.

This characteristic of a liquid to be compressed is best described by the reciprocal of compressibility (if you like its incompressibility) by what is termed the bulk

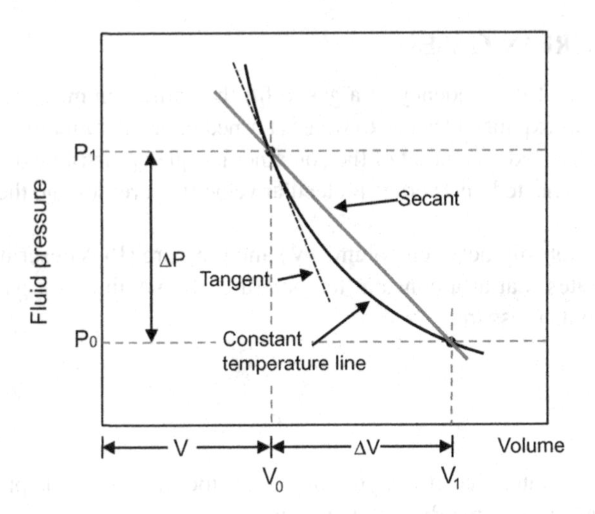

FIGURE 2.12 When fluids are under compression, the volume change is not linear.

modulus – the ratio of the very small decrease in volume resulting from a uniformly applied external pressure.

It is important to realise that fluids under compression do not follow Hooke's Law and that the relationship between pressure and that for a given temperature, volume change is not linear. Figure 2.12 illustrates the relationship between the fluid pressure and the corresponding volume at a constant temperature.

As the pressure increases, the bulk modulus of all fluids initially increases rapidly because of the decrease in the intermolecular gaps. However, as the pressure increases even further, molecules come in contact with each other and the rate of increase in the bulk modulus value is reduced.

Consequently, at a given pressure P_1, the bulk modulus may be defined as the slope of the tangent to the curve at P_1 called *tangent bulk modulus* (K_T):

$$K_T = V \cdot \frac{\Delta P}{\Delta V} \tag{2.21}$$

where

V = oil volume after compression (m^3)
ΔV = oil volume change (m^3)
ΔP = pressure change (Pa)

Because the tangent bulk modulus is only specified at a fixed pressure, a more useful value that is easier to determine and which is suitable for large pressure changes is to take an average value over a given range.

This is called the *secant bulk modulus* (K_S) and is given by the slope of a line cutting the curve at two points P_1 and V_1:

$$K_S = -V_0 \cdot \frac{\Delta P_1}{\Delta V_1} \tag{2.22}$$

where

V_0 = initial oil volume (m^3)

The minus sign indicates that for most normal fluids a negative change in volume results in an increase in the pressure.

Nonetheless, for mineral oil over a pressure range of up to about 830 bar (12,000 psi) the bulk modulus can be expressed as a linear function of pressure.

Consequently, at a given pressure P, the bulk modulus may be defined as either the slope of a line connecting P to the origin, which can be regarded as an average value of bulk modulus over the range from P_0 to P (called *secant bulk modulus*) or on the slope of the tangent to the curve at P (called tangent or instantaneous bulk modulus).

Although the tangent bulk modulus is more correct, since it is derived from the approximate equation of state for a fluid (liquid), the most frequently used value is the secant bulk modulus since it is easier to determine.

2.10.1 Effect of Pressure on Bulk Modulus

When a fluid is compressed it will normally undergo an increase in temperature which, in most fluids, will decrease the bulk modulus value. Consequently, it is normal to refer to the *isothermal bulk modulus* in which changes occur slowly enough to allow the system to continually adjust in such a manner that the temperature remains constant.

In reality, most hydraulic applications are concerned with rapidly moving systems in which the compression speed produces a rise in temperature. Consequently, the bulk modulus is termed *adiabatic* or *isentropic*. From the foregoing it seems obvious that unless the bulk modulus is given at some specific pressure and at a specific temperature it's an almost meaningless value. Unfortunately, this is exactly what happens in too many cases.

2.11 VISCOSITY

One of the most important primary properties of a fluid (liquid or gas) is its viscosity – its resistance to flow or to objects passing through it. Conceptually, viscosity might be thought of as the 'thickness' of a fluid. In essence, it is an internal frictional force between the different layers of the fluid as they move past one another. In a liquid, this is due to the cohesive forces between the molecules whilst in a gas it arises from collisions between the molecules. Thus, water is 'thin', having a low viscosity, while vegetable oil is 'thick' having a high viscosity.

If the fluid is regarded as a collection of moving plates, one on top of the other, then when a force is applied to the fluid, shearing occurs and the viscosity is a measure of the resistance offered by a layer between adjacent plates.

Figure 2.13 shows a thin layer of fluid sandwiched between two flat metal plates of area A – the lower plate being stationary and the upper plate moving with velocity v. The fluid directly in contact with each plate is held to the surface by the adhesive force between the molecules of the fluid and those of the plate. Thus the upper surface of the fluid moves at the same speed v as the upper plate whilst the fluid in contact with the stationary plate remains stationary. Since the stationary layer of the fluid retards

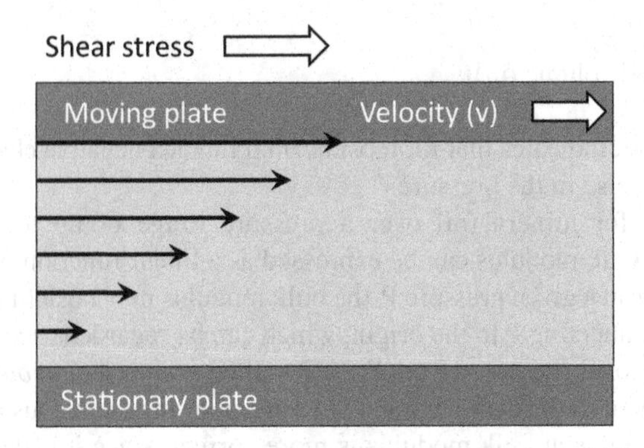

FIGURE 2.13 When a thin layer of fluid is sandwiched between two flat metal plates, shearing occurs and the upper surface of the fluid moves at the same speed as the upper plate whilst the fluid in contact with the stationary plate remains stationary.

the flow of the layer just above it and this layer, in turn, retards the flow of the next layer, the velocity varies from zero to v, as shown.

The relative force acting on the layers is called the *shear stress* (the force per unit area). In Figure 2.13, the fluid flows under the action of the shear stress due to the motion of the upper plate. It is also clear that the lower plate exerts an equal and opposite shear stress to satisfy a 'no-slip' condition at the lower stationary surface.

It follows, therefore, that at any point in the flow, the velocity at which the layers move relative to each other, referred to as the *shear rate*, is directly proportional to the shear stress (Figure 2.14).

Since:

$$\text{Shear stress} \propto \text{Shear rate} \tag{2.23}$$

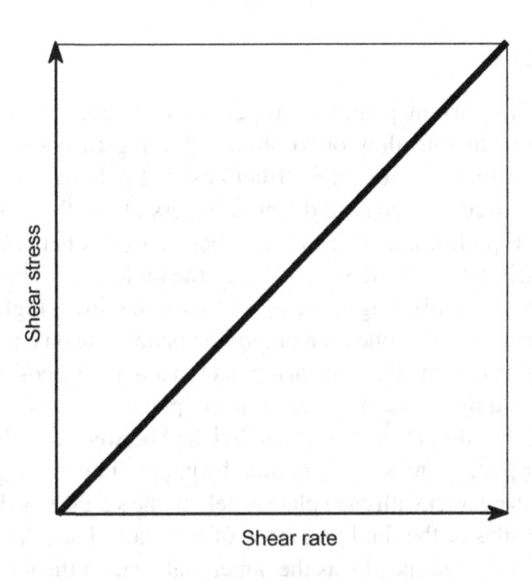

FIGURE 2.14 Shear rate is directly proportional to the shear stress.

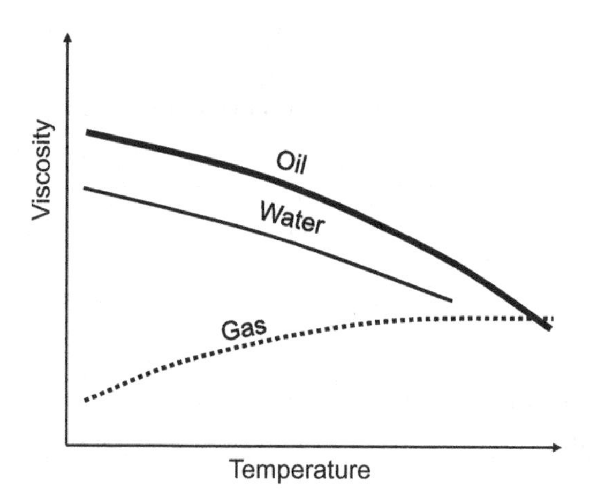

FIGURE 2.15 Viscosity of fluids is strongly dependent on the temperature.

we can also state:

$$\text{Shear stress} = \mu \cdot \text{Shear rate} \tag{2.24}$$

where

μ = dynamic viscosity – the ratio between the shear stress and shear rate

The SI unit of dynamic (absolute) viscosity is the pascal second (Pa s) defined as: *the shear stress (pressure) required to move the plate a distance, equal to the thickness of the layer between the plates, in one second.*

As shown in Figure 2.15, the viscosity of a fluid depends strongly on temperature and generally decreases when the temperature increases. Gases, however, show the opposite behaviour and the viscosity increases with increasing temperature.

Subsequently, Table 2.3 lists the viscosity of various fluids at specified temperatures – with the viscosity of liquids such as motor oil for example decreasing rapidly as temperature increases.

The viscosity of a fluid also depends on the pressure but, surprisingly, pressure has less effect on the viscosity of gases than on liquids.

A pressure increase from 0 to 70 bar (in air) results in only an approximate 5% increase in viscosity. However, with methanol for example a 0–15 bar increase results in a 10-fold increase in viscosity. Some liquids are more sensitive to changes in pressure than others.

In practice, the most commonly used unit for dynamic viscosity is the poise.

Based on metric centimetre–gram–second (cgs) units the poise is defined as: *force in dynes required to move a surface 1 cm² in area past a parallel surface at a speed of 1 cm/s with the surfaces separated by a fluid film of 1 cm thickness.*

And in turn, the dyne (Figure 2.16) is defined as: *the force required to accelerate a mass of one gram at a rate of one centimetre per second per second*:

1 dyne is equal to 10 μN (micro-newton)

TABLE 2.3
Comparison of the Viscosities of Various Fluids

Fluid	Temperature (°C)	Viscosity μ (Pa s)
Molasses	20	100
Glycerine	20	1.5
Engine oil (SAE 10)	50	0.02
Milk	20	5×10^{-3}
Blood	37	4×10^{-3}
Water	0	1.8×10^{-3}
Ethyl alcohol	20	1.2×10^{-3}
Water	20	1×10^{-3}
Water	100	0.3×10^{-3}
Air	20	0.018×10^{-3}
Water vapour	100	0.013×10^{-3}
Hydrogen	0	0.009×10^{-3}

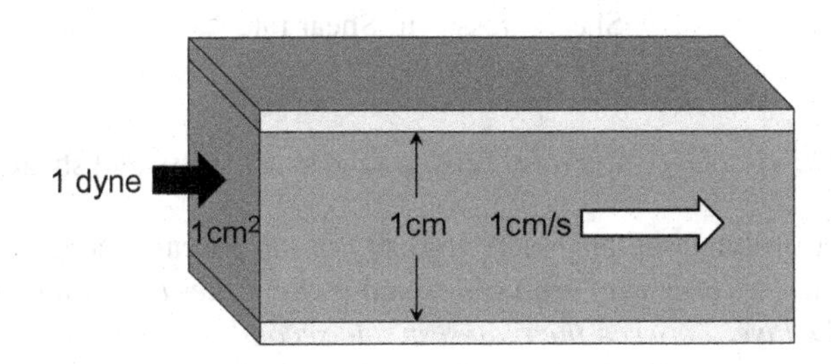

FIGURE 2.16 Dyne is defined as the force required to accelerate a mass of 1 gram at a rate of 1 cm/s/s.

In practice the poise is too large and we make use of centipoise (cP). Table 2.4 compares the dynamic viscosity expressed in centipoise with pascal second:

$$1 \text{ Pa s} = 1{,}000 \text{ cP} = 1 \text{ N s/m}^2 = 1 \text{ kg/(m s)}$$

In the oil industry, we tend not to use either of these measurements. The reason lies with a widely used practical method of measurement – the falling sphere viscometer. As illustrated in Figure 2.17 a liquid, at a controlled temperature, is held stationary in a vertical glass tube and a sphere, of known size and density, is allowed to descend through the liquid. The time it takes to pass two marks on the tube is measured and Stokes' Law is then used to calculate the viscosity of the fluid.

Because this measurement is also dependent on the density of the fluid, this unit of measurement is called the *kinematic viscosity*.

The unit of kinematic viscosity is the stokes, expressed in square centimetres per second (cm^2/s) and is usually given the Greek letter nu (ν). The more customary unit is the centistoke (cSt).

TABLE 2.4
Comparison of Dynamic Viscosity Expressed in Centipoise with Pascal Seconds

Fluid	Dynamic Viscosity (cP)	Dynamic Viscosity (Pa s)
Water (5°C)	1.519	1.519×10^{-3}
Water (20.2°C)	1.0	1.0×10^{-3}
Water (30°C)	0.798	0.798×10^{-3}
Blood (37°C)	4.0	4.0×10^{-3}
Light crude	0.5	0.5×10^{-3}
Heavy crude	>2,000	>2

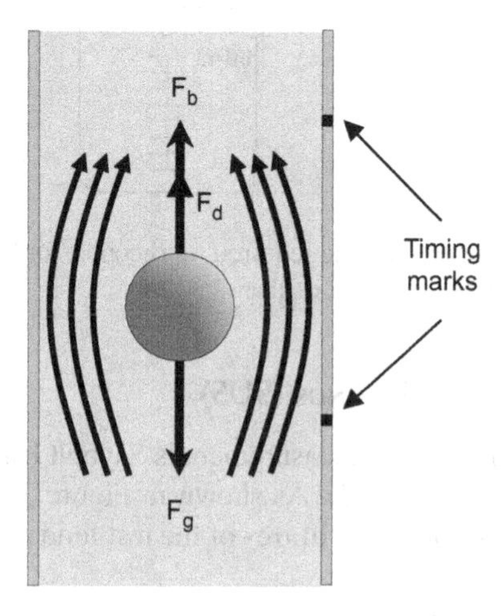

FIGURE 2.17 In the falling sphere viscometer, a sphere is allowed to descend through the liquid. The time it takes to pass two marks on the tube is measured and Stokes' Law is then used to calculate the viscosity of the fluid.

Kinematic viscosity can be related to dynamic viscosity by:

$$\nu = \frac{\text{Dynamic viscosity}}{\text{Density}} = \frac{\mu}{\rho} \tag{2.25}$$

Kinematic viscosity is generally measured in centistokes (cSt) where:

$$1\,\text{m}^2/\text{s} = 10^6\ \text{cSt} \tag{2.26}$$

To obtain centistokes (cSt) from centipoise (cP) it is therefore necessary to divide by the density. Since most hydrocarbons have a density of about 0.85–0.9 the centistoke value will be about 10–15% higher than the centipoise value.

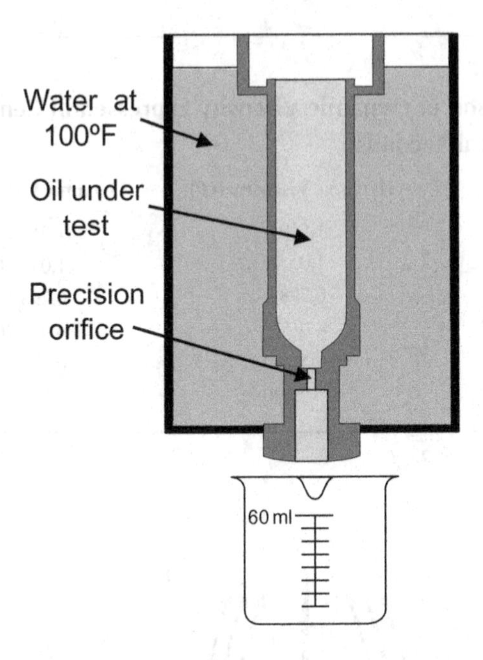

FIGURE 2.18 SUS is based on the time required, in seconds, for 60 millilitres of the tested fluid at 100°F to pass through a standard orifice.

2.11.1 Saybolt Universal Seconds (SUS)

Another frequently used viscosity measurement is Saybolt Universal Seconds (SUS) or Saybolt Seconds Universal (SSU). As shown in Figure 2.18 this is based on the time required, in seconds, for 60 millilitres of the test fluid at 100°F to pass through a standard orifice.

For SSU values greater than 100:

$$\mathrm{cSt} = 0.22 \cdot \mathrm{SSU} - \left(\frac{135}{\mathrm{SSU}}\right) \tag{2.27}$$

Table 2.5 compares some SSU values with centipoise.

2.12 NON-NEWTONIAN FLUIDS

Most fluids used in engineering systems exhibit what is termed a Newtonian behaviour in that, for a given value of pressure and temperature, the shear stress is directly proportional to the shear rate. Thus, if the shear stress is plotted against the shear rate the result is a straight line passing through the origin (Figure 2.19).

Many fluids, however, do not exhibit this behaviour. Examples include: tar, grease, printers' ink, colloidal suspensions, hydrocarbon compounds with long-chain molecules, and polymer solutions. In addition, some fluids, called viscoelastic fluids, do not immediately return to a condition of zero shear rate when stress is removed.

TABLE 2.5
Comparison of some SSU Values with Centipoise

cP = mPa s	SSU
10	60
50	233
100	463
160	741
200	927
260	1,204
300	1,390
360	1,668
400	1,853
460	2,131
500	2,316

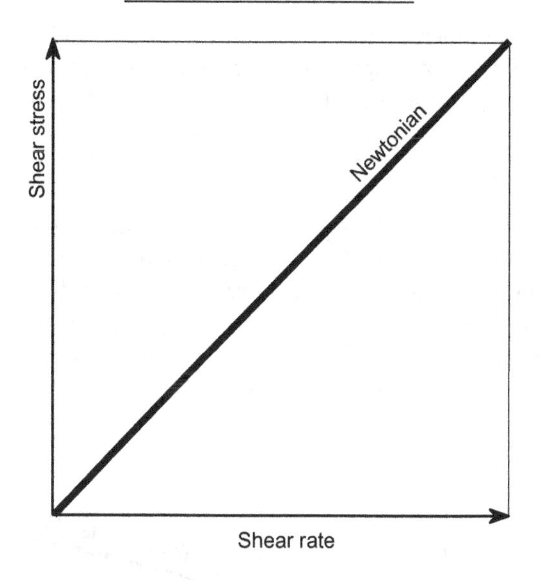

FIGURE 2.19 In a Newtonian fluid the shear stress, plotted against shear rate, results in a straight line passing through the origin.

2.12.1 Ideal Plastic

The so-called *ideal plastics* or *Bingham fluids* exhibit a linear relationship between the shear stress and shear rate. However, such substances only flow after a definite yield point has been exceeded (Figure 2.20).

When at rest, these materials possess sufficient rigidity to resist shear stresses smaller than the yield stress. Once exceeded, however, this rigidity is overcome and the material flows in much the same manner as a Newtonian fluid. A common

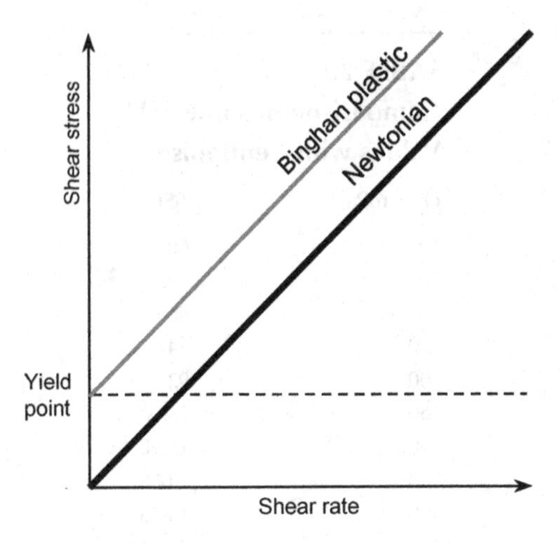

FIGURE 2.20 Bingham fluids again exhibit a linear relationship between the shear stress and shear rate – but only flow after a definite yield point has been exceeded.

example is toothpaste which will not exude until pressure is applied and then pushed out as a solid plug. Other examples of materials exhibiting this type of behaviour include: tar, chewing gum, grease, slurries, sewage slugs, and drilling muds.

2.12.2 Pseudoplastic

A pseudoplastic substance, such as printer's ink, is characterised by polymers and hydrocarbons that possess long-chain molecules and suspensions of asymmetric particles. Although exhibiting a zero yield stress, the relationship between the shear stress and shear rate is non-linear and the viscosity decreases as the shear stress increases (Figure 2.21). The shear force thus causes it to go from thick like honey to

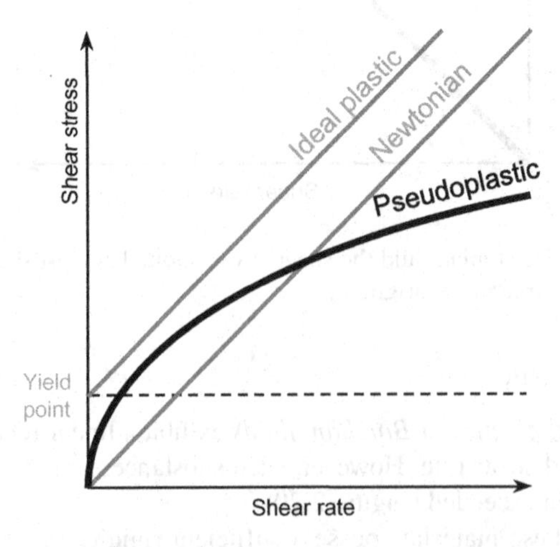

FIGURE 2.21 Although a pseudoplastic exhibits zero yield stress, the relationship between the shear stress and shear rate is non-linear and the viscosity decreases as the shear stress increases.

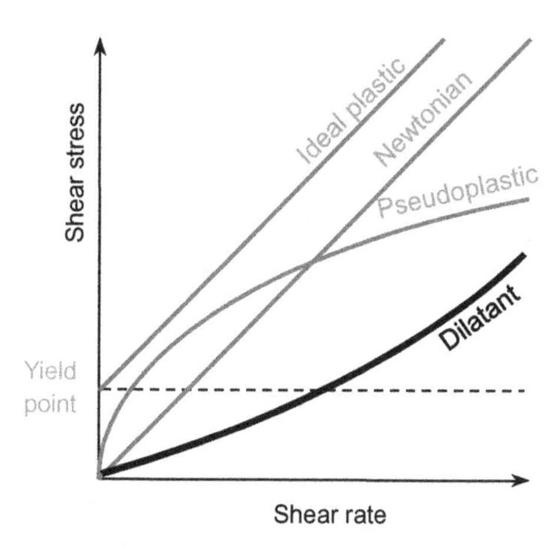

FIGURE 2.22 Dilatant materials also exhibit a non-linear relationship between the shear stress and shear rate and a zero yield stress – with the viscosity increasing as the shear stress increases.

flowing like water. One example is paint. When modern paints are applied the shear created by the brush or roller will allow them to thin and wet out the surface evenly.

2.12.3 Dilatant

Dilatant materials also exhibit a non-linear relationship between the shear stress and shear rate and a zero yield stress. However, in this case, the viscosity increases as the shear stress increases (Figure 2.22).

This type of behaviour is found in highly concentrated suspensions of solid particles. At low rates of shear, the liquid lubricates the relative motion of adjacent particles, thereby maintaining relatively low stress levels. As the shear rate increases, the effectiveness of this lubrication is reduced and the shear stresses are increased.

3 Measurement Considerations

3.1 INTRODUCTION

What do we actually mean by flow measurement? Do we want to measure the flow rate in terms of its velocity – metres (or feet) per second? Or are we more interested in how much in terms of capacity for example litres (or gallons) per second? Or are we more interested in how much in terms of its mass kilograms (or pounds) per second?

Although there are a couple of technologies that measure the flow directly in terms of capacity, most tend to measure the flow rate in terms of its velocity and then infer the volumetric flow rate (capacity) or the mass flow rate.

3.2 VOLUMETRIC FLOW RATE

The measurement of flow in terms of volume is termed the *volumetric flow rate* (*Q*) and represents the total volume of fluid flowing through a pipe per unit of time. This is usually expressed in in a number of ways for example:

- Litres per second (L/s)
- Gallons per minute (g/min)
- Cubic metres per hour (m^3/h)

The calculation of the volumetric flow rate is most frequently achieved by measuring the mean velocity of a fluid as it travels through a pipe of known cross-sectional area A (Figure 3.1):

$$Q = v \cdot A \tag{3.1}$$

Figure 3.2 illustrates the range of volumetric flow rates according to the flow velocity and pipe diameter.

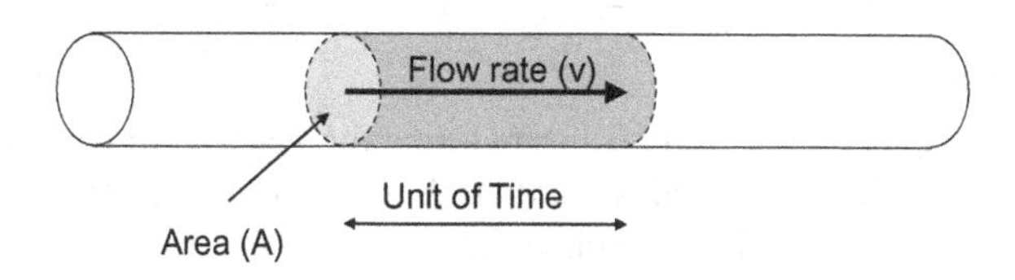

FIGURE 3.1 Volumetric flow rate, Q, represents the total volume of fluid flowing through a pipe per unit of time.

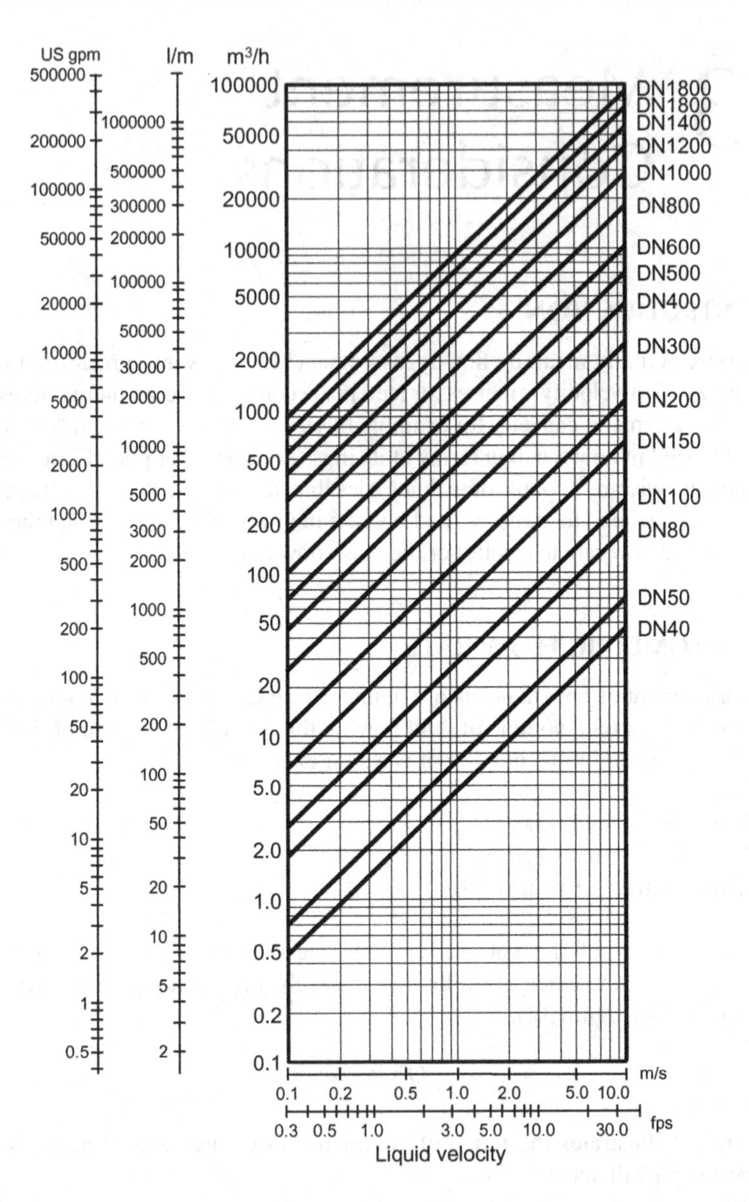

FIGURE 3.2 Different volumetric flow rates according to the flow velocity and pipe diameter.

PIPE SIZES

Pipes are rated according to their normal pipe diameter and given according to a preferred series – referred to as their *Nominal Pipe Size* (NPS) or their *Nominal Bore* (NB) – both based on inches.

The European equivalent to NPS is the *Nominal Diameter* (DN) measured in millimetres (Table 3.1).

TABLE 3.1
Preferred Series of ANSI and DN Pipe Sizes

Pipe Diameter (in) (ANSI – NPS)	Pipe Diameter (mm) (DN)	Pipe Diameter (in) (ANSI – NPS)	Pipe Diameter (mm) (DN)
0.5	15	8	200
0.75	20	10	250
1	25	12	300
1.5	40	14	350
2	50	16	400
3	80	24	600
4	100	36	900
6	150	48	1,200

It should be noted that, in this case, 'nominal' means just that – in other words: approximate or 'not necessarily corresponding exactly to the real value'.

3.3 MASS FLOW RATE

Most chemical reactions are largely based on their mass relationship and, consequently, in order to control the process more accurately, it is often desirable to measure the mass flow of the product.

The *mass flow rate*, Q_m, gives the total mass of fluid flowing at any instant in time. The knowledge of volume flow rate, Q and the fluid density, ρ (Figure 3.3) determines the mass flow rate from:

$$Q_m = Q \cdot \rho \tag{3.2}$$

or:

$$Q_m = v \cdot A \cdot \rho \tag{3.3}$$

Some flowmeters, such as Coriolis meters, measure the mass flow directly. However, in many cases, mass flow is determined by measuring the volumetric flow and the density and then calculating the mass flow as shown above. Sometimes the

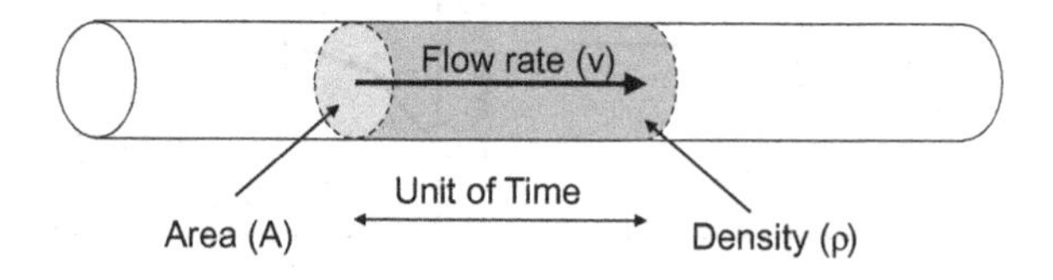

FIGURE 3.3 Mass flow rate, Q_m, represents the total mass of fluid flowing through a pipe per unit of time.

density is inferred from the measurement of the pressure and temperature of the fluid. This type of measurement is referred to as the inferred method of measuring the mass flow.

3.4 TOTALISATION

Totalisation is simply a running total of how much of fluid (liquid, gas, or steam) has flowed through the metering system in a given time – allowing the fluid to be dispensed in set quantities.

The flow rate itself may be constant during this period (Figure 3.4a) or it may vary (Figure 3.4b).

In any event, totalisation entails integrating the flow rate between two timed intervals: t_1 and t_2:

$$\text{Total}_{t_2-t_1} = \int_{t_1}^{t_2} Q_{(t)} \cdot dt \tag{3.4}$$

In essence there are two basic metering systems: those that measure the flow rate in fixed discrete packages (either directly or inferred) or measure the quantity on a continuous (analogue) basis.

Positive displacement meters for example provide a pulse output for each measured volume of the product. Consequently, totalisation entails counting the total number of pulses generated during the timing period. Since this is not a continuous measurement, there will be a resolution issue – how much of the medium is contained in each discrete package – with the resolution determined by the 'width' of each sample (Figure 3.5).

Many meters, such as turbines generate pulses according to the inferred flow rate. Consequently, resolution will be based on the number of pulses that can be generated during a single revolution of the turbine.

The measurement of flow rates from metering systems based on head loss, ultrasonic, and electromagnetic technology provide a continuous analogue output and totalisation therefore requires either some form of A to D conversion to generate a pulse train or some form of electronic integration.

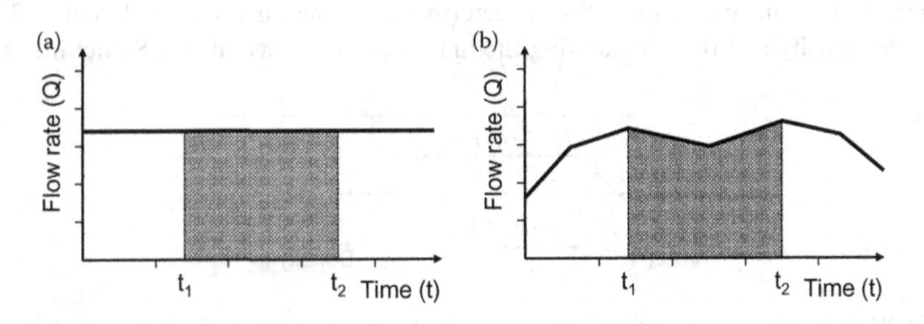

FIGURE 3.4 During the timing period the flow rate (a) may be constant or (b) may vary.

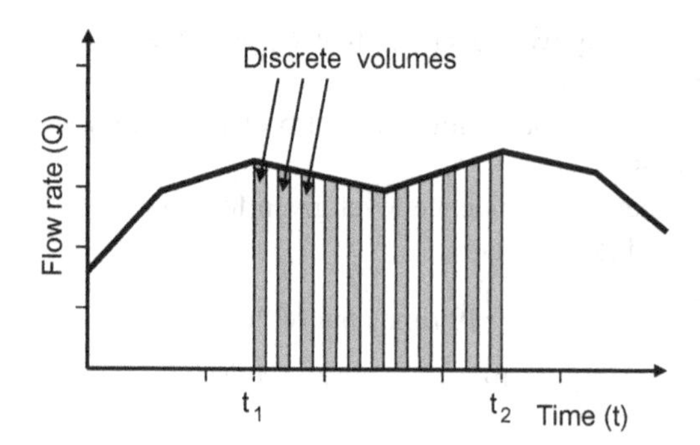

FIGURE 3.5 Totalisation entails counting the total number of the pulses generated during the timing period – with the resolution determined by the 'width' of each sample.

3.5 ACCURACY REVIEW

Any measurement tells us about the properties of something with the result normally given in two parts: a number and a unit of measurement. For example to the question: 'How hot is it?' the answer might be: 'It's 65°C'.

But how sure are we that the temperature really is 65°C?

One term generally used to describe how close an agreement there is between the result of the measurement and its true value is *accuracy* – or, as some would claim, *inaccuracy*.

It is generally accepted that the term accuracy refers to the truthfulness of the instrument. An instrument having a high accuracy gets closer to a true reading than an instrument having low accuracy.

However, if accuracy refers to the agreement between a measurement and the true or correct value, we have to ask ourselves, can the true value ever be known?

The fact is that the true value can never truly be known and we can only estimate it. Consequently, many standards authorities say that accuracy should only be used as a *qualitative* term and that no numerical value should be attached to it.

Accordingly we can say that the measurement is: 'fairly accurate' or 'highly accurate', or 'not very accurate' – but we shouldn't put a figure to it. And although virtually all manufacturers make use of the term accuracy as a *quantitative* figure, that doesn't make it any less incorrect!

So what about the term *error*? Whilst we describe accuracy as the closeness of agreement between a measured value and the true value, error is the difference between a measurement and the true value. But again, since we still don't know what the actual true value is they are both qualitative and we still shouldn't put a figure to it.

If we cannot put a figure to the terms accuracy or error, what should we use? Since there is always a margin of doubt about any measurement, we should rather make use of the term *uncertainty* – the doubt that exists about the result of any measurement.

But, we also need to ask: 'How big is the margin?' and 'How bad is the doubt?'

This result could be written as:

65°C ± 5°C, at a level of confidence of 95%.

Expressed another way we are saying that we are 95% sure that the temperature lies between 60°C and 70°C.

The *confidence level* is thus a statement of probability and we shouldn't state an uncertainty without it.

The generally accepted confidence level given for flow is 95%. This means that, on average, we should expect that:

19 times out of 20

$$19/20 = 95/100 = 95\%$$

the reading of the meter will fall within the bracket specified (e.g. ±1% of the actual calibrated value).

If a flowrate of 50 ± 1 L/s was reported at a 50% confidence level we would only expect repeated measurements to lie within the range 49 and 51 L/s half the time – with half the results lying outside this interval.

3.6 ACCURACY

Despite the aforementioned problems associated with the term accuracy, it still continues to be used by the majority of manufacturers and, by a more or less common agreement, is generally used to express the maximum deviation between the meter's indication and the true value of the process being measured.

In flow measurement accuracy may be quoted in two ways:

- Relative accuracy – the proportion of the error to the whole value (i.e. 100 L/min ± 1%)
- Absolute accuracy – the quantity of the error (i.e. 100 ± 1 L/min)

Another common problem associated with accuracy is that it may also be subject to what can only be described as 'specmanship' – an attempt to make the specification appear better than it really is.

To illustrate this problem, consider three flowmeters:

- One with a claimed accuracy of ±1% of span
- One with a claimed accuracy of ±1% of a reading
- One with a claimed accuracy of ±1% of Upper Range Limit (URL)

The URL is defined as the highest flowrate that a meter *can* be adjusted to measure – sometimes referred to as the Full Scale Deflection (FSD). Another term, the Upper Range Value (URV) is defined as the highest flowrate that the meter *is* adjusted to measure. In this example, each meter has a URL of 100 L/min, and is calibrated 0 to 50 L/min.*

For the *percentage of span* instrument, the absolute error is determined at the 100% span reading, and then used to determine the accuracy at lower flow rates. Since the

* In reality, few meters will actually measure down to a zero flow rate and, indeed, usually have some form of low flow cut-off limit below which the readings are not considered to be reliable.

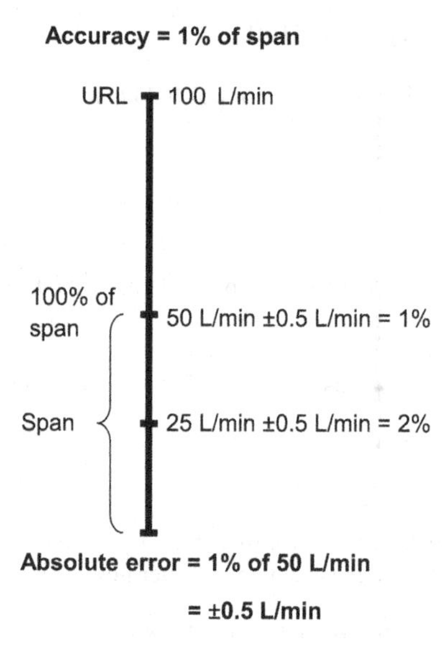

FIGURE 3.6 In the percentage of span instrument, the absolute error is determined at the 100% span reading, and then used to determine the accuracy at lower flow rates.

span is 50 L/min the absolute error would be ±1% of 50 = ±0.5 L/min. Consequently, the accuracy of the meter at 50 L/min would be 50 ± 0.5 L/min = ±1%. And at 25 L/min the accuracy at would be 25 ± 0.5 L/min = ±2% (Figure 3.6).

For the *percentage of reading* instrument, the absolute error is determined with the actual reading, and varies with the flow rate. The absolute error at 50 L/min is ±1% of 50, or ±0.5 L/min. The absolute error at 25 L/min is ±1% of 25, or ±0.25 L/min. This means the meter has a constant accuracy of ±1% at all readings (Figure 3.7).

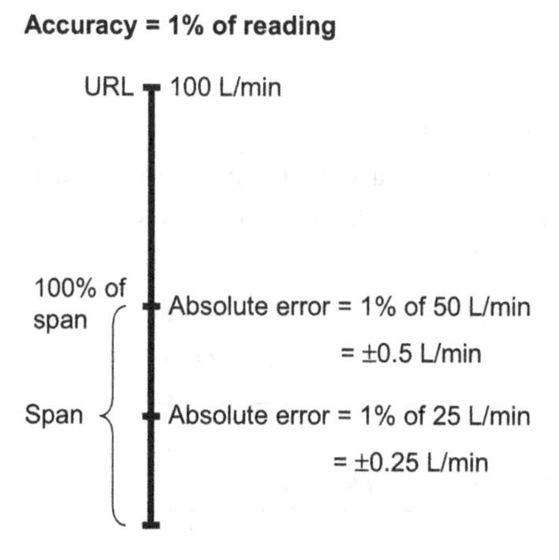

FIGURE 3.7 In the percentage of reading instrument, the absolute error is determined at the actual reading, and varies with the flow rate.

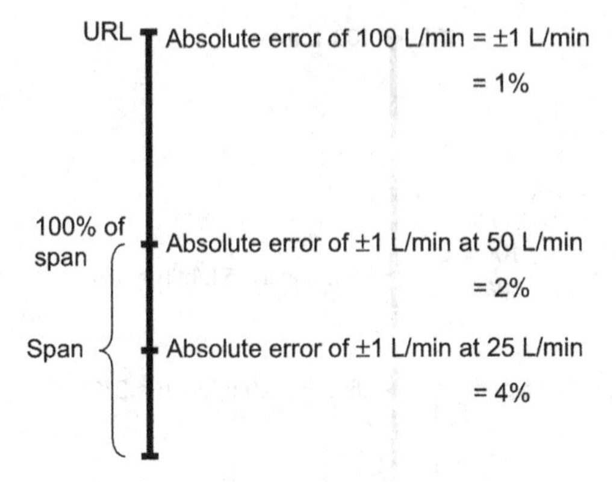

FIGURE 3.8 In the percentage of URL instrument, the absolute error is determined at the URL and then used to determine the accuracy at lower flow rates.

For the *percentage of URL* instrument, the absolute error is determined at the URL and then used to determine the accuracy at lower flow rates. The absolute error would be ±1% of 100, or ±1 L/min. The accuracy of the meter at 50 L/min would be 50 ± 1 L/min, or ±2%. The accuracy at 25 L/min would be 25 L/min at ±1 L/min, or ±4% (Figure 3.8).

In the above example, all three meters would have the same accuracy, ±1%, when calibrated at the URL, 100 L/min.

3.7 FLOW RANGE AND RANGEABILITY

Whilst there is considerable confusion regarding basis terminology, nowhere is this more evident than in the difference between the terms flow range, turndown ratio, span, and rangeability.

3.7.1 Flow Range

The flow range is simply the difference between the maximum and minimum flow rate over which a meter produces acceptable performance within the basic accuracy specification of the meter. This is illustrated in Figure 3.9.

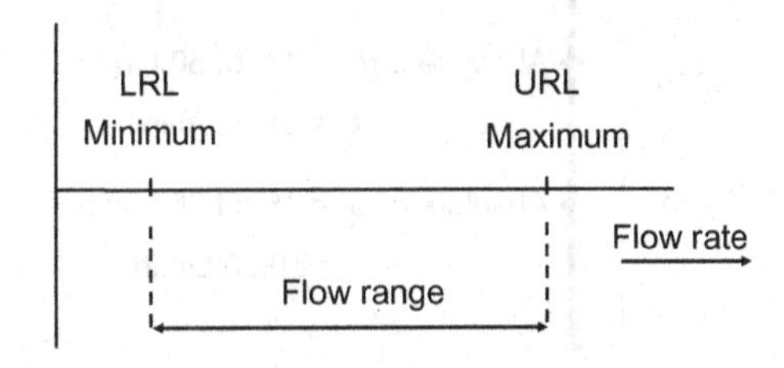

FIGURE 3.9 Flow range is the difference between the maximum and minimum flow rates over which a meter produces acceptable performance within the basic accuracy specification of the meter.

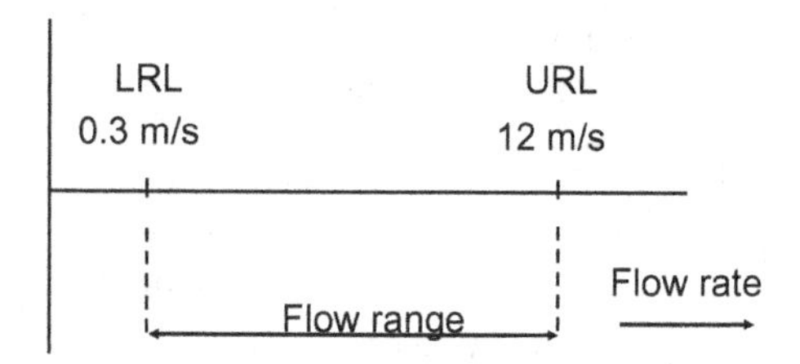

FIGURE 3.10 Measuring range of a magnetic flow meter might be 0.3–12 m/s within an accuracy of 0.3%.

For example, the measuring range of a magnetic flow meter might be 0.3–12 m/s within an accuracy of 0.3% (Figure 3.10).

For flowmeters that exhibit a minimum flowrate, the flow range is thus the interval from the minimum flow rate to the maximum flowrate. If the meter does not exhibit a minimum flow rate (very unlikely), the flow range is the interval from zero flow to maximum flow.

The flow range is often (erroneously) referred to as rangeability.

3.7.2 Turndown Ratio

The turndown ratio is the ratio of the maximum flow rate to the minimum flow rate for a measuring range that is within a stated accuracy. In the example of the magnetic flow meter, given above, the measuring range extends from 0.3 to 12 m/s. Consequently, the turndown ratio is 40:1 – within an accuracy of 0.3. If the measuring range were extended from 0.2 to 12 m/s, within a reduced accuracy of 0.5% the turndown ratio is 60:1. It is, therefore, meaningless to express the turndown ratio without a specified accuracy.

3.7.3 Span

The term span relates to the flowmeter output signals and is the difference between the upper and lower range values assigned to the output signal.

Again, in the example of the magnetic flowmeter having a 4–20 mA analogue output the upper and lower range values might be assigned as:

Lower range value: 4 mA = 3 m/s

Upper range value: 20 mA = 8 m/s

The span is thus the difference between the two values, that is 8 – 3 = 5 m/s (Figure 3.11). The minimum span is the lowest flowrate able to produce a full-scale output and the maximum span is equal to the maximum range of the sensor.

3.7.4 Rangeability

The term rangeability is often used very loosely. By definition rangeability is a measure of how much the range of an instrument can be adjusted and is defined as the ratio of the maximum flow range (maximum span) and the minimum span.

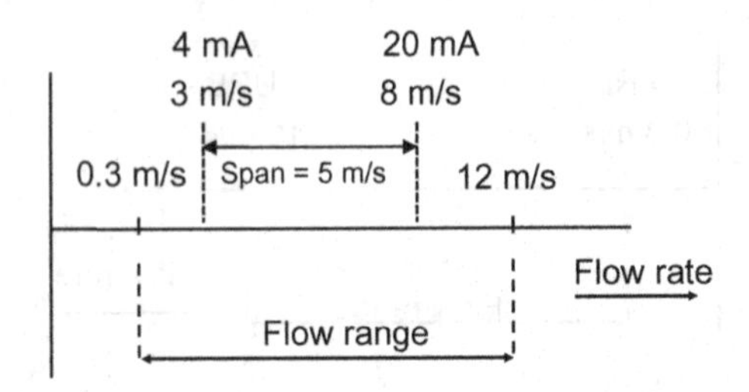

FIGURE 3.11 Span relates to the difference between the upper and lower range values assigned to the output signal.

However, it is also often confused with the turndown ratio and users should be careful as to what is actually meant when the terms are used.

3.8 FURTHER COMMONLY USED DEFINITIONS

Assume a marksman was firing at a target (Figure 3.12). This illustrates a number of commonly used definitions.

Although not hitting anywhere near the bull's-eye, the grouping was quite good. In other words, they were inaccurate but had high repeatability – if you like a high degree of precision.

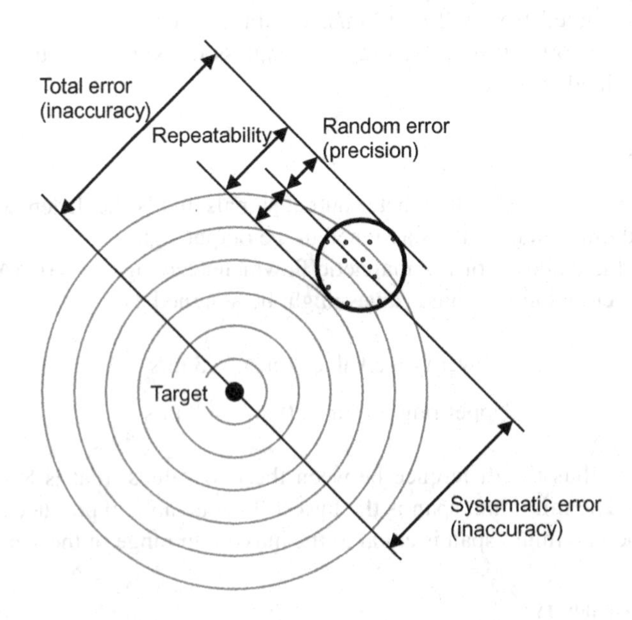

FIGURE 3.12 Marksman firing at a target illustrating a number of commonly used definitions.

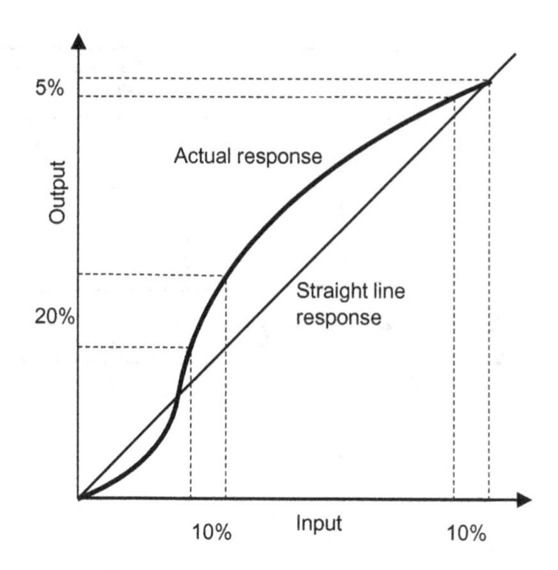

FIGURE 3.13 A 10% input near the top of the scale results in an output of only 5%. However, when lower down the scale, the same 10% input results in an output of 20%.

3.8.1 Repeatability

Repeatability is thus defined as how close a second measurement is to the first, under the same operating conditions and the same input. This measurement can also be an indication of instrument consistency.

In some cases, repeatability is considered more important than accuracy.

3.8.2 Linearity

Linearity is a term that may be used for instruments that give a reading approximately proportional to the true reading over their specified range. Linearity refers to the closeness within which the meter achieves a truly linear or proportional response – and is usually defined by stating the maximum deviation or nonconformity for example ±1% of flow rate, within which the response lies over a stated range.

The biggest danger associated with nonlinearity is that assumptions may be made judged on a single measurement. As illustrated in Figure 3.13, a 10% input near the top of the scale results in an output of only 5%. However, when lowering down the scale, the same 10% input results in an output of 20%.

3.9 DATA SPECIFICATIONS

Figure 3.14 illustrates a typical flowmeter envelope of a meter having a 10:1 turndown with an uncertainty of ±1% of reading at a confidence level of 95%.

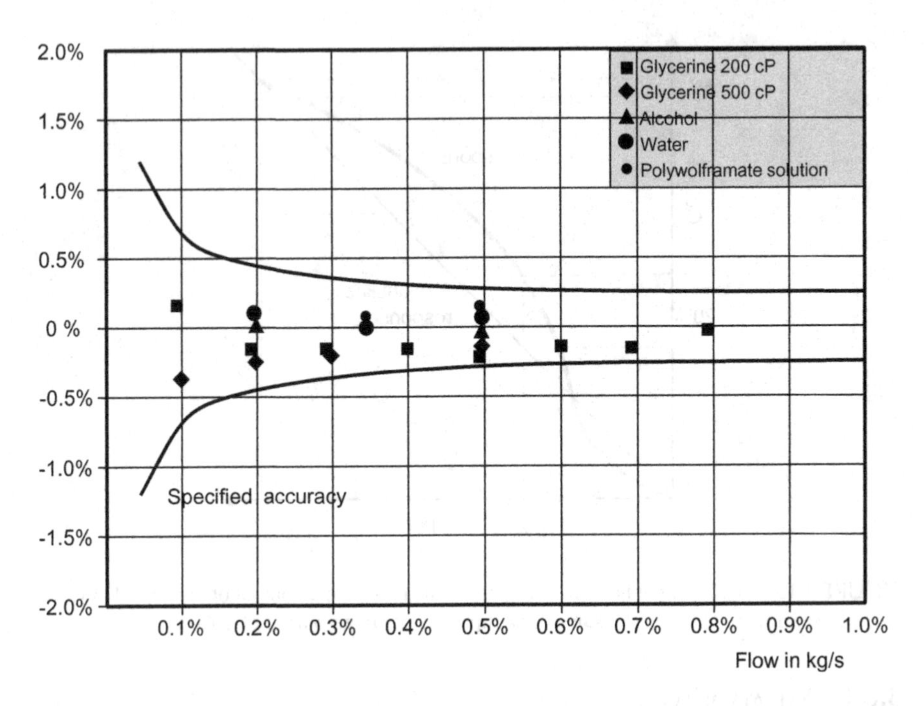

FIGURE 3.14 Typical flowmeter envelope of a meter having a 10:1 turndown with an uncertainty of ±1% of reading at a confidence level of 95%.

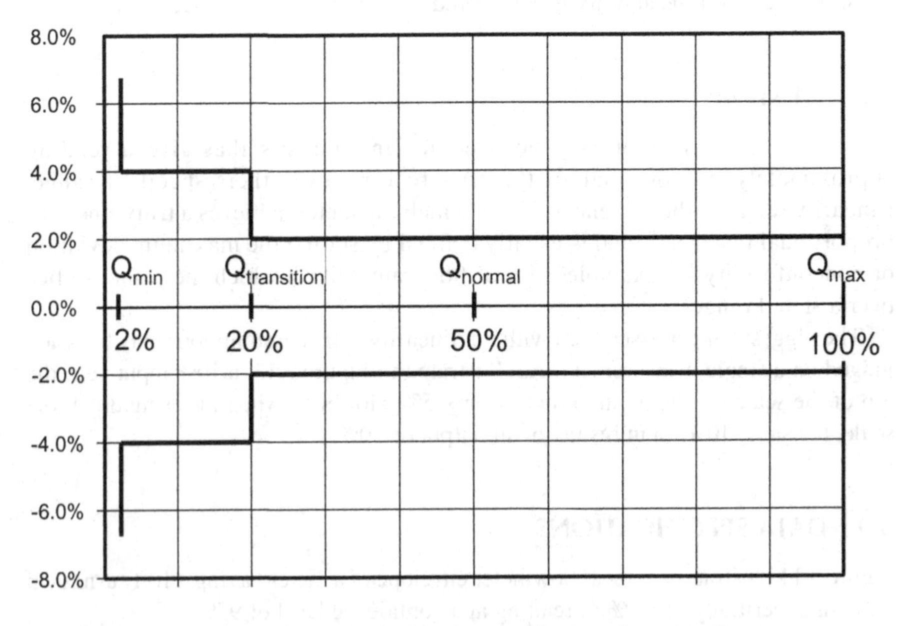

FIGURE 3.15 Envelope commonly used in the specification of water and gas meters.

The envelope also includes an indication of the uncertainty spread for a variety of different media. This is a reasonable performance for a flowmeter and would probably satisfy many requirements in industry.

Figure 3.15 shows another type of envelope that is particularly common in the specification of water and gas meters. Here, the meter has an uncertainty factor of ±2% of rate from full flow down to 20%. Below this value, the uncertainty is ±4% of flow rate, down to 2% of range.

In practice, a meter might have more steps in its envelope.

4 Flow Profile and Conditioning

4.1 INTRODUCTION

One of the most important fluid characteristics affecting the flow measurement is the shape of the velocity profile in the direction of flow.

In a frictionless pipe in which there is no retardation at the pipe walls, a flat 'ideal' velocity profile would result (Figure 4.1) in which all the fluid particles move at the same velocity.

4.2 LAMINAR FLOW PROFILE

We have already seen, however, that real fluids do not 'slip' at a solid boundary but are held to the surface by the adhesive force between the fluid molecules and those of the pipe. Consequently, at the fluid/pipe boundary, there is no relative motion between the fluid and the solid.

At low flow rates, the fluid particles move in straight lines in a laminar manner – with each fluid layer flowing smoothly past adjacent layers with no mixing between the fluid particles in the various layers. As a result the flow velocity increases from zero, at the pipe walls, to a maximum value at the centre of the pipe and a velocity gradient exists across the pipe. The shape of a fully developed velocity profile for such a *laminar flow* is parabolic, as shown in Figure 4.2, with the velocity at the centre equal to twice the mean flow velocity. Clearly, if not corrected for, this concentration of velocity at the centre of the pipe can compromise the flow computation.

4.3 TURBULENT FLOW PROFILE

One of the earliest investigators into fluid flow was Osborne Reynolds (1842–1912) who conducted a number of experiments using what is now termed a Reynolds instrument – a device that injects ink into the flow stream (Figure 4.3).

For a given pipe and liquid, as the flow rate increases, the laminar path of an individual particle of fluid is disturbed and is no longer straight. This is called the transitional stage (Figure 4.4).

As the velocity increases further the individual paths start to intertwine and cross each other in a disorderly manner so that thorough mixing of the fluid takes place. This is termed *turbulent flow*.

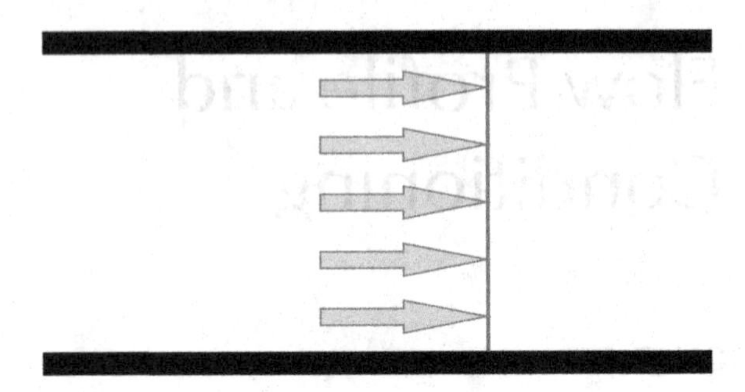

FIGURE 4.1 Flat 'ideal' velocity profile.

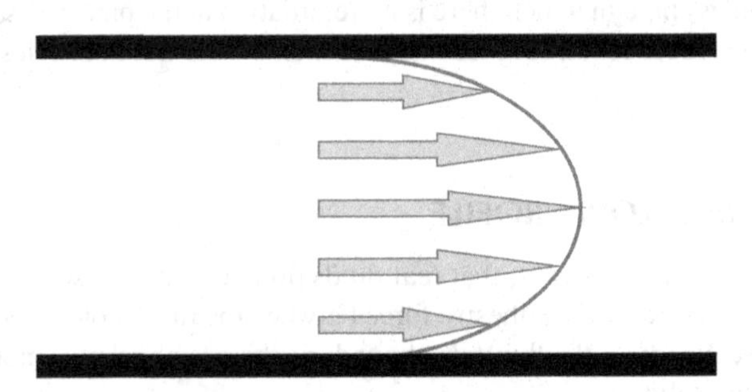

FIGURE 4.2 Laminar 'parabolic' velocity profile.

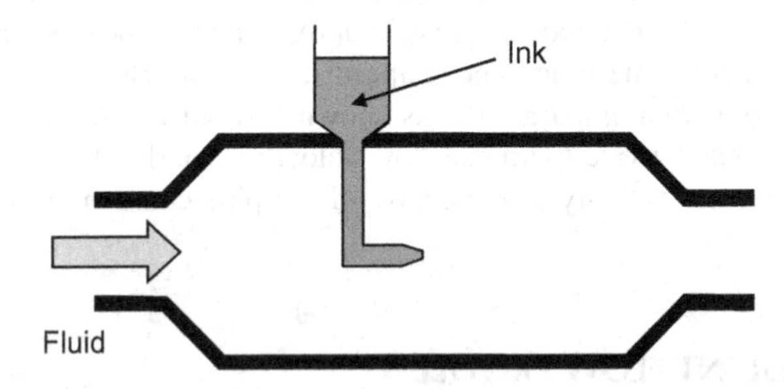

FIGURE 4.3 Reynolds's instrument injects ink into the flow stream in order to observe the flow regime. (Courtesy: Emerson.)

Since the flow velocity is almost constant in all of the pipe cross-section, the velocity profile for turbulent flow is flatter than for laminar flow and thus closer approximates the 'ideal' or 'one dimensional' flow (Figure 4.5).

The transition from a flat profile through laminar to turbulent is also illustrated in Figure 4.6 which shows how the influence of the boundary layer on the flow profile increases until it reaches the centre of the pipeline flow – at which point the flow profile is said to be *fully developed* or *turbulent*.

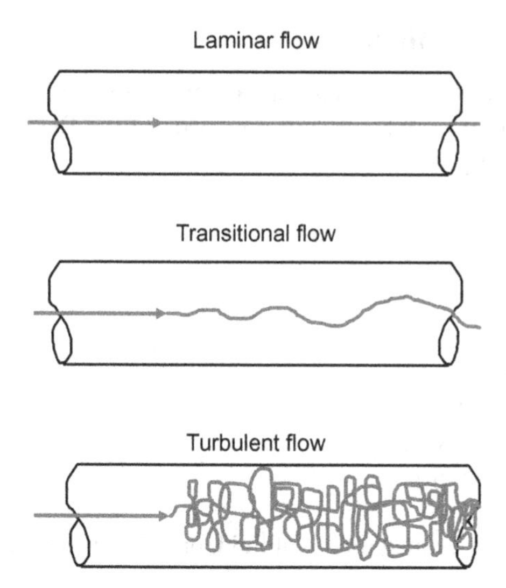

FIGURE 4.4 Transition from laminar through to turbulent flow.

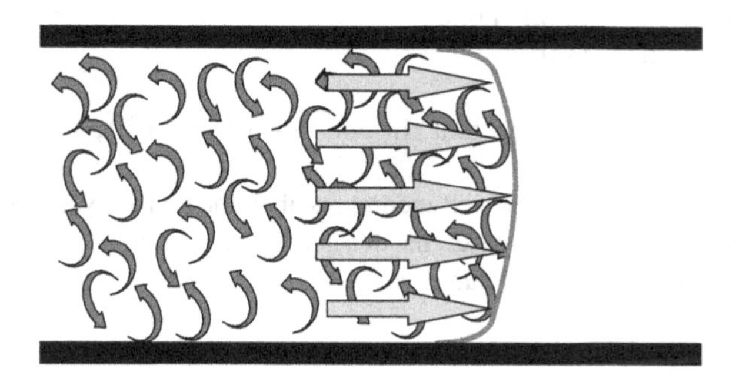

FIGURE 4.5 Turbulent velocity profile.

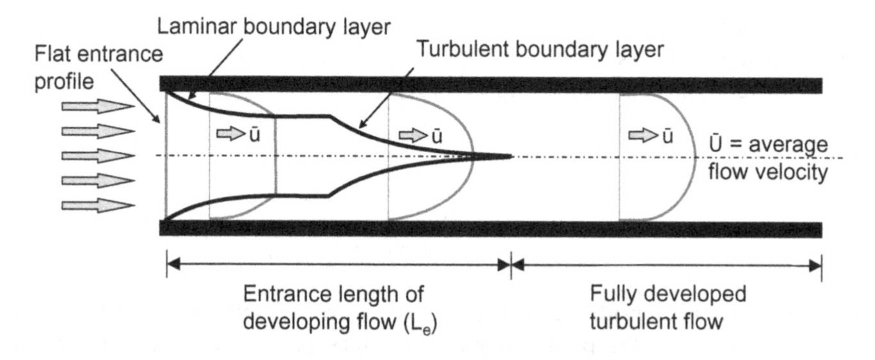

FIGURE 4.6 Influence of the boundary layer on the flow profile.

4.4 REYNOLDS NUMBER

► The onset of turbulence is often abrupt and to be able to predict the type of flow present in a pipe, for any application, use is made of the Reynolds number, Re – a dimensionless number given by:

$$Re = \frac{\rho \cdot v \cdot d}{\mu} \tag{4.1}$$

where

ρ = density of fluid (kg/m³)
μ = viscosity of fluid (Pa s)
v = mean flow velocity (m/s)
d = diameter of pipe (m)

Irrespective of the pipe diameter, type of fluid, or velocity, Reynolds showed that the flow is

Laminar: $Re < 2{,}000$
Transitional: $Re = 2{,}000–4{,}000$
Turbulent: $Re > 4{,}000$

From the foregoing it is seen that, in addition to viscosity, Re also depends on density.

Since most liquids are pretty well incompressible, the density varies only slightly with temperature. However, for gases, the density depends strongly on the temperature and pressure in which (for ideal gas):

$$PV = m \cdot R \cdot T \tag{4.2}$$

where

P = pressure (Pa)
V = volume of the gas (m³)
T = temperature (K)
m = number of moles
R = universal gas constant (8.315 J/mol K)

since:

$$\rho = \frac{m}{V} \frac{P}{R \cdot T} \tag{4.3}$$

Most gases may be considered ideal at room temperature and low pressures. Both, laminar and turbulent flow profiles require time and space to develop. At an entrance to a pipe, the profile may be very flat – even at low Re. And it may stay laminar, for a short time, even at high Re. ◄

4.5 DISTURBED FLOW PROFILE

The data supplied by most manufacturers is based on steady flow conditions and installation in long straight pipes both upstream and downstream of the meter. In practice, most meter installations rarely meet these idealised requirements – with bends, elbows, valves, T-junctions, pumps, and other discontinuities all producing disturbances that have an adverse effect on meter accuracy.

Such disturbed flow, which should not be confused with turbulent flow, gives rise to a number of effects that include:

Swirl – fluid rotation about the pipe axis
Vortices – areas of swirling motion with high local velocity which are often caused by separation or a sudden enlargement in pipe area
Asymmetrical profile – see Figure 4.7
Symmetrical profile with high core velocity – caused by a sudden reduction in pipe area

Ultimately the flow profile will be restored by the natural mixing action of the fluid particles as the fluid moves through the pipe. However, the effect of such disturbances can have an important bearing on accuracy for as much as 40 pipe diameters upstream of the measuring device. Figure 4.8 shows the ongoing disturbance in a pipe following a simple elbow.

Both swirl and distortion of the flow profile can occur – either separately or together. Research has shown that swirl can persist for distances of up to 100 pipe diameters (100D) from a discontinuity whilst in excess of 150 pipe diameters (150D) may be required for a fully developed flow profile to form.

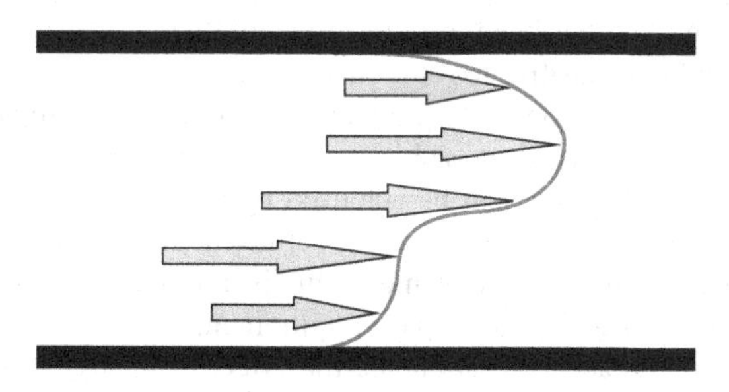

FIGURE 4.7 Asymmetric flow profile due to disturbed flow.

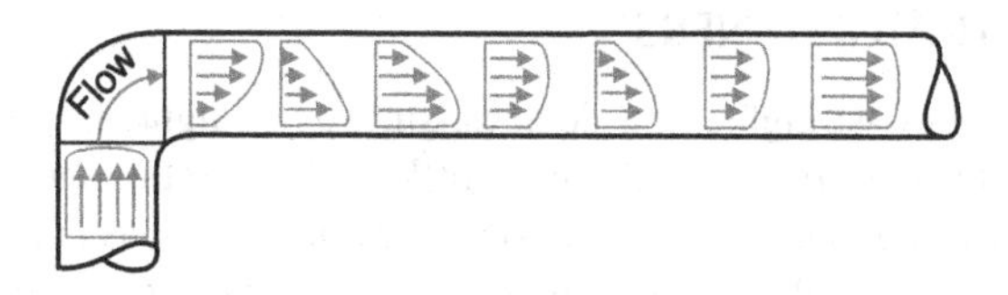

FIGURE 4.8 Ongoing disturbance in a pipe following a simple elbow.

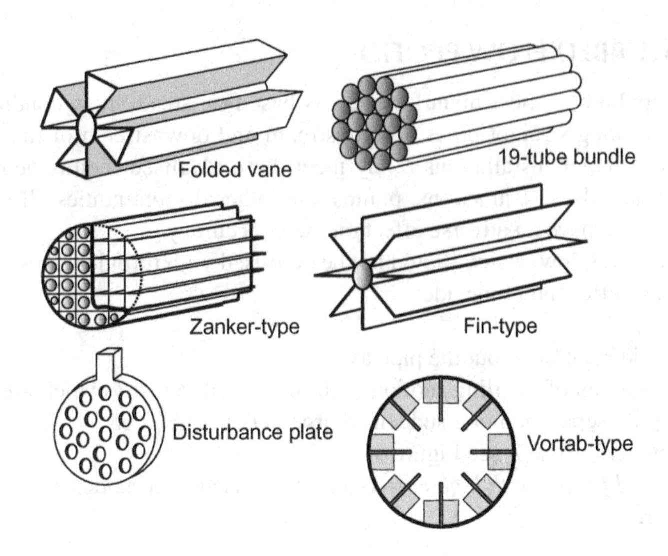

FIGURE 4.9 Some commonly used straighteners and conditioners.

4.6 FLOW CONDITIONING

The effect of disturbed flow on the measurement accuracy of the flow metering device depends very much on the metering technology being employed. For positive displacement and Coriolis mass flowmeters, the measurement is totally independent of the flow profile.

However, an inferential meter, such as the orifice plate, requires that the output signal should represent the average fluid velocity occurring over the complete area of the meter tube. On the other hand, rotating-type inferential meters, such as turbine meters, are particularly sensitive to rotational swirl.

Consequently, most flowmeters require an axi-symmetrical, fully developed flow profile, with a zero swirl, in order to achieve an acceptable accuracy and consistency. And while this desired profile can be achieved by ensuring that there are at least 25–40 pipe diameters of straight run piping before the flow element and about 4–5 pipe diameters downstream of the element, this is not always practical in terms of higher installation costs and greater space requirements.

An alternative approach is to attenuate the effect of upstream disturbances by the use of flow straighteners or flow conditioners. Figure 4.9 illustrates some of the commonly used straighteners/conditioners.

4.7 FLOW STRAIGHTENERS

Generally, whilst flow-straightening devices, such as straightening vanes and tube bundles, eliminate swirl, they actually have little or no effect on the velocity profile.

Assuming the straightener has equally sized entrance holes, as in the case of a tube bundle, the distorted profile entering the device will have exactly the same profile once it exits the straightener. And although the friction due to the wall roughness of

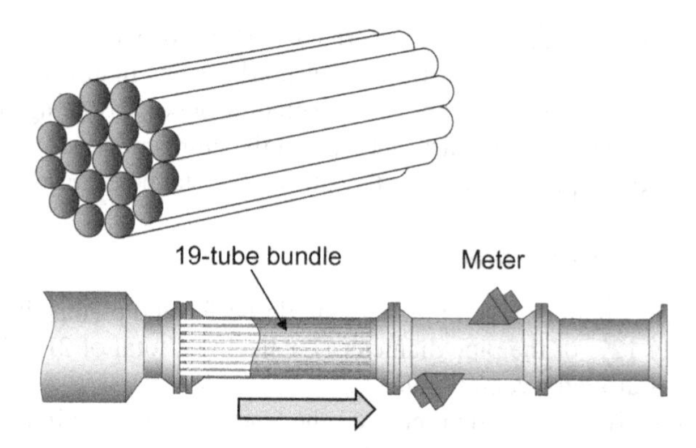

FIGURE 4.10 Majority of gas flow installations in North America still make use of the 19-tube bundle flow conditioner.

the individual tubes would eventually restore the developed flow profile, if they were long enough, they are typically only two to three pipe diameters (2D to 3D) long. Consequently, once the fluid leaves the pipe bundle and the flows recombine, they tend to produce a single highly distorted profile and the tube bundle itself becomes an installation effect.

Another assumption is that because a tube bundle has a higher open area than a typical disturbance plate, it would have a lower pressure drop. In reality, because the flow passes over both the inside and outside surfaces of the tubes, a tube bundle has a significantly higher surface area than a disturbance plate. Typically, a 25 mm thick disturbance plate has a surface area of around 830 cm^2 compared with an AGA3 19-tube bundle having a surface area of more than 1,800 cm^2.

Despite the fact that numerous studies (Karnick et al., 1994; Morrow and Park, 1992, and others) are now available providing performance results indicating less than acceptable meter performance when using the conventional 19-tube bundle, the majority of gas flow installations in North America still make use of the 19-tube bundle flow conditioner (Figure 4.10). Indeed, general indications are that the tube bundle will cause an orifice installation to over-register flow values up to 1.5% when the tube bundle is 1 pipe diameter to approximately 11 pipe diameters from the orifice plate. And slight under-registration of flows occurs for distances between approximately 15 and 25 pipe diameters.

4.8 FLOW CONDITIONING PLATES

Most of the modern solutions to flow conditioning centres on conditioning plates – often used in combination with a straightening technology and expansion chambers.

A typical conditioning plate comprises a solid metal disc, usually 25–50 mm thick, having a large number of circular passages arranged in a specific geometric pattern. A graded flow resistance is achieved by varying the number, spacing and size of the holes – designed to ensure a fully developed profile is produced downstream of the conditioner.

Some of the earlier designs of what are termed 'perforated plates' can be found in the proposals from Laws, Spearman (NEL), and Mitsubishi Heavy Industries (MHI), which focus on such parameters as overall porosity, the grading of porosity along the radius, the wetted perimeter, the perforation distribution, and the number and size of holes in the plate. In this regard, the porosity refers to the ratio of the total hole area to the plate area.

4.8.1 Laws Flow Conditioner

The Laws device (Figure 4.11) comprises a single-perforated plate, with a thickness of approximately 12% of the pipe diameter (D), having a total of 21 holes arranged in circular arrays around a central hole.

The apertures (d1–d3) are chamfered on the leading edge – with their diameters specified in terms of the plate diameter (D):

$d1 = 0.192D$
$d2 = 0.1693D$
$d3 = 0.1462D$

Here, the ratio of the total hole area to the plate area (the porosity) may thus be calculated at just on 51.5% and a pressure loss of about 0.8–2 velocity heads – the head loss of a pipe is the same as that produced in a straight pipe having a length equal to the pipes of the original systems.

4.8.2 Spearman (NEL)

The National Engineering Laboratory (NEL) flow conditioner (Figure 4.12), introduced by E. P. Spearman, comprise a single-perforated plate with 16 holes in an outer ring, eight holes in an inner ring and four holes arranged in a square at the centre.

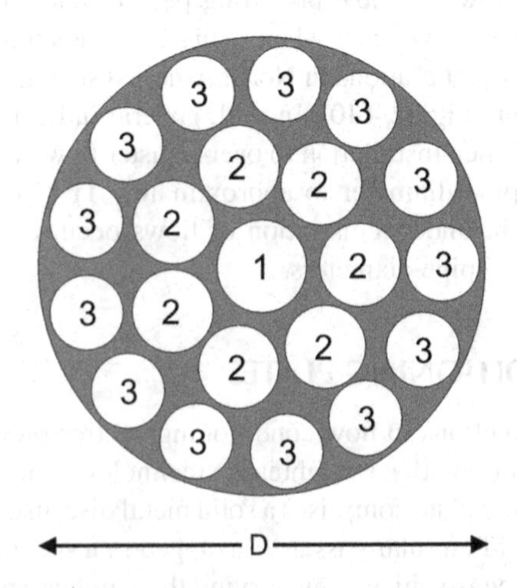

FIGURE 4.11 Laws conditioning plate comprises a single-perforated plate, having a total of 21 holes arranged in circular arrays around a central hole.

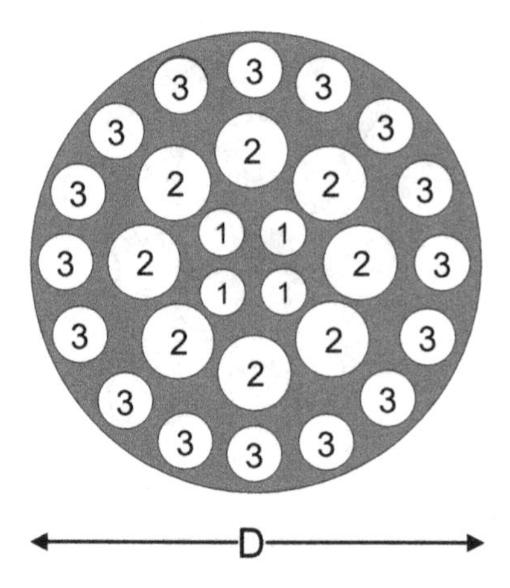

FIGURE 4.12 Spearman (NEL) conditioning plate comprises a single-perforated plate, having a total of 28 holes.

The diameters of the plate apertures (d1–d3) are again specified in terms of the plate diameter (D):

d1 = 0.1D
d2 = 0.16D
d3 = 0.12D

giving a porosity of 47.5% and a pressure loss of about 2.9 velocity heads. There is limited scope for reducing the pressure loss by increasing the plate porosity, without the central four holes interfering with the holes in the inner ring.

4.8.3 Mitsubishi Heavy Industries (MHI) Conditioner

The MHI conditioner, (Figure 4.13) comprises a single-perforated plate with 35 holes positioned so that the resistance is graded across the conditioner to allow more flow through the centre than at the edge – assisting in the production of a fully developed flow profile.

The hole size and plate thickness are both 13% of the pipe diameter (D) – giving a porosity of approximately 59%. Depending on whether a chamfer is used, the head loss coefficient is in the range of 1–1.7.

4.8.4 Zanker Flow Conditioner Plate

The Zanker flow conditioner plate (Figure 4.14) comprises 32 bored holes arranged in a symmetrical circular pattern having a thickness of 0.12D.

The diameters of the plate apertures (d1–d5) are specified in terms of the plate diameter (D):

d1 = 0.141D four holes
d2 = 0.139D eight holes

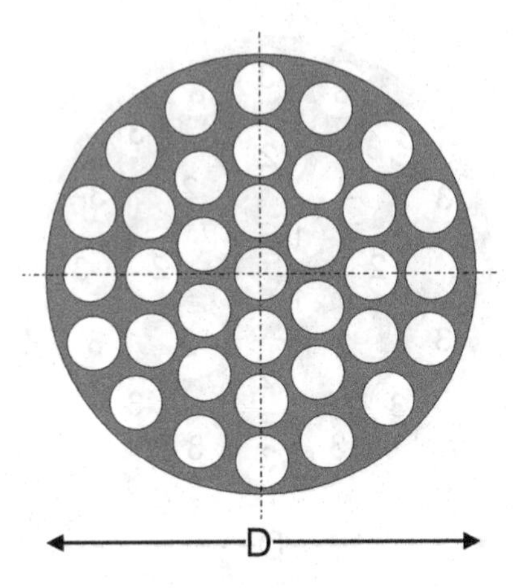

FIGURE 4.13 MHI conditioner comprises a single-perforated plate with 35 equally sized holes.

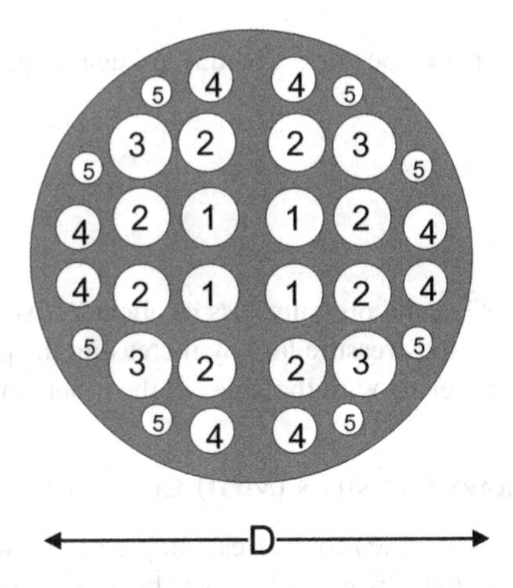

FIGURE 4.14 Zanker flow conditioner plate comprises 32 bored holes arranged in a symmetrical circular pattern.

d3 = 0.136D four holes
d4 = 0.110D eight holes
d5 = 0.077D four holes

giving a porosity of approximately 43% and a head loss coefficient around 2.

4.9 FLOW CONDITIONERS

All of the foregoing devices condition the flow with varying degrees of success. Indeed, research indicates that a flow-conditioning plate still requires downstream lengths of 20D or more to develop a fully developed flow regime. Furthermore, swirl may still be an issue. Consequently, flow-conditioning plates are often combined with flow-straightening

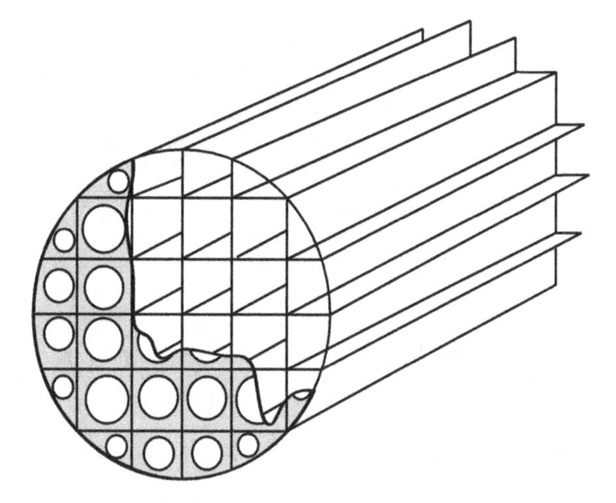

FIGURE 4.15 In the Zanker flow conditioner the flow plate is followed by an egg-box honeycomb arrangement.

devices in order to produce satisfactory results. Whilst many of these are now in the public domain as a result of patent expiration (e.g. Zanker, Sprenkle), many of the modern systems are proprietary in nature (e.g. Vortab, CPA, Gallagher, Emerson et al).

4.9.1 Zanker Flow Conditioner

In the Zanker flow conditioner, the Zanker flow plate is associated with an egg-box honeycomb arrangement (Figure 4.15) in which the plate itself is followed by a number of channels (one for each hole) formed by the intersection of a number of thin plates. Whilst thus providing protection against swirl, the pressure loss coefficient increases approximately to 5.

4.9.2 Sprenkle Flow Conditioner

The Sprenkle conditioner (Figure 4.16) comprises three perforated plates in series held together by studs – with 0.1D spacing between successive plates. The holes are normally chamfered at 45° on the upstream side to reduce the pressure loss.

Typically, the hole diameter is 0.05D (the same as the plate thickness) and the number of holes optimised to provide a porosity of at least 40%.

The pressure loss coefficient is approximately 11 with an inlet bevel or 14 without.

4.10 PROPRIETARY FLOW CONDITIONERS

4.10.1 Vortab Conditioner

Many of the flow-conditioning devices discussed previously incur significant pressure losses – typically up to 5 or more head loss coefficients. This has both power and cost implications that can deter users from the use of such devices. A major feature of the Vortab-type flow conditioner (Figure 4.17) is a head loss coefficient of only

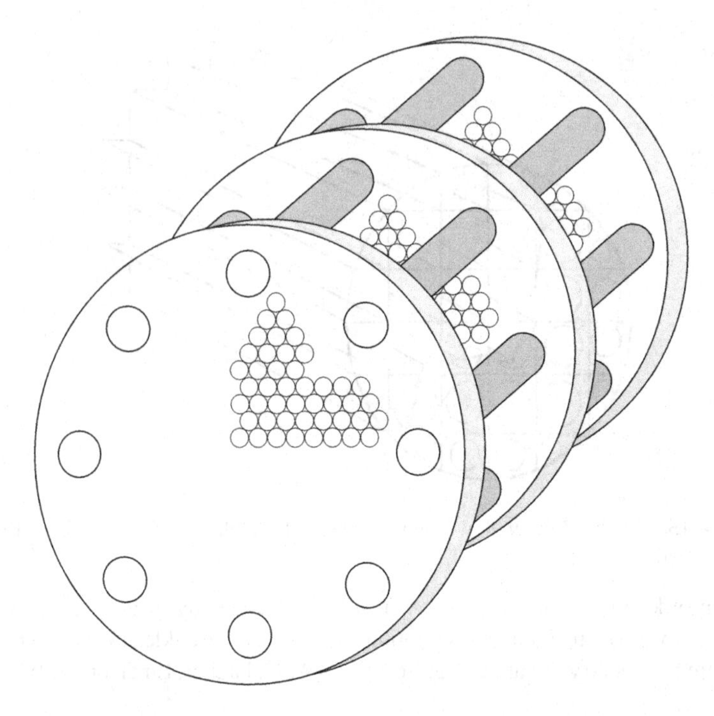

FIGURE 4.16 Sprenkle conditioner comprises three perforated plates in series held together by studs.

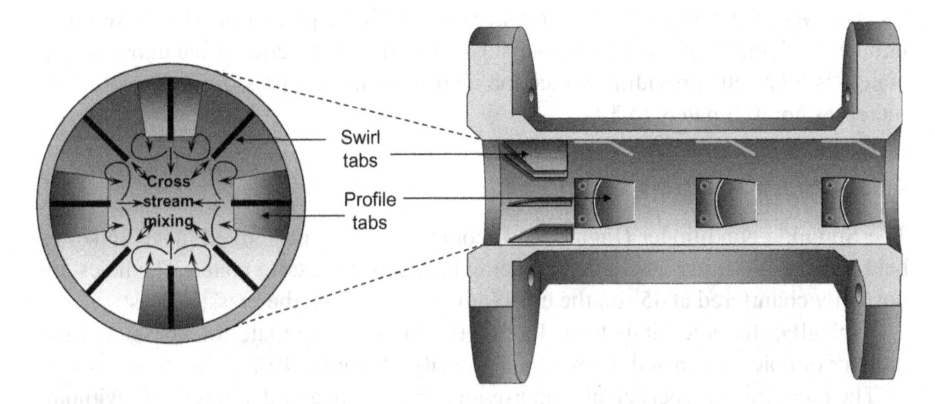

FIGURE 4.17 Vortab-type short-run flow conditioner uses a series of radial and inclined traverse tabs to generate a series of vortices. (Courtesy: Vortab Company.)

0.7. Although available in several different styles (insertion, meter run, short run, and elbow) the basic elements are similar in that they use a series of radial and inclined traverse tabs to generate a series of vortices. The inlet comprises eight swirl reduction tabs uniformly spaced around of the pipe's circumference that produce essentially non-swirling flow.

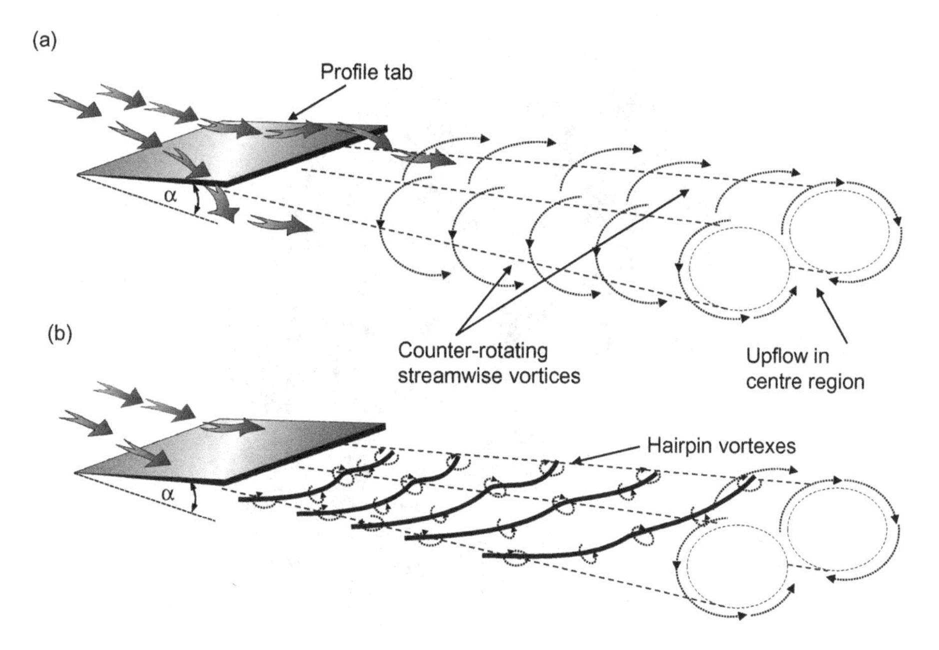

FIGURE 4.18 Generation of (a) counter-rotating stream and (b) hairpin vortices.

The swirl reduction tabs are followed by a series of profile tabs that produce counter-rotating streamwise vortices (Figure 4.18a). In addition the flowing media is pulled towards the pipe wall by the sharp pressure gradient generated by the profile tabs.

This generates an upflow in the centre region of the counter-rotating streamwise vortices that spiral downstream. A further effect is due to the high-shear region at the edge of the tabs which shed transient hairpin-like vortices (Figure 4.18b) that merge to produce a homogeneous fully developed velocity profile in three pipe diameters.

4.10.2 CPA 50E Flow Conditioner

From a commercial standpoint the CPA 50E flow is one of the two generally accepted flow conditioner brands used throughout the Unites States.

The conditioner comprises a steel plate, 0.125D–0.15D in thickness, having a central circular hole and two rings of circular holes in concentric circles around it (Figure 4.19). This design produces a fully developed profile downstream of the conditioner – with the holes accounting for approximately 50% of the plate area. The result is that the velocity profile even a short distance downstream is fully developed. The thickness of the plate also eliminates swirl.

4.10.3 Gallagher Flow Conditioner

The Gallagher flow conditioner comprises an anti-swirl device, a settling chamber and a profile device mounted sequentially in the pipe.

FIGURE 4.19 CPA 50E flow conditioner comprises a steel plate, 0.125D–0.15D in thickness, having a central circular hole and two rings of circular holes in concentric circles around it. (Courtesy: Canada Pipeline Accessories.)

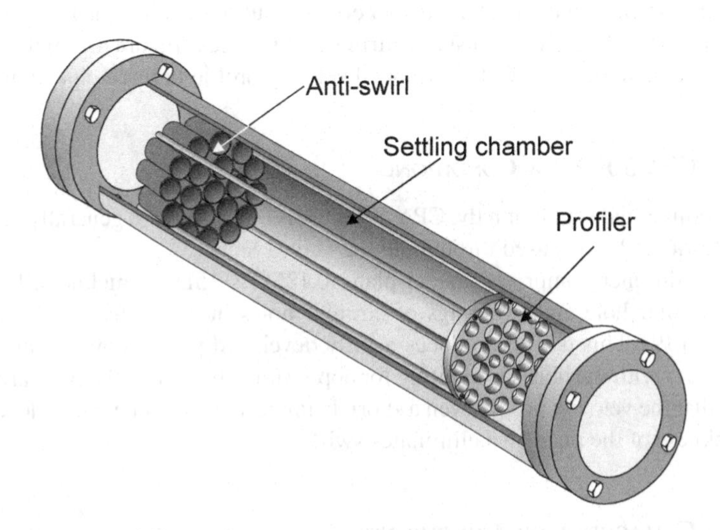

FIGURE 4.20 Gallagher flow conditioner comprises an anti-swirl device, a settling chamber and a profile device mounted sequentially in the pipe. (Courtesy: Savant Measurement Corporation.)

The anti-swirl device may take the form of a short length of tube bundle (Figure 4.20) or anti-swirl vanes. In applications where the swirl level is less than 15°, the anti-swirl device may be omitted with no loss in performance.

The settling chamber, or isolation chamber, minimises any possible interaction between the anti-swirl device and profile device.

The profile device consists of a flat steel plate with a number of circular holes arranged in concentric circles.

5 Positive Displacement Meters

5.1 INTRODUCTION

Positive displacement meters (sometimes referred to as direct-volumetric totalisers) all operate on the general principle where defined volumes of the medium are separated from the flow stream and moved from the inlet to the outlet in discrete packages.

Totalising the number of packages provides the total volume passed and the total volume passed in a given time provides the flow rate for example litres/min. Because they pass a known quantity, they are ideal for certain fluid batches, blending and custody transfer applications. They give very accurate information and are generally used for production and accounting purposes.

5.2 SLIDING VANE

Used extensively in the petroleum industry for gasoline and crude oil metering, the sliding vane meter comprises a rotor assembly fitted with four blades in opposing pairs – with each pair mounted on rigid spacing rods (Figure 5.1). The rotor is mounted on a shaft that is eccentric to the centre of the meter chamber.

As the liquid enters the measuring chamber, the pressure on the exposed portion of the vane causes the rotor to turn. As the blades rotate, they are guided onto the measuring crescent to form an efficient seal. The fluid trapped between the two vanes provides an accurate measure of the volume.

This process is repeated, without pulsations, as the vanes move around the measuring chamber – with 'packets' of fluid trapped and passed to the outlet manifold as discrete known quantities of fluid.

A mechanical counter register or electronic pulse counter is attached to the shaft of the rotor so that the flow volume is directly proportional to the shaft rotation.

Close tolerances and carefully machined profiles of the casing ensure that the blades are guided smoothly through the measuring crescent to give high performance.

5.2.1 Advantages of the Sliding Vane Meter Include

- Suitable for accurately measuring small volumes
- High accuracy of ±0.2%
- High repeatability of ±0.05%
- Turndown ratio of 20:1
- Suitable for high-temperature service, up to 180°C
- Pressures up to 7 MPa
- Not affected by viscosity

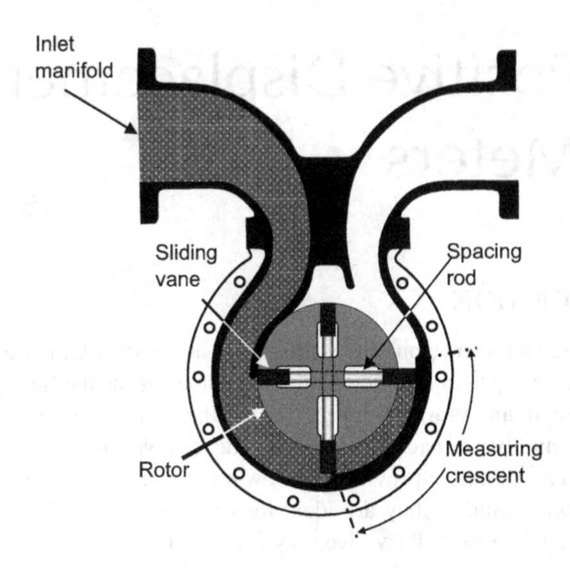

FIGURE 5.1 Sliding vane positive displacement meter comprising a rotor assembly fitted with four blades in opposing pairs – with each pair mounted on a rigid-spacing rod. (Courtesy: Avery-Hardoll.)

5.2.2 Disadvantages Include

- Suitable for clean liquids only
- Limitations due to leakage
- High unrecoverable pressure loss

5.3 ROTARY VANE

Although there are several versions of this type of meter, the technology is typified by the Smith Meter rotary vane meter. As illustrated in Figure 5.2, the housing contains a rotor that revolves and carries evenly spaced blades. As the liquid flows through the meter, the rotor and blades (vanes) revolve about a fixed cam – causing the blades to move outwards.

This successive movement of the blades forms a measuring chamber of precise volume between two of the blades, the rotor, the housing, the bottom, and the top covers – with a continuous series of closed chambers produced for each rotor revolution.

Neither blades nor the rotor is in contact with the stationary walls of the measuring chamber. The liquid within the small gap between the blades and the chamber wall forms a capillary seal. However, a small amount of liquid flow (called 'slippage') can occur through these clearances – determined by the differential pressure across the meter and the medium viscosity. Consequently, as shown in Table 5.1, there is a trade-off between the minimum flow rate and the viscosity – with the minimum flow rate decreasing as viscosity increases.

The flow is virtually undisturbed while it is being metered – thus energy is not wasted by unnecessary hydraulic bending of the liquid.

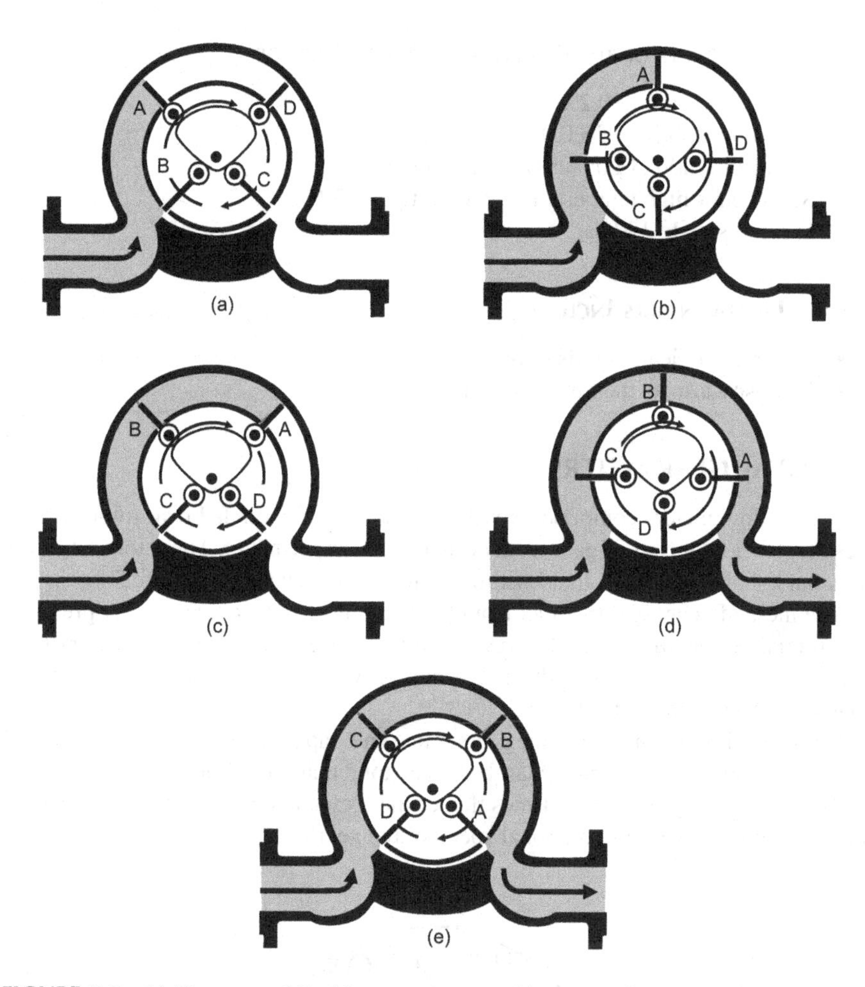

FIGURE 5.2 (a) Unmeasured liquid enters meter – with rotor and vanes turning clockwise. Vanes A and D are fully extended forming the measuring chamber; while vanes B and C are retracted. (b) After the rotor and vanes have made a further one-eighth revolution: Vane A is fully extended; Vane B is partially extended; Vane C is fully retracted; Vane D is partially retracted. (c) A quarter revolution has now been made: Vane A is still extended and Vane B is now fully extended. An exact and known volume of new liquid is now in the measuring chamber. (d) One-eighth revolution later, the measured liquid is moving out of the meter. A second measuring chamber is being created between Vanes B and C. Vane A has begun to retract and Vane C is beginning to extend. (e) In three-eighths of a revolution, one measured segment of fluid has been passed and a second segment is being created. This cycle is repeated as long as liquid flows.

TABLE 5.1
Trade-off between the Minimum Flow Rate and Viscosity

Viscosity (cP)	400	100	20	5	1	0.5
Minimum flow rate (L/min)	0.3	1.25	6	25	60	100

5.3.1 Advantages of the Rotary Vane Meter Include

- High accuracy of ±0.2%
- High repeatability of ±0.02%
- Turndown ratio of 20:1 (dependent on viscosity)
- Suitable for high temperature service, up to 95°C
- Pressures up to 150 bar

5.3.2 Disadvantages Include

- Suitable for clean liquids only
- Viscosity affects the turndown ratio

5.4 OVAL GEAR METERS

Oval gear flow meters comprise two identical precision moulded oval rotors which mesh together by means of gear teeth around the gear perimeter. The rotors rotate on stationary shafts which are fixed within the measuring chamber (Figure 5.3).

The meshed gears seal the inlet from the outlet flow, developing a slight pressure differential across the meter that results in the movement of the oval rotors (see Figure 5.4a–d). This alternate driving action provides a smooth rotation of almost constant torque without dead spots.

With flow through the meter, the gears rotate and trap precise quantities of liquid in the crescent-shaped measuring chambers. The total quantity of flow for one rotation of the pair of oval gears is four times that of the crescent-shaped gap and the rate of flow is proportional to the rotational speed of the gears.

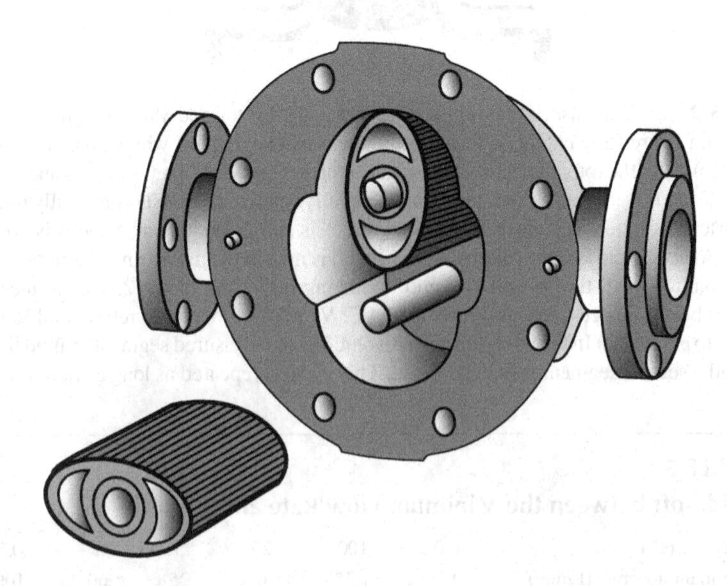

FIGURE 5.3 Construction of the oval gear meter. (Courtesy: Emerson.)

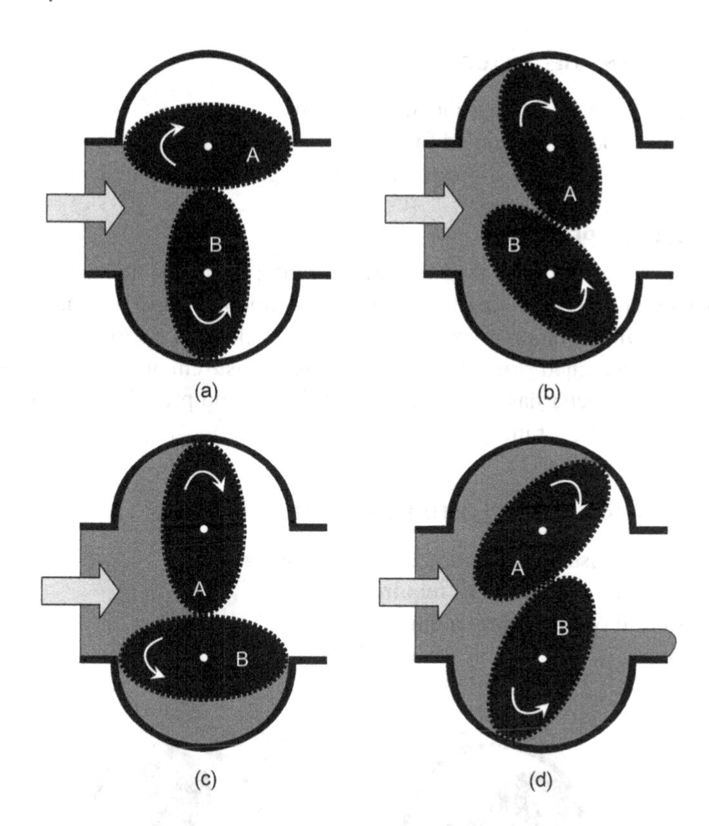

FIGURE 5.4 (a) Gear A receives torque from the pressure difference. The net torque on Gear B is zero. (b) Gear A drives Gear B. (c) Gear B traps a defined quantity of fluid. The net torque on Gear A is zero. Gear B receives torque from the pressure difference. (d) Gear B drives Gear A. Defined quantity of fluid passes to the outlet.

Because the amount of slippage between the oval gears and the measuring chamber wall is minimal, the meter is essentially unaffected by the changes in the viscosity and lubricity of the liquids.

The major disadvantage of this meter is that the alternate driving action is not constant and, as a result, the meter introduces pulsations into the flow.

Newer designs of this type of meter use servomotors to drive the gears. These eliminate the pressure drop across the meter and the force required to drive the gear. This applies mainly to smaller-sized meters and significantly increases the accuracy at low flows.

5.4.1 Advantages of the Oval Gear Meter Include

- High accuracy of ±0.25%
- High repeatability of ±0.05%
- Low-pressure drop of less than 20 kPa
- High operating pressures, up to 10 MPa
- High temperatures, up to 300°C
- Wide range of materials of construction

5.4.2 Disadvantages Include

- Pulsations caused by alternate drive action
- Generally not recommended for water or low viscosity fluids because of the increased risk of fluid slippage between the gears and chamber walls

5.5 LOBED IMPELLER

Similar in operation to the oval meter, the lobed impeller type meter (Figure 5.5) is a non-contact meter comprising two high precision lobed impellers that are geared externally and which rotate in opposite directions within the enclosure.

Essentially designed for high-volume gas measurement up to 1,000 m^3/h (30,000 ft^3/h), the meter has a repeatability of ±0.1% and provides accuracies of up to ±1% over a 10:1 flow range.

5.5.1 Advantages of the Lobed Impeller Meter Include

- Excellent measuring accuracy for gas measurements
- No inlet and outlet sections required
- No external power supply required
- Low-pressure drop – typically 0.7 kPa (0.1 psi)

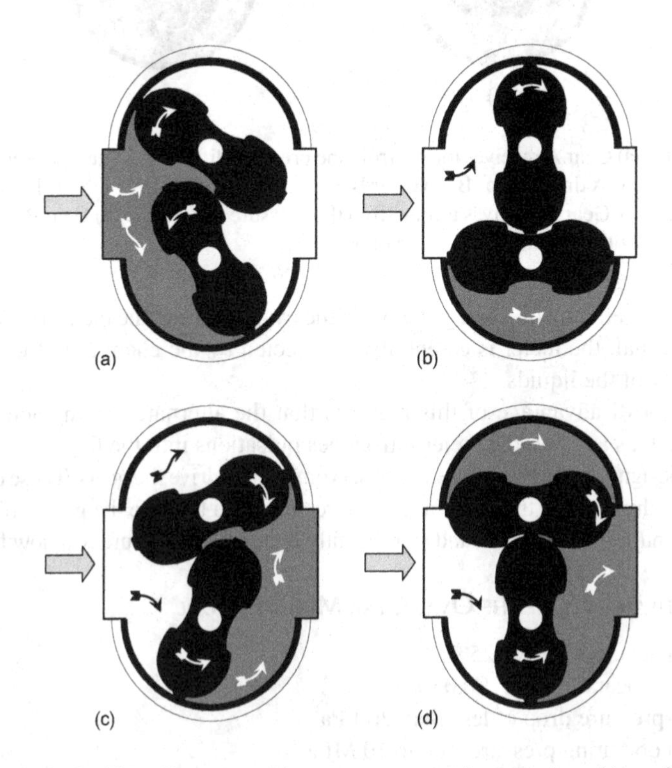

FIGURE 5.5 Lobed impeller meter: (a) gas enters measuring chamber, (b) precise quantity of gas trapped by lower impeller, (c) gas passed on to exit, (d) precise quantity of gas trapped by upper impeller. (Courtesy: Roots-Dresser.)

5.5.2 Disadvantages Include

- Pulsations caused by alternate drive action
- Temperature of process medium limited to about 60°C
- Wear due to moving parts

5.6 OSCILLATING PISTON

Widely used in the water industry, the oscillating or the rotating piston meter consists of a stainless-steel housing and a rotating piston as shown in Figure 5.6. The only moving part in the measuring chamber is the oscillation piston which moves in a circular motion.

To obtain an oscillating motion, the movement of the piston is restricted in two ways. First, the piston is slotted vertically to accommodate a partition plate which is fixed to the chamber. This plate prevents the piston from spinning around its central axis and also acts as a seal between the inlet and outlet ports of the chamber. Second, the piston has a centre vertical pin which confines the piston's movement to a circular track which is part of the chamber.

The differential pressure across the meter causes the piston to sweep the chamber wall in the direction of flow – displacing the liquid from the inlet to the outlet port in a continuous stream.

The openings for filling and discharging are located in its base and thus in Figure 5.7a, areas 1 and 3 are both receiving liquid from the inlet port (A) and area 2 is discharging through the outlet port (B).

In Figure 5.7b, the piston has advanced and area 1, which is connected to the inlet port, has enlarged; and area 2, which is connected to the outlet port, has decreased, while area 4, is about to move into position to discharge through the outlet port.

In Figure 5.7c, area 1 is still admitting liquid from the inlet port, while areas 2 and 3 are discharging through the outlet port. In this manner known discrete quantities of the medium have been swept from the inlet port to the outlet port.

The movement, and therefore the rotation, of the hub is normally sensed by a single magnet in the central spindle. Alternatively, a series of external magnets arranged allow the rotation to be detected through the meter wall thereby increasing the meter's resolution.

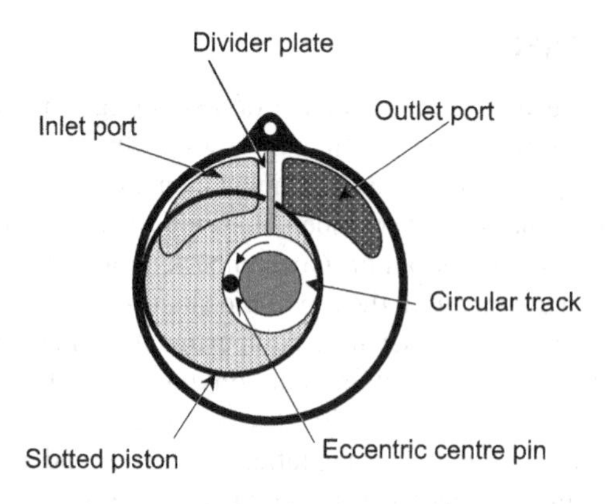

FIGURE 5.6 Basic layout of the oscillating or rotating piston meter.

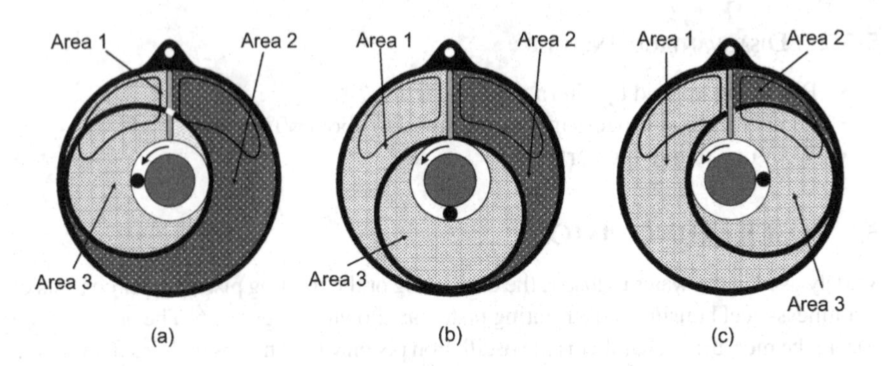

FIGURE 5.7 Oscillating or rotating piston meter showing the principle of operation.

The rotating piston meter is particularly suitable for accurately measuring small volumes.

5.6.1 Advantages of the Rotating Piston Include

- Accuracy of ±0.5%
- Performance largely unaffected by viscosity (from heating oil to paste)
- Insensitive to mounting orientation
- Measurement of low and high viscosity liquids
- Only one moving part (oscillating piston)
- No stagnate chambers to accumulate contaminants or stale products
- Extremely compact and comparatively light weight
- High turndown ratio for example 100:1 for water – increasing for higher viscosities for example 3,000:1 at 250 cSt

5.6.2 The Main Disadvantages of the Oscillating Piston Meter are

- Depending on the choice of materials, the piston may be subject to rapid wear
- Can produce pulsations

5.7 NUTATING DISC

The term nutation is derived from the action of a spinning top whose axis starts to wobble and describe a circular path as the top slows down.

In a nutating disc-type meter the displacement element is a disc that is pivoted in the centre of a circular measuring chamber (Figure 5.8). The lower face of the disc is always in contact with the bottom of the chamber on one side, and the upper face of the disc is always in contact with the top of the chamber on the opposite side. The chamber is therefore divided into separate compartments of known volume.

Liquid enters through the inlet connection on one side of the meter and leaves through an outlet on the other side – successively filling and emptying the compartments and moving the disc in a nutating motion around a centre pivot. A pin attached to the disc's pivot point drives the counter gear train.

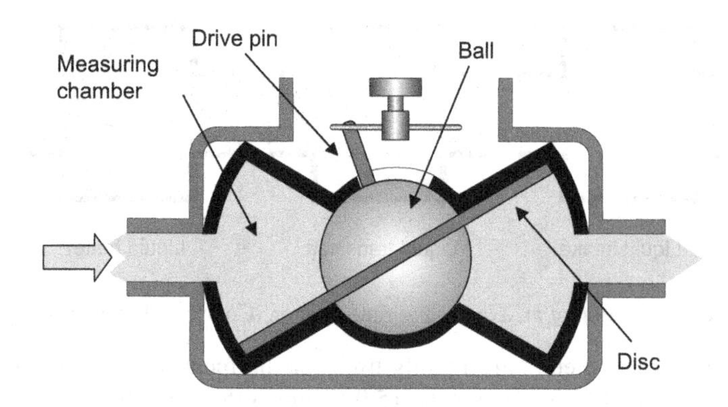

FIGURE 5.8 Nutating disc meter in which the displacement element is a disc pivoted in the centre of a circular measuring chamber.

Although there are inherently more leakage paths in this design, the nutating disc meter is also characterised by its simplicity and low cost.

It tends to be used where longer meter life, rather than high performance, is required for example domestic water service. The meter is also suitable for use under high temperatures and pressures.

5.8 AXIAL-FLUTED ROTOR METERS

The axial-fluted rotor meter (Figure 5.9) makes use of two spiral-fluted rotors working within the same measuring chamber. Axial flow of medium causes the pair of spindles to rotate and form volumetrically defined measuring chambers that are a measure of the delivered volumetric flow in a uniform, non-pulsating manner. Rotation is sensed by inductive proximity switches.

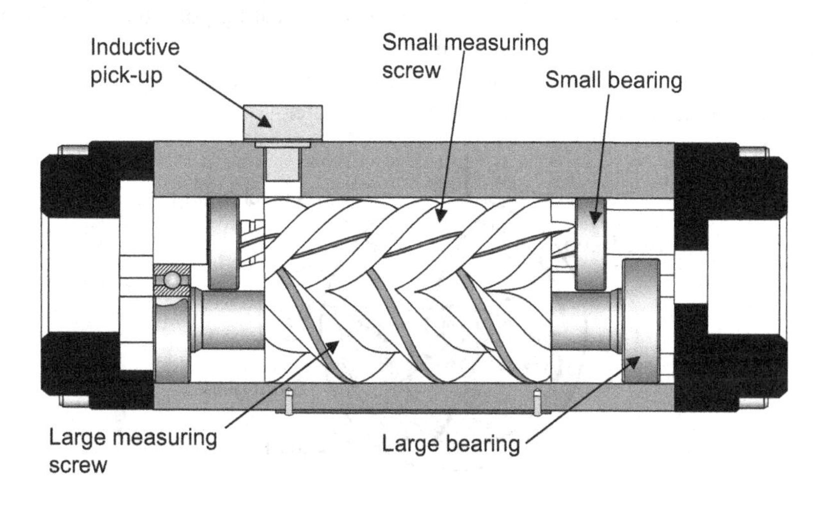

FIGURE 5.9 Physical construction of a typical axial radial-fluted meter. (Courtesy: Kobold.)

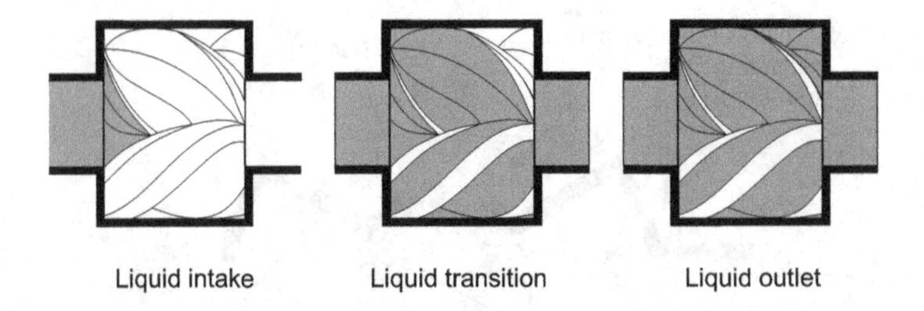

FIGURE 5.10 Operation of an axial radial-fluted meter. (Courtesy: Emerson.)

Axial-fluted rotor meters are mainly used for the measurement or batching of highly viscous non-abrasive media up to 5,000 mm^2/s (5×10^9 cSt).

As the product enters the intake of the measuring unit chamber (Figure 5.10), the two rotors divide the volume being measured into segments; momentarily separating each segment from the flowing inlet stream and then returning them to the outlet of the measuring unit chamber.

5.9 WET-TYPE GAS METERS

The wet-type gas meter (Figure 5.11) comprises a gas-tight casing containing a measuring drum, with four separate compartments, mounted on a spindle that is free to revolve. The casing is filled to approximately 60% of its volume with water or light oil.

Under normal operation, the gas passes through the measuring drum so that each compartment of the drum must, in turn, be emptied of water and filled with gas – thus forcing the drum to rotate. In an alternative arrangement, the gas is introduced into the space above the water in the outer casing and then passes through the drum to the outlet of the meter.

The calibration of the measuring drum (i.e. the quantity of gas passed for each revolution) is determined by the height of the water in the casing. Consequently, the normal calibration point is shown by a water-level indicating point that is visible in the sight box located on the side of the meter casing.

The spindle on which the measuring drum is mounted is connected through gears to record the quantity of gas passing through the meter.

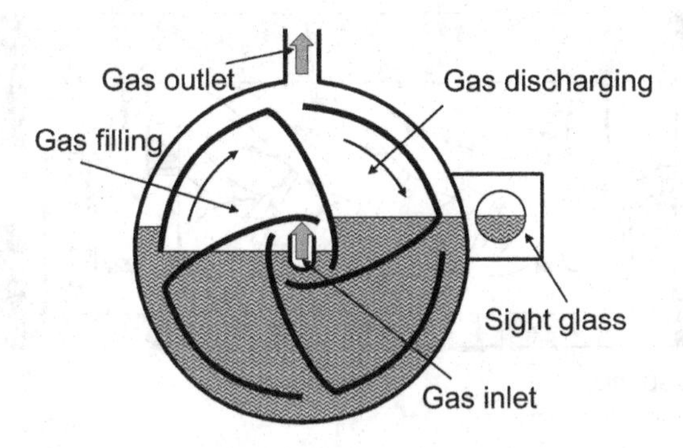

FIGURE 5.11 The wet-type gas meter. (Courtesy: Alexander Wright division of GH Zeal Ltd.)

Such meters are available in capacities ranging in size from 5 to 9,000 L/h with an accuracy down to ±0.5%.

5.10 GENERAL SUMMARY

Because of their high accuracy, positive displacement meters are used extensively in liquid custody transfer applications where duty is applicable on such commodities as petrol, wines, and spirits.

In use, some of the following application limitations should be noted:

- Owing to the mechanical contact between the component parts, wear and tear is a problem. In general, therefore, positive displacement meters are primarily suited for clean, lubricating, and non-abrasive applications.
- In some cases, filters (down to 10 μm) may be required to filter debris and clean the fluid before the meter. Such filters require regular maintenance. If regular maintenance is not carried out, the added pressure drop may also need to be considered.
- Their working life also depends on the nature of the fluid being measured, especially with regards to solids build-up and the media temperature.
- Positive displacement meters are an obstruction to the flow path and consequently produce an unrecoverable pressure loss.
- Because many positive displacement meters have the same operating mechanisms as pumps, they may be driven by a motor and used as dosing or metering pumps.
- One of the drawbacks of the positive displacement meter is its high differential pressure loss. This, however, may be reduced by measuring the differential pressure across the meter and then driving it with a motor that is controlled by a feedback system.
- Positive displacement meters are limited at both high and low viscosities. Errors can occur due to the leakage (slippage) around the gears or pistons. Slippage may be reduced by using viscous fluids which have the ability to seal the small clearances. However, if the fluid is too viscous then it can coat the inner chambers of the meter and reduce the volume passed – causing reading errors. Thus, whilst low viscosities limit the use at low flows (due to increased slippage), high viscosities limit the use at high flows due to the high pressure loss.
- If slippage does occur, and is calibrated for, it can change with temperature as the viscosity varies.
- Positive displacement meters can be damaged by over-speeding.
- In certain cases (e.g. the oval gear meter) positive displacement meters give rise to pulsations. This may inhibit the use of this type of meter in certain applications.
- Positive displacement meters are primarily used for low volume applications and are limited when high volume measurement is required.

6 Head Loss Meters

6.1 INTRODUCTION

Head loss meters, also known as differential pressure meters, encompass a wide variety of meter types that includes: orifice plates, Venturi tubes, nozzles, Dall tubes, target meters, Pitot tubes, and variable area meters. Indeed, the measurement of flow using differential pressure is still the most widely used technology.

One of the features of the head loss meter is that flow can be accurately determined from the differential pressure, accurately measurable dimensions of the primary device, and the properties of the fluid. Thus, an important advantage of differential type meters over other instruments is that they do not always require direct flow calibration. In addition, they offer excellent reliability, reasonable performance, and modest cost.

Another advantage of orifice plates in particular, is that they can be used in liquid or gas applications with little change.

6.2 BASIC THEORY

Differential pressure flow rate meters are based on a physical phenomenon in which a restriction in the flow line creates a pressure drop that bears a relationship to the flow rate.

► This physical phenomenon is based on two well-known equations: the *Equation of continuity* and *Bernoulli's equation.*

6.2.1 Equation of Continuity

Consider the pipe in Figure 6.1 that rapidly converges from its nominal size to a smaller size followed by a short parallel sided throat before slowly expanding to its full size again. Further, assume that a fluid of density ρ flowing in the pipe of area A_1, has a mean velocity v_1 at a line pressure P_1. It then flows through the restriction of area A_2, where the mean velocity increases to v_2 and the pressure falls to P_2.

The ratio of the diameters of the restriction (d) to the inside diameter (ID) (D) of the pipe is called the *beta ratio* (β), that is:

$$\frac{d}{D} = \beta \tag{6.1}$$

The equation of continuity states that for an incompressible fluid the volume flow rate, Q, must be constant. Very simply, this indicates that when a liquid flows through a restriction, then in order to allow the same amount of liquid to pass (to achieve a constant flow rate) the velocity must increase (Figure 6.2).

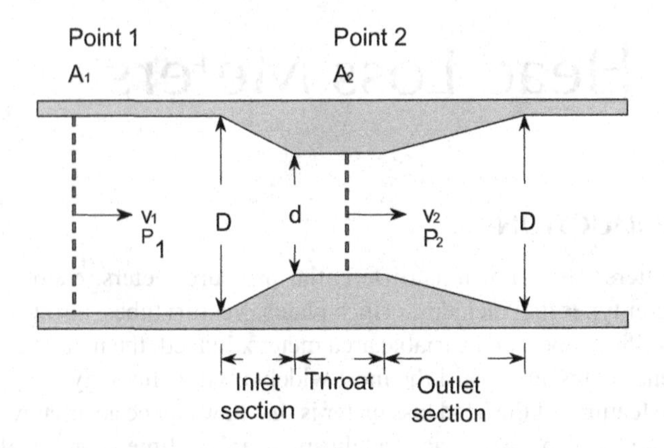

FIGURE 6.1 Basic definition of terms.

Mathematically:

$$Q = v_1 \cdot A_1 = v_2 \cdot A_2 \tag{6.2}$$

where v_1 and v_2 and A_1 and A_2 are the velocities and cross-sectional areas of the pipe at points 1 and 2, respectively.

6.2.2 Bernoulli's Equation

In its simplest form, Bernoulli's equation states that under steady fully developed flow (turbulent) conditions, the total energy (kinetic + pressure + gravitational) per unit mass of an ideal incompressible fluid (i.e. one having a constant density and zero viscosity) remains constant along a flow line:

$$\frac{v^2}{2} + \frac{P}{\rho} + gz \tag{6.3}$$

where

v = the velocity at a point in the streamline
P = the pressure at that point

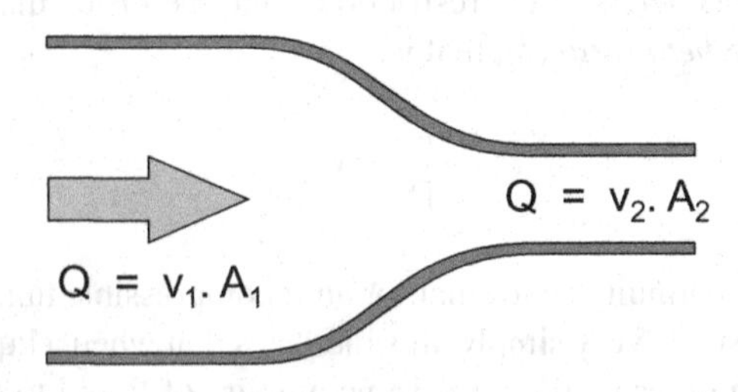

FIGURE 6.2 To allow the same amount of liquid to pass the velocity must increase, that is $Q = v_1 \cdot A_1 = v_2 \cdot A_2$.

ρ = the fluid density
g = the acceleration due to gravity
z = the level of the point above some arbitrary horizontal reference plane with the positive z-direction in the direction opposite to the gravitational acceleration
k = constant

In the restricted section of the flow stream, the kinetic energy (dynamic pressure) increases due to the increase in velocity and the potential energy (static pressure) decreases.

Relating this to the conservation of energy at two points in the fluid flow then:

$$\frac{v_1^2}{2} + \frac{P_1}{\rho} = \frac{v_2^2}{2} + \frac{P_2}{\rho} \tag{6.4}$$

Multiplying through by ρ gives:

$$\frac{1}{2} \cdot \rho \cdot v_1^2 + P_1 = \frac{1}{2} \cdot \rho \cdot v_2^2 + P_2 \tag{6.5}$$

rearranging:

$$P_1 - P_2 = \Delta P = \frac{1}{2} \cdot \rho \cdot v_2^2 - \frac{1}{2} \cdot \rho \cdot v_1^2 \tag{6.6}$$

Now from the continuity equation (6.2) we can derive:

$$v_1 = \frac{Q}{A_1} \tag{6.7}$$

and:

$$v_2 = \frac{Q}{A_2} \tag{6.8}$$

substituting in (6.6):

$$\Delta P = \frac{1}{2} \cdot \rho \cdot \left(\frac{Q}{A_2}\right)^2 - \frac{1}{2} \cdot \rho \cdot \left(\frac{Q}{A_1}\right)^2 \tag{6.9}$$

Solving for Q:

$$Q = A_2 \sqrt{\frac{2 \cdot \Delta P}{\rho \cdot \left[1 - (A_2/A_1)\right]^2}} \tag{6.10}$$

Since it is more convenient to work in terms of the diameters of the restriction (d) and the ID (D) of the pipe we can substitute for:

$$A_1 = \frac{\pi \cdot D^2}{4} \tag{6.11}$$

and:

$$A_2 = \frac{\pi \cdot d^2}{4} \tag{6.12}$$

$$Q = \frac{\pi}{4} \cdot d^2 \sqrt{\frac{2 \cdot \Delta P}{\rho \cdot (1 - (d^4/D^4))}} \tag{6.13}$$

The term:

$$\sqrt{\frac{1}{1 - (d/D)^4}} \tag{6.14}$$

is called the *Velocity of Approach Factor* (E_v)* and by substituting from (6.1):

$$E_v = \sqrt{\frac{1}{1 - \beta^4}} \tag{6.15}$$

Now, substituting in (6.13) we have:

$$Q = E_v \cdot d^2 \sqrt{\frac{2 \cdot \Delta P}{\rho}} \tag{6.16}$$

6.2.3 Discharge Coefficient (C_D)

Unfortunately, Equation 6.16 only applies to a fully developed inviscid flow regime. To take into account the effects of viscosity and turbulence a term called the *discharge coefficient* (C_d) has been introduced. This term, also known as the *coefficient of discharge*, is simply a measure of the actual flow rate compared with the theoretical flow rate:

$$C_d = \frac{\text{Actual flow rate}}{\text{Theoretical flow rate}} \tag{6.17}$$

In practice C_d is a function of the design of the restriction (orifice plate, Venturi, nozzle, etc.), the beta ratio (β), the Reynolds number (Re), the location of the pressure taps, and the friction due to pipe roughness – all of which marginally reduce the flow rate (Q) by a factor of between 0.6 and 0.9.

* Clearly although E_v is a constant, determined by the beta ratio, its value will vary with temperature.

Over a period of many years an extensive range of tests have been carried out to characterise the C_d, over a wide range of possible flow conditions and covering most meter types. The analysis of this huge body of empirically derived data has allowed a curve to be determined that best fits these thousands of collected data points (Figure 6.3).

The uncertainty for this variable is determined using the Standard Estimate of the Error (SEE) – the standard deviation of the data sample referenced to the calculated (curve) values.

Subsequently several equations have been developed to fit the data curve and predict the discharge coefficient of any geometrically similar primary element. Currently forming the basis of ISO 5167, the Reader-Harris/Gallagher (1998) Equation, serves as a calibration constant so that primary elements of similar construction do not need to each be calibrated in a laboratory. This requires that each device is manufactured to a strict manufacturing standard with reference texts and standards listing typical values and tolerances for C_d under certain flows in standard installations:

$$
\begin{aligned}
C_d = {} & 0.5961 + 0.0261\beta^2 - 0.216\beta^8 + 0.000521\left(\frac{10^6\beta}{\mathrm{Re}}\right)^{0.7} \\
& + (0.0188 + 0.0063A)\left(\frac{10^6}{\mathrm{Re}}\right)\beta^{3.5} \\
& + (0.043 + 0.08\varepsilon^{-10L_1} - 0.123\varepsilon^{-7L_1})(1 - 0.11A)\frac{\beta^4}{1-\beta^4} \\
& - 0.031\left(\frac{2L_2}{1-\beta} - 0.8\left(\frac{2L_2}{1-\beta}\right)^{1.1}\right)\beta^{1.3} + M_2
\end{aligned}
\tag{6.18}
$$

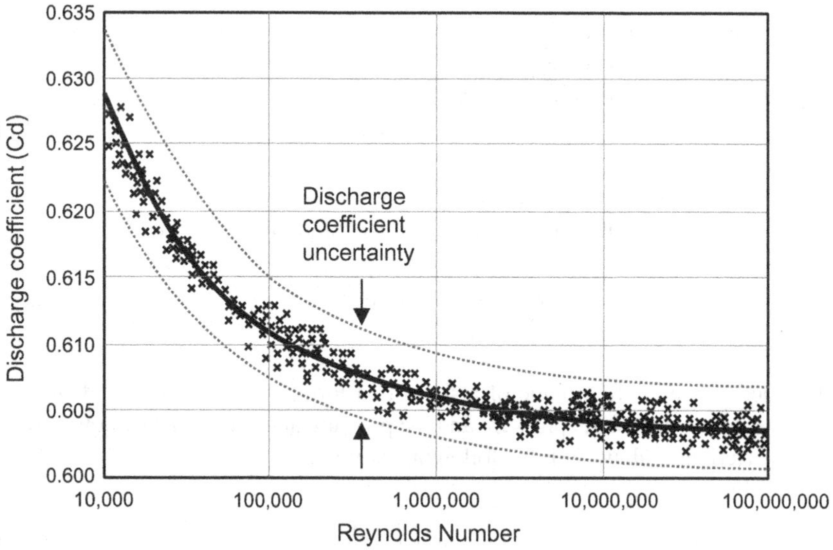

FIGURE 6.3 A best-fit curve that best characterises the C_d, over a wide range of possible flow conditions and covering most meter types.

where:

$$A = \left(\frac{19{,}000\beta}{Re}\right)^{0.8} \tag{6.19}$$

and M_2 may be one of two values – dependent on the internal diameter (D) of the pipe:
If $D \geq 71.12$ mm (2.8 in.) then:

$$M = 0 \tag{6.20}$$

If $D < 71.12$ mm (2.8 in.) then:

$$M = 0.011(0.75 - \beta)\left(2.8 - \frac{D}{0.0254}\right) \tag{6.21}$$

L_1 and L_2 are functions determined by the tap type:

For 1″ flange taps: $L_1 = L_2 = 0.0254/d$ (m)
For corner taps: $L_1 = L_2 = 0$
For D and D/2 taps: $L_1 = 1$; $L_2 = 0.47$

and for $D < 71.12$ mm (2.8 in) the following should be added arithmetically:

$+0.9(0.75 - \beta)(2.8 - D/25.4)$ where D is in mm

Referring back to Figure 6.3, the best uncertainties associated with Equation 6.18 is:

$(0.7 - \beta)\%$ for $0.1 \leq \beta < 0.2$
0.5% for $0.2 \leq \beta < 0.6$
$(1.667\beta - 0.5)\%$ for $0.6 \leq \beta < 0.75$

◀

Incorporating both the discharge coefficient (C_d) and the velocity of approach factor (E_v), the full equation for an incompressible fluid thus becomes:

$$Q = C_d \cdot E_v \cdot d^2 \sqrt{\frac{2 \cdot \Delta P}{\rho}} \tag{6.22}$$

In practice use is often made of a simplified formula that relates the difference between the upstream static pressure and the pressure at or immediately downstream of the restriction, to flow, with the following expression:

$$Q = kC_d \sqrt{\frac{2\Delta P}{\rho}} \tag{6.23}$$

where k is a lumped constant.

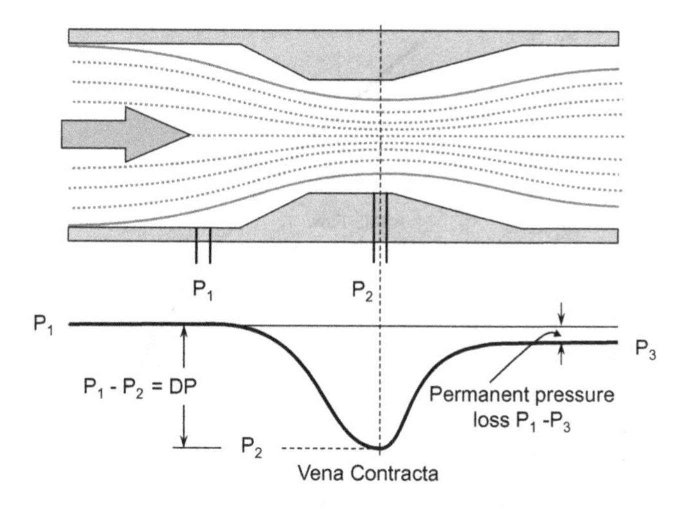

FIGURE 6.4 Defining the 'head' loss.

The foregoing formulae highlight three major limitations that are applicable to all differential pressure systems:

- The square root relationship between differential pressure (ΔP) and flow (Q) severely limits the turndown ratio of such techniques to a maximum of 5:1 or less; and
- If density (ρ) is not constant, it must be known or measured. In practice the effect of density changes is not significant in the majority of liquid flow applications and needs only to be taken into account in the measurement of gas flow.
- All differential pressure meters create a permanent pressure loss (Figure 6.4). This 'head' loss depends on the type of meter and the square of the volume flow.

The point at which the minimum cross-sectional area of the flow stream reaches a minimum is known as the *vena contracta* (from the Latin *contracted vein*) and occurs (typically) at some 0.35–0.85 pipe diameter downstream from the narrowest point of the constriction (Figure 6.5).

6.2.4 Gas Flow

Vapour or gas flow through a restriction differs from liquid flow in that as a gas flows through a restriction there is a decrease in pressure which results in the expansion of the gas and a decrease in density.

Thus, for the mass flow to remain constant, the velocity must increase to compensate for the lower density. The result is that the formula for gas flow is modified by the addition of the term *gas expansion factor* (Y_1). With a lowered density, the velocity

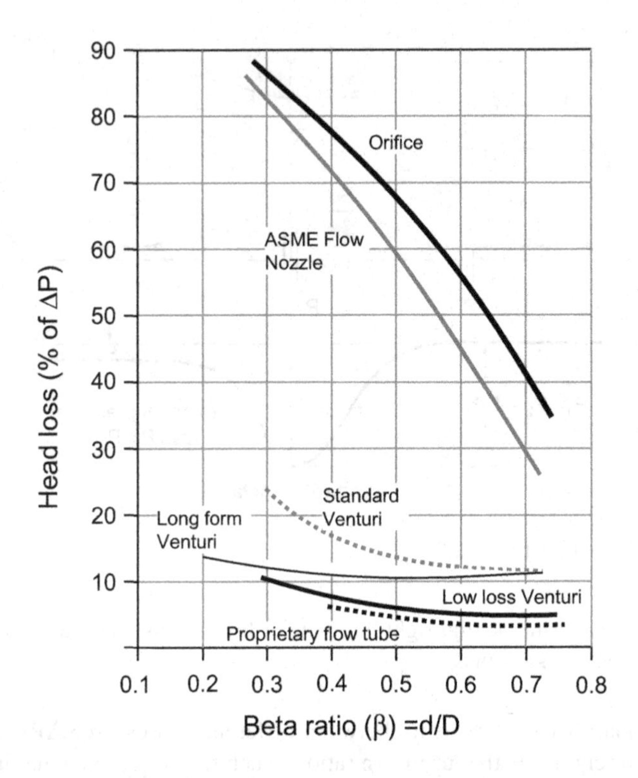

FIGURE 6.5 The permanent 'head' loss for various measurement techniques. The orifice plate produces the most drop whilst the low-loss tube causes the least. (Courtesy: Emerson.)

will be slightly higher than predicted by the theoretical flow equation ($Y_1 = 1$ for a liquid):

$$Q = C_d \cdot E_v \cdot Y_1 \cdot d^2 \sqrt{\frac{2 \cdot \Delta P}{\rho}} \tag{6.24}$$

Like the discharge coefficient, the gas expansion factor is also derived empirically.

The gas expansion factor is based on the determination of density at the upstream of the restriction. Tables and graphs are available for the expansion factor as a function of the pressure ratio across the restriction and the specific heat of the gas (BS 1042). Alternatively, the expansion factor may be calculated by standard equations listed in BS 1042. The mass flow rate for both liquids and gases is found by multiplying the theoretical mass flow equation by the expansion factor and the appropriate discharge coefficient.

6.3 ORIFICE PLATE

The orifice plate is the simplest and most widely used differential pressure flow measuring element and generally comprises a metal plate with a concentric round hole

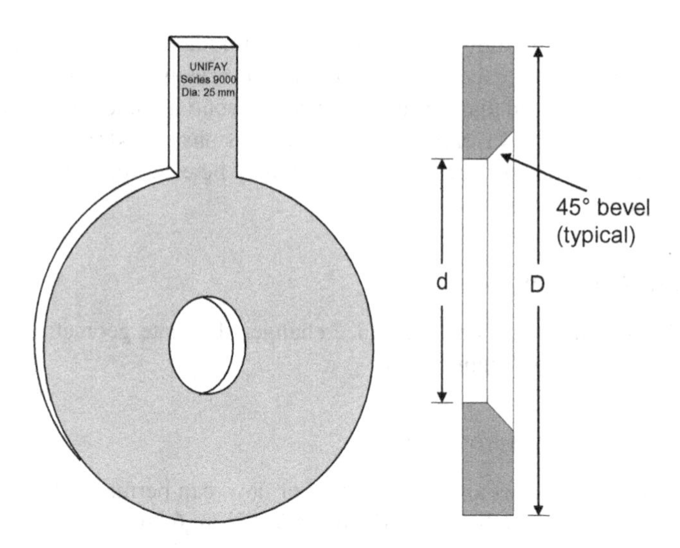

FIGURE 6.6 Concentric orifice plate with integral metal tab.

(orifice) through which the liquid flows (Figure 6.6). An integral metal tab facilitates installation and carries details of the plate size, thickness, serial number, etc. The plate, usually manufactured from stainless steel, Monel or phosphor bronze should be of sufficient thickness to withstand buckling (3–6 mm). The orifice features a sharp square upstream edge and, unless a thin plate is used, a bevelled downstream edge.

A major advantage of the orifice plate is that it is easily fitted between adjacent flanges that allow it to be easily changed or inspected (Figure 6.7).

Although a correctly installed new plate may have an uncertainty of 0.6%, the vast majority of orifice meters measure flow only to an accuracy of about ±2–3%.

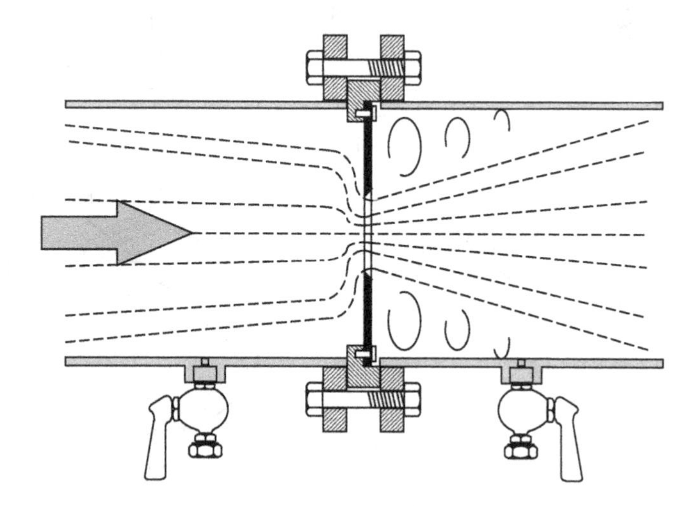

FIGURE 6.7 Orifice plate fitted between adjacent flanges.

This uncertainty is due mainly to errors in temperature and pressure measurement, variations in ambient and process conditions, and the effects of upstream pipework.

It is commonly assumed that, since the orifice is essentially fixed, its performance does not change with time. In reality the orifice dimensions are extremely critical and even for a new plate, the measurement accuracy may be even further compromised due to a wide number of factors.

6.3.1 Plate Contamination

Debris and lubricant on the upstream face changes the plate geometry producing errors up to 15% or more (Figure 6.8a).

6.3.2 Buckled or Warped Plate

Deformation of the plate, as a result of excess of flow, can permanently enlarge the orifice leading to a negative bias of –6% to –9% (Figure 6.8b).

6.3.3 Reversed Plates

Orifice plates are often mistakenly installed in the reverse direction to the flow. If, installed backwards, with the sharp orifice edge on the exit or downstream side of the plate, negative biases of between –15% and –20% may be experienced – dependent on the beta ratio (Figure 6.8c).

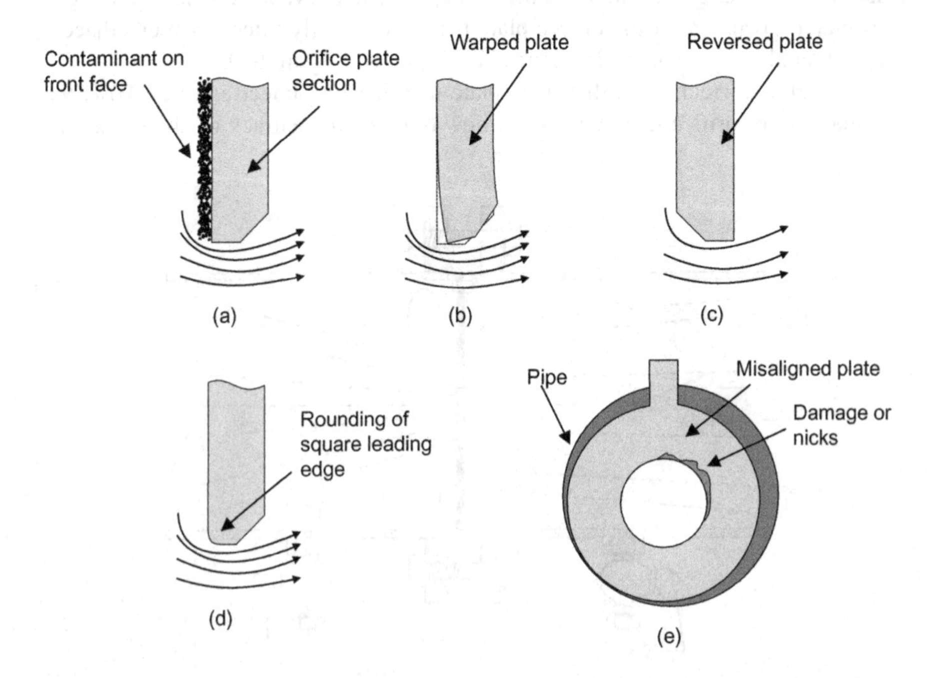

FIGURE 6.8 Errors incurred as a result of wear and contamination on the orifice plate.

6.3.4 Worn-Leading Orifice Edge

A rounding or chamfering of the sharp-leading edge of the orifice hole, due to erosion or corrosion, of as little as 0.5 mm (0.02 in), can produce an approximate negative bias of –5%. Increasing the radius to 1.3 mm (0.05 in) has produced negative bias errors of approximately 13% (Figure 6.8d).

6.3.5 Damage or Nicks in the Plate

Errors of up to 10% or higher are largely determined by the beta ratio and the extent of the nick (Figure 6.8e).

6.3.6 Debris Trapped at the Orifice

Debris lodged in the orifice changes the effective geometry and creates a positive bias on the flow rate prediction.

6.3.7 Plate Misalignment

Orifices that are not centred in the pipe can produce misalignment errors of up to 3% (Figure 6.8e).

6.3.8 Blocked Impulse Tubes

Blocking, or partial blocking, of the impulse tubes or tapping points can produce varying degrees of positive or negative errors – dependent on the amount of blockage and whether it occurs in the upstream or downstream tapping.

6.3.9 Saturated ΔP Transmitter

Because of their limited turndown ratio the differential pressure delivered across the orifice plate commonly exceeds the transmitter's range and the transmitter is said to be 'saturated'. Consequently the transmitter's output is limited at this lower value and will thus a produce a negative error.

Note: It should be appreciated that none of these error-producing factors would be apparent to the operator. Indeed, one of the major problems associated with flow measurement using an orifice plate (or any other head loss device) is that operators generally assumed that if they have a reading, the system is working.

Fittings that allow the plate to be periodically removed and inspected provide a good indication of the meter performance. However, these are only 'spot checks' that still leave the operator blind to any problems that might occur between such checks.

That is not to say that diagnostic information is not available. One such diagnostic system ('Prognosis®') is illustrated in Figure 6.9. In addition to the conventional upstream (high pressure P_1) and downstream (low pressure P_2) tappings used to measure the differential pressure, a third downstream tapping (P_3) is provided. This

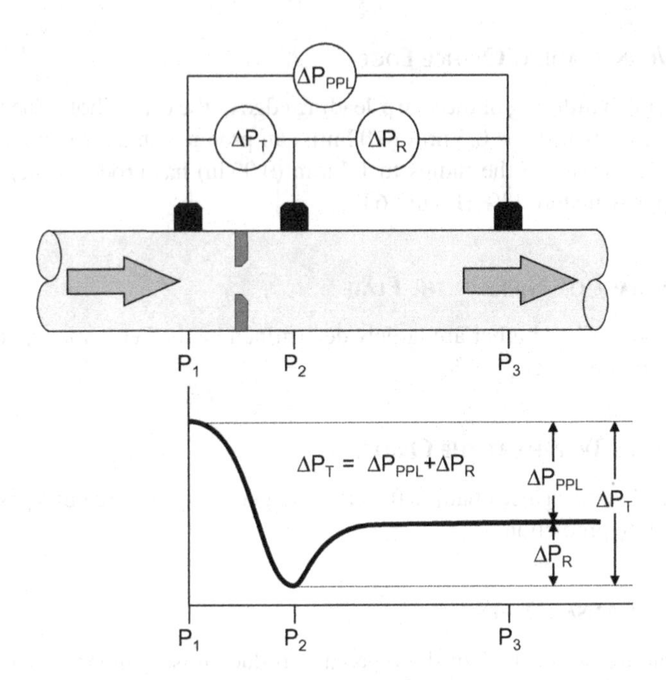

FIGURE 6.9 Diagnostic system in which, in addition to the conventional differential pressure tappings at P_1 and P_2, a third downstream tapping (P_3) is provided. (Courtesy: Swinton Technology.)

allows both the total head loss or 'Permanent Pressure Loss' (ΔP_{PPL}) as well as the 'Recovered' differential pressure (ΔP_R) to be measured.

Software manipulation of these three measurements allows three sets of ratios to be determined that are set characteristics of the meter. In turn these are used to produce seven validation results that are then plotted as four points on a live graphic display (Figure 6.10).

Points inside the box indicate that the meter is performing correctly, whereas points outside of the box indicate a problem and raises an alarm.

Note: One of the four points is ΔP_T and ΔP_T, inf. This latter value (ΔP_T, inf) is the inferred value of the differential pressure (ΔP_T) calculated from the addition of ΔP_{PPL} and ΔP_R (see Figure 6.9).

6.4 QUADRANT EDGE ORIFICE PLATE

An adaptation of the sharp, square edge is the quadrant edge orifice plate (also called quarter circle and round edge). As shown in Figure 6.11 this has a concentric opening with a rounded upstream edge that produces a coefficient of discharge that is practically constant for Reynolds numbers from 300 to 25,000, and is therefore useful for use with high viscosity fluids or at low flow rates.

The radius of the edge is a function of the diameters of both the pipe and the orifice. In a specific installation this radius may be so small as to be impractical to

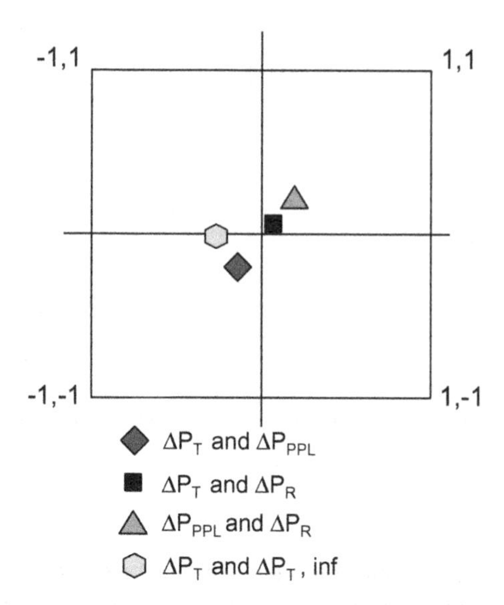

FIGURE 6.10 Software manipulation allows three sets of ratios to be determined that are used to produce seven validation results that are then plotted as four points on a live graphic display. (Courtesy: Swinton Technology.)

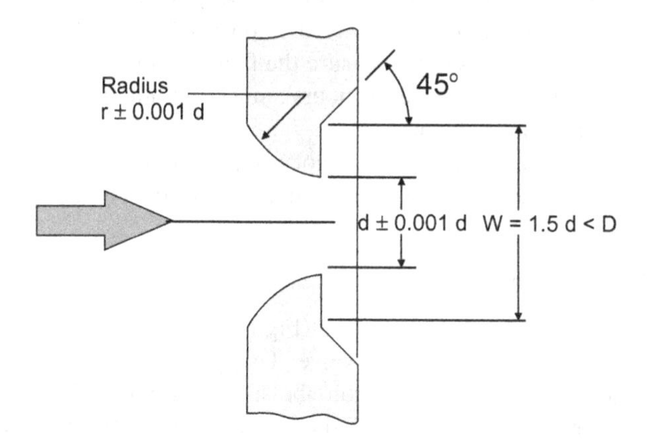

FIGURE 6.11 The quadrant edge orifice plate with a rounded upstream edge.

manufacture or it can be so large that it practically becomes a flow nozzle. As a result, on some installations, it may be necessary to change maximum differentials or even pipe sizes to obtain a workable solution for the plate thickness and its radius.

6.5 ORIFICE PLATE CONFIGURATIONS

Although the concentric orifice (Figure 6.12a) is the most frequently used, other plate configurations are used.

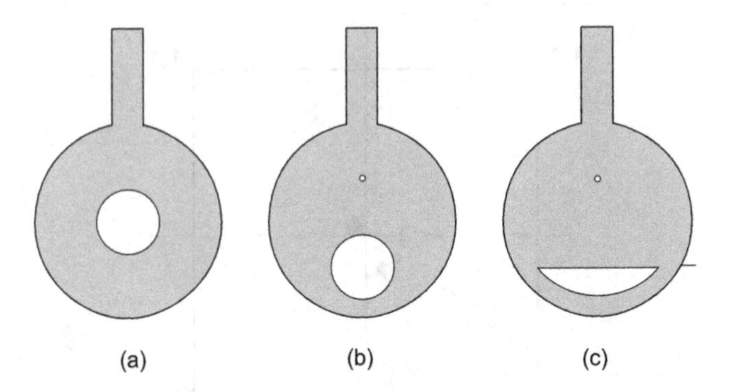

FIGURE 6.12 Various types of orifice plate configurations: (a) concentric, (b) eccentric, and (c) segmental.

6.5.1 Eccentric

In the eccentric bore orifice plate (Figure 6.12b), the orifice is offset from the centre and is usually set at the bottom of the pipe bore. This configuration is mainly used in applications where the fluid contains heavy solids that might become trapped and accumulate on the back of the plate. With the orifice set at the bottom, these solids are allowed to pass. A small vent hole is usually drilled in the top of the plate to allow gas, which is often associated with liquid flow, to pass. It should be noted, however, that the vent hole adds an unknown flow error and runs the risk of plugging.

Eccentric plates are also used to measure the flow of vapours or gases that carry small amounts of liquids (condensed vapours), since the liquids will carry through the opening at the bottom of the pipe.

The coefficients for eccentric plates are not as reproducible as those for concentric plates, and in general, the error can be 3–5 times greater than on concentric plates.

6.5.2 Segmental Orifice Plates

The opening in a segmental orifice plate (Figure 6.12c) is a circular segment – comparable to a partially opened gate valve. This plate is generally employed for measuring liquids or gases that carry non-abrasive impurities, which are normally heavier than the flowing media such as light slurries, or exceptionally dirty gases.

6.6 TAPPING POINTS

The measurement of differential pressure requires that the pipe is 'tapped' at suitable upstream (high pressure) and downstream (low pressure) points. The exact positioning of these taps is largely determined by the application and desired accuracy.

6.6.1 Vena Contracta Tapping

Because of the fluid inertia, its cross-sectional area continues to decrease after the fluid has passed through the orifice. Thus its maximum velocity (and lowest pressure)

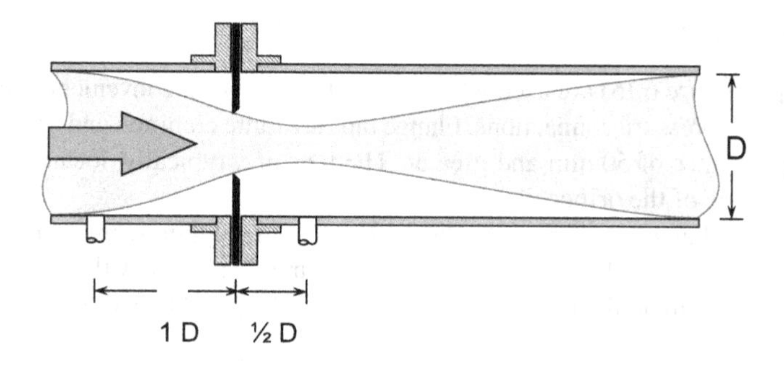

FIGURE 6.13 For maximum differential pressure the high pressure tap is located 1 pipe diameter upstream and the low pressure tap at the vena contracta – about ½-pipe diameter downstream.

is at some point downstream of the orifice – at the vena contracta. On standard concentric orifice plates these taps are designed to obtain the maximum differential pressure and are normally located one pipe diameter upstream and at the vena contracta – about ½-pipe diameter downstream (Figure 6.13).

The main disadvantage of using the vena contracta tapping point is that the exact location depends on the flow rate and on the orifice size – an expensive undertaking if the orifice plate size has to be changed.

Vena contracta taps should not be used for pipe sizes under 150 mm diameter because of interference between the flange and the downstream tap.

6.6.2 Pipe Taps

Pipe taps (Figure 6.14) are a compromise solution and are located 2½ pipe diameter upstream and 8 pipe diameters downstream. Whilst not producing the maximum available differential pressure, pipe taps are far less dependent on the flow rate and orifice size.

Pipe taps are used typically in the existing installations, where radius and vena contracta taps cannot be used. They are also used in applications of greatly varying flow since the measurement is not affected by the flow rate or orifice size. Since pipe taps do not measure the maximum available pressure, accuracy is reduced.

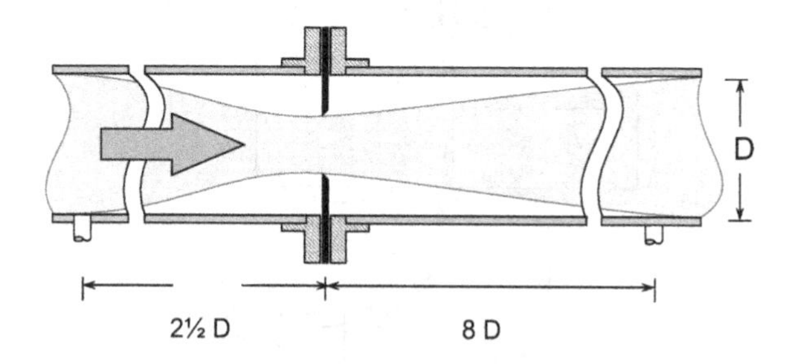

FIGURE 6.14 Pipe taps are far less dependent on the flow rate and orifice size and are located 2½ pipe diameter upstream and 8 pipe diameter downstream.

6.6.3 Flange Taps

Flange taps (Figure 6.15) are used when it is undesirable or inconvenient to drill and tap the pipe for pressure connections. Flange taps are quite common and are generally used for pipe sizes of 50 mm and greater. The taps are, typically, located 25.4 mm (1 in) either side of the orifice plate.

Usually, the flanges, incorporating the drilled pressure tappings, are supplied by the manufacturer (Figure 6.16). With the taps thus accurately placed by the manufacturer the need to recalculate the tapping point, when the plate is changed, is eliminated.

6.6.4 Corner Taps

For pipe diameters less than 50 mm, the vena contracta starts to become close to and, possibly, forward of the downstream tapping point. Consequently, use is made of *corner taps* – an adaptation of the flange tap (Figure 6.17) in which the tappings are made to each face of the orifice plate. The taps are located in the corner formed by the pipe wall and the orifice plate on both the upstream and downstream sides and require the use of special flanges or orifice holding rings.

6.7 SPECIAL ORIFICE PLATE ASSEMBLIES

In many industries, particularly oil and gas, there is a need to regularly inspect or change the orifice plate under operating conditions without interrupting the flow. This is normally achieved using a bypass pipe and isolation valves (Figure 6.18).

Whilst standard orifice flanges are widely used to hold orifice plates, their removal and replacement, necessitating spreading of the flanges, is time consuming and runs the risk of leakage and spillage.

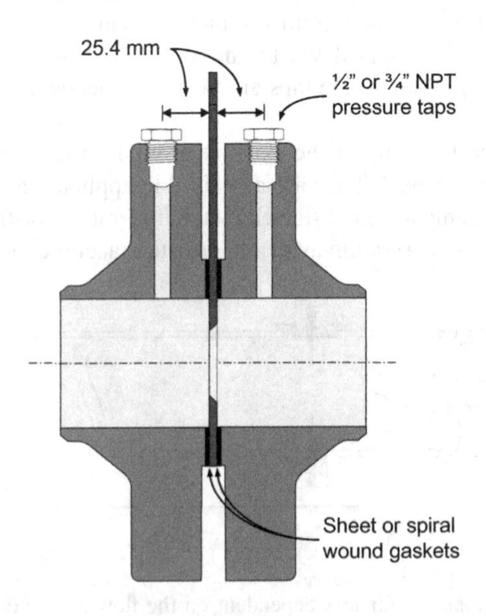

FIGURE 6.15 Flange taps are located 25.4 mm (1 in) either side of the orifice plate.

FIGURE 6.16 Typical flange and plate assembly supplied by the manufacturer, incorporating the drilled pressure tappings.

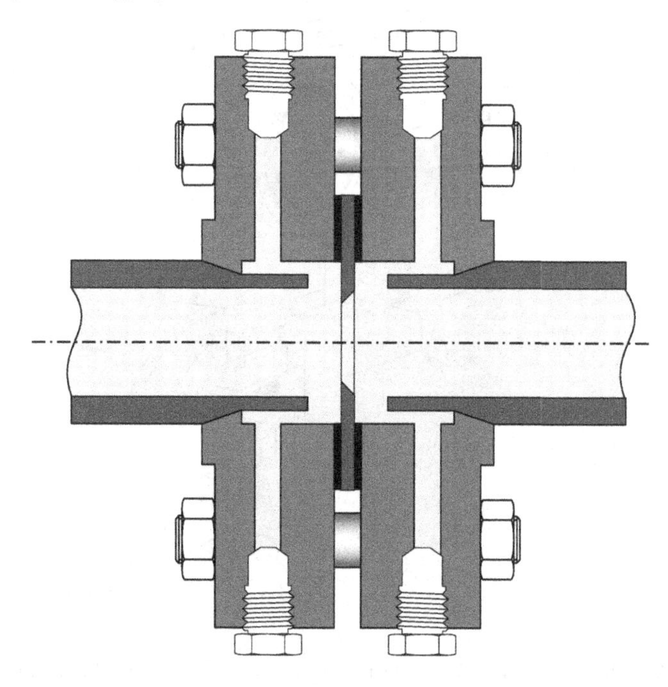

FIGURE 6.17 Corner tap is made to each face of the orifice plate.

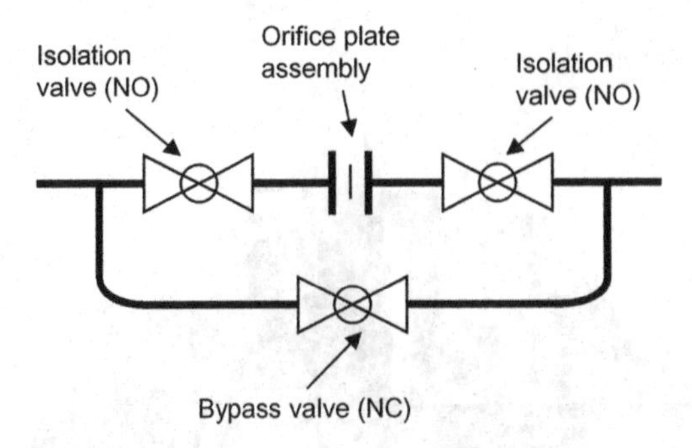

FIGURE 6.18 A bypass pipe and isolation valves are used to gain access to the orifice plate without interrupting the flow.

One solution lies with a *single chamber orifice* fitting as illustrated in Figure 6.19. Here the orifice plate is removed by first loosening the top set of screws, sliding the clamping bar out, and then lifting out the sealing bar carrier and orifice plate as one unit.

In custody and fiscal transfer applications a bypass system is illegal. Consequently, use may be made of a dual chamber system typified by the Daniel Senior. Figure 6.20a shows such an arrangement operating under normal flow conditions while Figure 6.20b illustrates a typical ratchet plate and orifice assembly.

Removal of the orifice plate is carried out in four distinct stages.

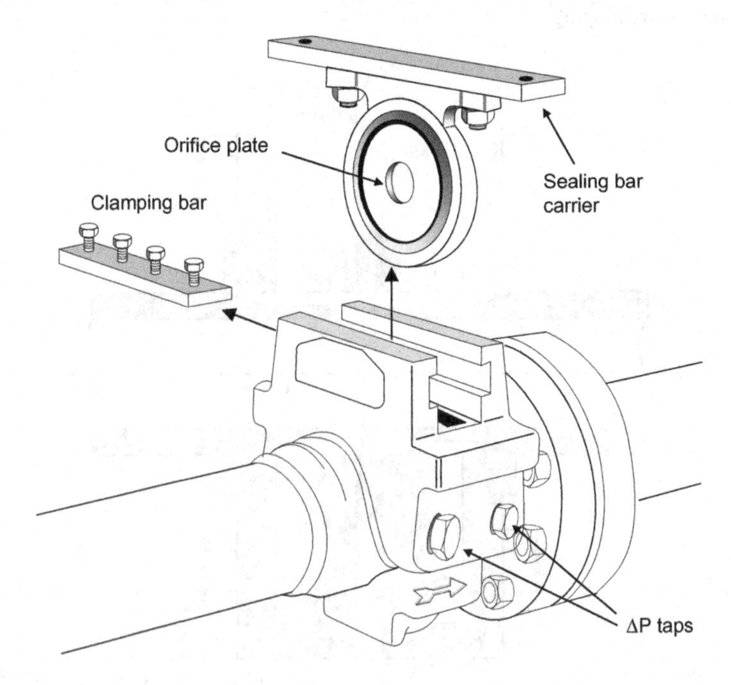

FIGURE 6.19 Single Chamber Orifice fitting allows the orifice plate to be removed quickly and safely.

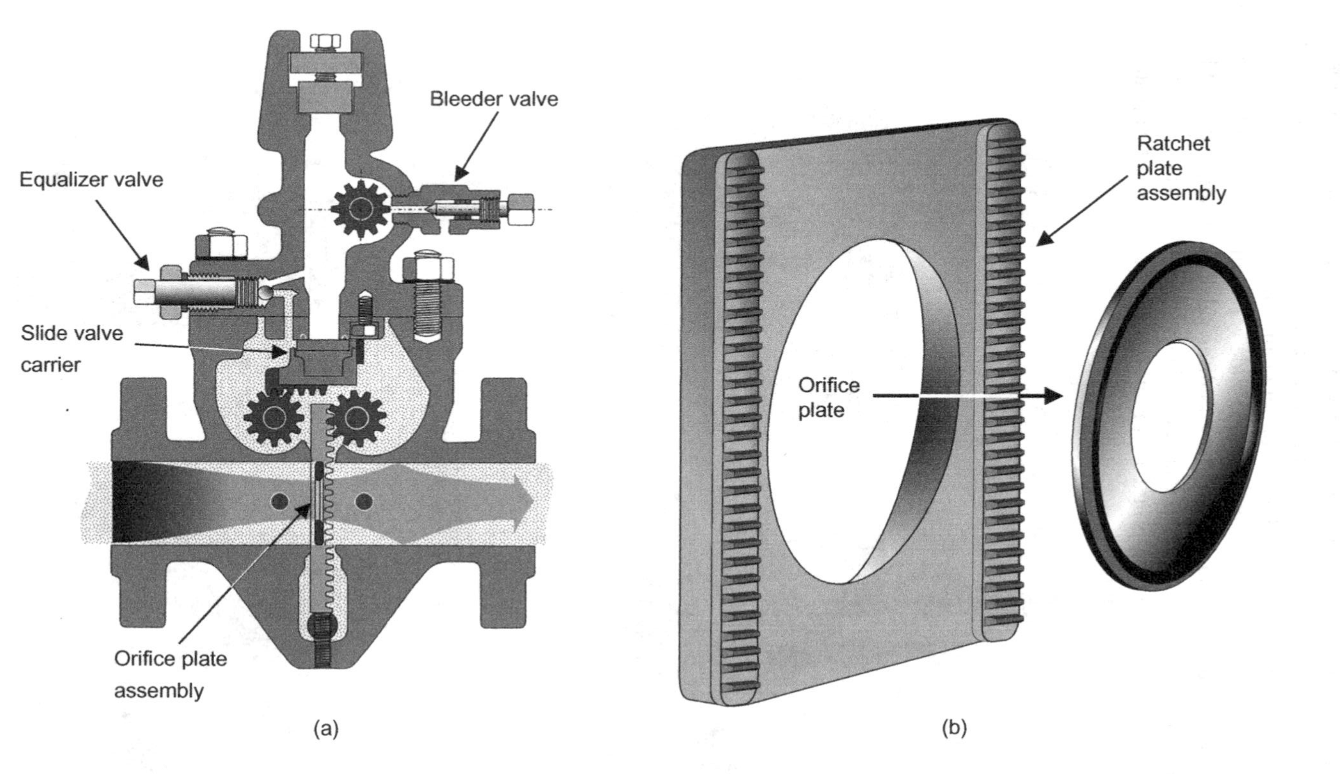

FIGURE 6.20 Daniel Senior dual chamber system: (a) operation under normal flow conditions and (b) a typical ratchet plate and orifice assembly. (Courtesy: Emerson.)

In the first stage (Figure 6.21a), the upper chamber pressure is equalised with that of the lower chamber by opening the equaliser valve. Next (Figure 6.21b), the slide valve carrier, which acts as an isolation mechanism, moves to the side – allowing the orifice plate assembly to be raised to the upper chamber.

In the penultimate stage (Figure 6.21c), the slide valve carrier is closed – isolating the upper and lower chambers. Next, the equaliser valve is closed and the bleeder valve is opened – allowing the upper chamber to be brought back to ambient pressure. Finally (Figure 6.21d), the clamping bar is removed allowing the orifice plate to be removed from the top chamber.

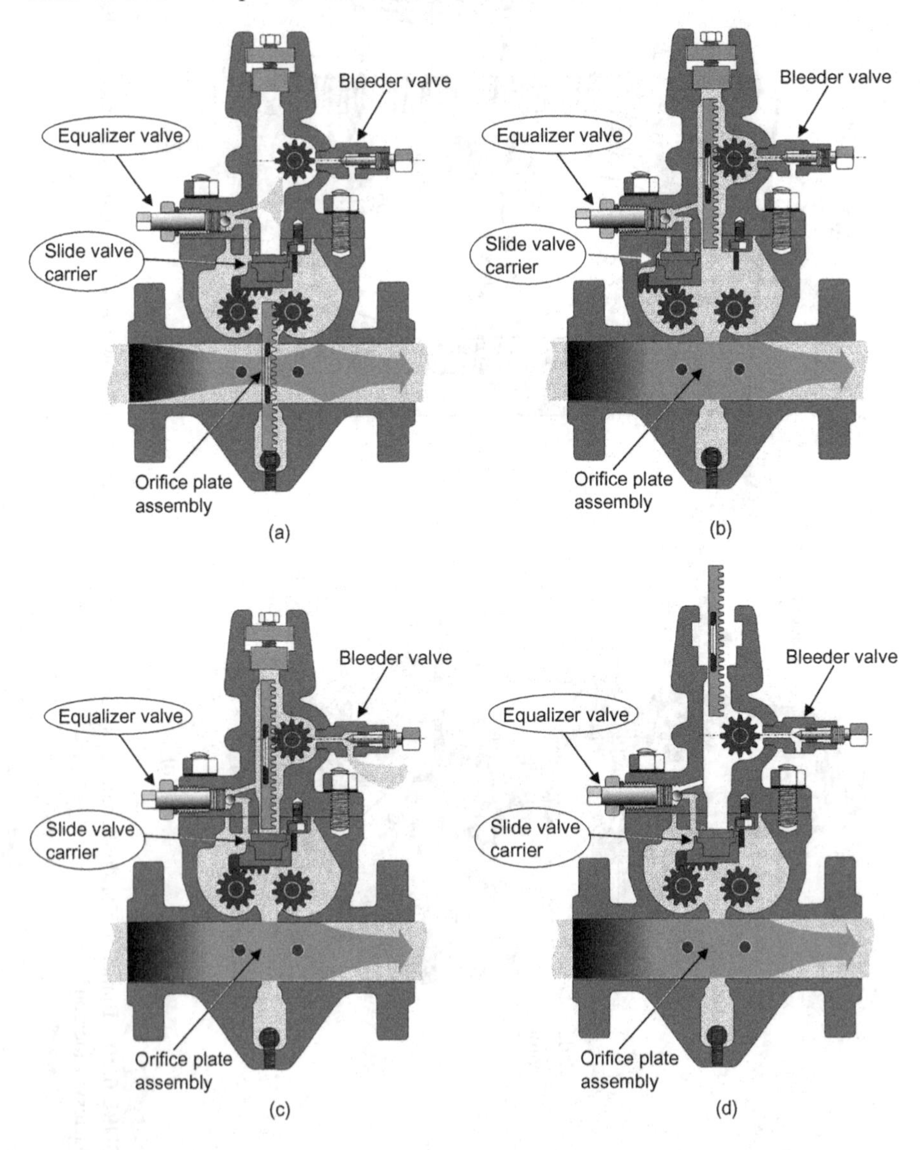

FIGURE 6.21 The four stages involved in removing the orifice plate assembly.

6.8 ORIFICE PLATE SIZING

Whilst use can be made of the formulae detailed in Equations 6.19 and 6.20, the modern practice is to make use of any one of a number of software sizing programs that are available from a number of sources. Several, are detailed below:

http://www.flowcalcs.com/
http://www.osti.gov/energycitations/prod…
http://www.farrisengineering.com/Product…
http://www.efunda.com/formulae/fluids/calc_orifice_flowmeter.cfm
http://www.flowmeterdirectory.com/flowmeter_orifice_calc.html

6.9 ORIFICE PLATES – GENERAL

At the beginning of this chapter, it was stated that an important feature of differential type meters is that flow can be determined directly – without the need for calibration. This is particularly true for the orifice plate where there is a comprehensive range of standard designs that require no calibration.

6.9.1 Advantages

- Simple construction
- Inexpensive
- Robust
- Internationally recognised range of standard orifice plates and sizes that obviate the need for calibration
- Easily fitted between flanges
- No moving parts
- Large range of sizes and opening ratios
- Suitable for most gases and liquids as well as steam
- Price does not increase dramatically with size
- Well understood and proven

The advantages listed above would normally be listed in most textbooks on the subject of orifice plates. However, a few observations regarding some of these 'advantages' are in order.

'*Expensive*' and '*inexpensive*' are, of course, relative terms. Certainly the primary element, the orifice plate itself, is relatively inexpensive compared with other flow measuring systems. However, as shown in Figure 6.22 the orifice plate is only one part of a number of ancillary components that include the flange plate assembly, the isolation valves, the impulse tubing, the valve manifold, and the differential pressure transmitter.

Designing, purchasing, installing, and commissioning an orifice plate–based flow measuring system can thus be a far more expensive proposal than first envisaged.

Well understood and proven often has a negative connotation in that many technically challenged instrumentation personnel would rather follow well-established instrumentation solutions even if, as is often the case, they are extremely outdated.

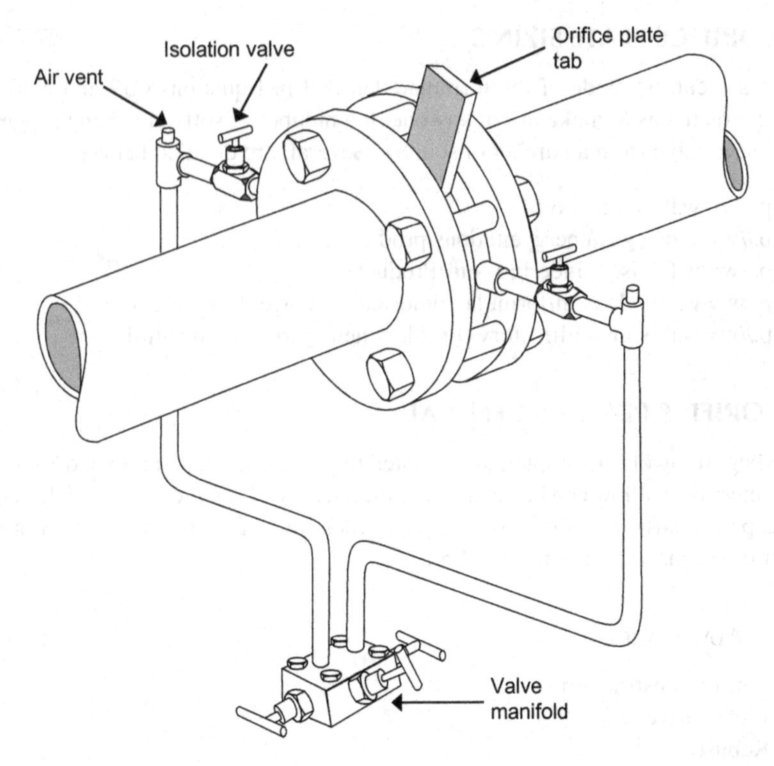

FIGURE 6.22 The orifice plate is only one of a number of ancillary components that includes the flange plate assembly, the isolation valves, the impulse tubing, the valve manifold, and the differential pressure transmitter (not shown).

6.9.2 Disadvantages

- High permanent pressure loss
- Inaccuracy, typically 2–3%
- Low turndown ratio, typically from 3 to 4:1
- Only suitable for clean non-abrasive fluids
- Accuracy is affected by density, pressure, and viscosity fluctuations
- Erosion and physical damage to the restriction compromises measurement accuracy without the operator being aware of the fact
- Requires a homogeneous single phase liquid
- Viscosity limits measuring range
- Requires straight pipe runs to ensure accuracy is maintained
- Pipeline must be full (typically for liquids)
- Output is not linearly related to flow rate
- Multiple potential leakage points
- Always gives a reading irrespective of the damage to plate
- Only provides a valid reading for fully developed flow profile
- Requires bypass to replace the plate under pressure

Again, a few observations regarding some of these 'disadvantages' are in order.

Low turndown ratio is a direct result for the need for square root extraction which severely limits the range over which the instrument can operate.

Slurries and dirty fluids cannot be measured with standard concentric orifice plates. However, half-circle or eccentric bores can be used for these applications but with the limitation of increasing the error by up to 3–5 times.

Straight run piping requirements, both before and after the orifice plate element, are rarely met – often through ignorance; often through a 'we'll probably get away with it' attitude; but, more often, because of the piping layout was designed well ahead of the instrumentation requirements.

Without flow-straightening, a typical installation requires 25–40 pipe diameter of straight run piping before the element and about 4–5 pipe diameter downstream of the element. These requirements vary quite considerably according to the upstream (and downstream) discontinuities and the beta ratio. Typically:

β-ratio of 0.5: 25 pipe diameter upstream and (25D) and 4 pipe diameter downstream (4D).
β-ratio of 0.7: 40D upstream and 5D downstream

The requirements for custody transfer applications are considerably more onerous:

The ASME (MFC 3M) requires up to 54D upstream and 5D downstream
AGA (Report Number 3) specifies up to 95D upstream and 4.2D downstream
ISO 5167 specifies up to 60D upstream and 7D downstream

The requirements for the API (RP550) are no less rigorous and specify the type of upstream and downstream disturbance (e.g. valve, elbow, double elbow, etc.) as illustrated in Figures 6.23 and 6.24.

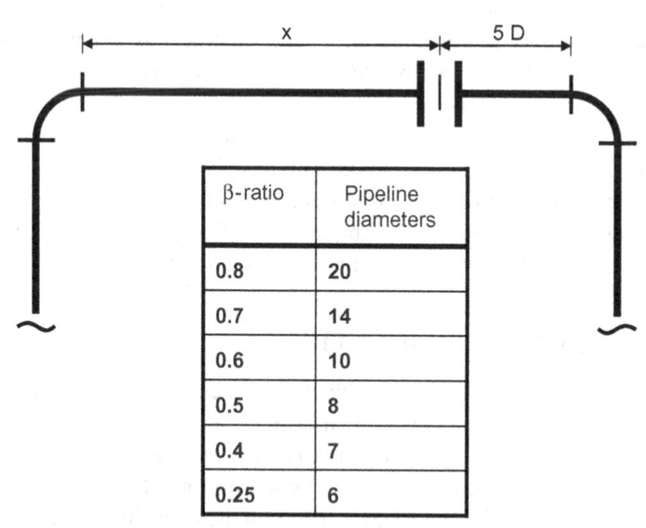

β-ratio	Pipeline diameters
0.8	20
0.7	14
0.6	10
0.5	8
0.4	7
0.25	6

FIGURE 6.23 API (RP550) straight-pipe run lengths for a single upstream and downstream elbow, for differing beta ratios.

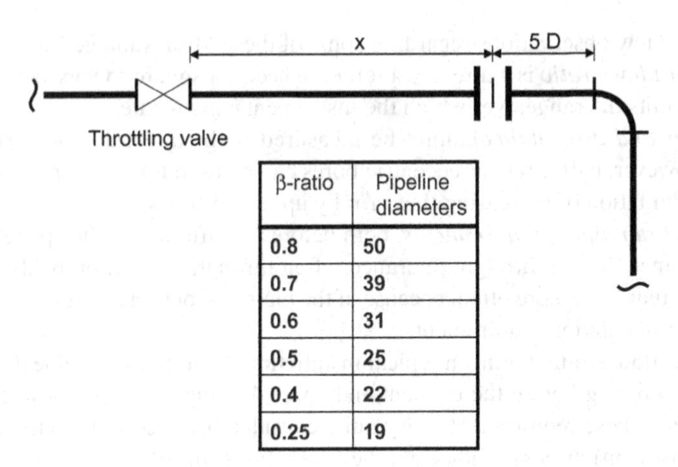

β-ratio	Pipeline diameters
0.8	50
0.7	39
0.6	31
0.5	25
0.4	22
0.25	19

FIGURE 6.24 API (RP550) straight-pipe run lengths for an upstream valve and a single downstream elbow.

Multiple leakage points. Figure 6.25 clearly illustrates the potential for multiple leakage points. Whilst many of these can be eliminated using continuous welded impulse tubing, the risks associated with blockages are increased.

6.10 ORIFICE PLATE THICKNESS

As the differential pressure across the orifice increases, the plate tends to deform elastically and, beyond a certain point, the deformation results in a shift in the meter characteristics and an increase in the measurement uncertainty.

The thickness of an orifice plate should thus be sufficient to ensure that the deflection does not exceed certain limits. The thickness is generally determined according to the guidelines given by ISO-5167; ISA-RP-3.2; API-2530; and ASME-MFC-3M. These are shown in Table 6.1.

In addition, the AGA-3 Appendix-2-F provides guidelines for using high differential for measurement of natural gas. This maximum limit is dependent upon the thickness, diameter, and beta ratio. For a given line size, there is always a maximum allowable differential pressure on the plate for example for 50 DN pipe, the maximum allowable ΔP is 1,000 MPa in 2.5 bar with a minimum thickness of 3.2 mm.

6.11 CONDITIONING ORIFICE PLATE

The 'Conditioning Orifice Plate', introduced by Emerson, is a differential pressure producer that differs from the conventional orifice plate by using four equally spaced holes (Figure 6.26) whose sum of the area is equivalent to the area of a bore in a conventional plate. This causes the flow to condition itself as it is forced through the four holes – eliminating swirl and irregular flow profiles and removing the requirement for a flow conditioner.

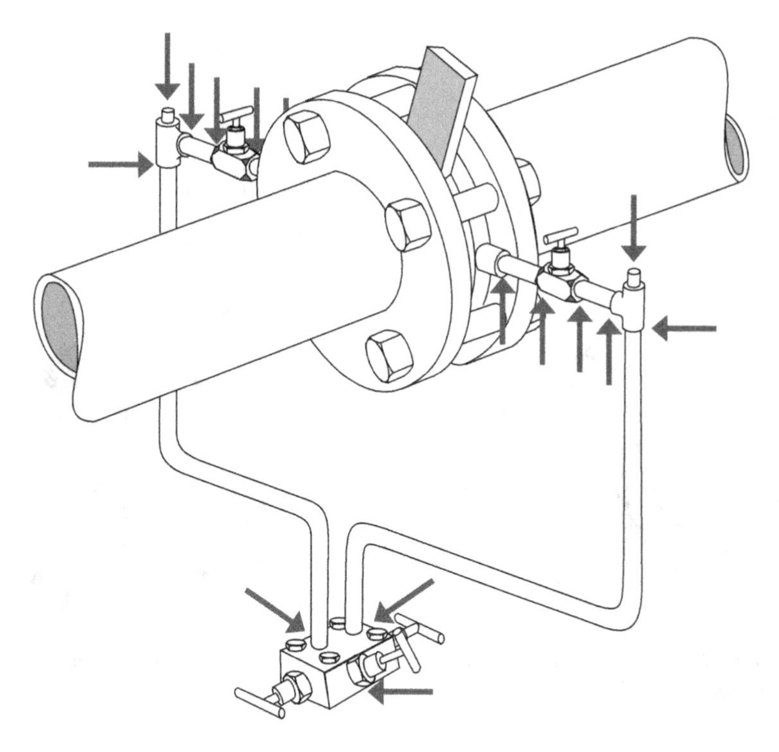

FIGURE 6.25 Potential for multiple leakage points.

Straight-run requirements are reduced to only 2 upstream and 2 downstream pipe diameters. Furthermore, the discharge coefficient (C_d) uncertainty is reduced to ±0.5% (for $\beta = 0.4$).

The *Conditioning Orifice Plate* is designed with 2 standard bore sizes, one for high flow rates and one for low flow rates, having bores equal to β of 0.4 and 0.65.

6.12 SEGMENTAL WEDGE METER

The segmental wedge element has a V-shaped restriction cast or welded into a flanged meter body that creates a differential pressure. The restriction is characterised by the

TABLE 6.1
Orifice Plate Thickness According to Pipeline Size

Line Size (DN)	Thickness in mm
<150	3.18
>200 to <400	6.125
>450	9.53

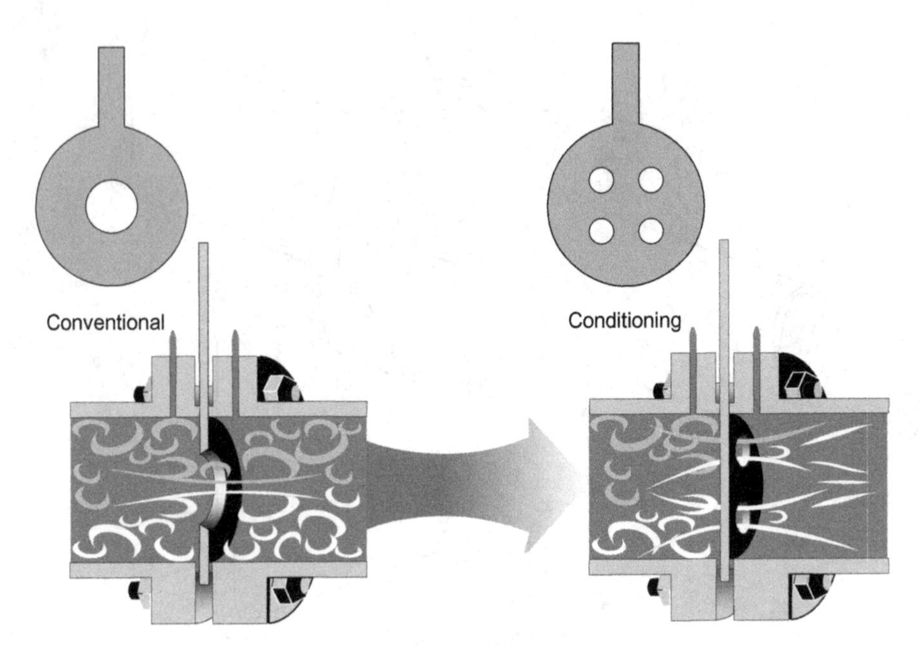

FIGURE 6.26 Conventional concentric orifice plate compared with Conditioning Orifice Plate. (Courtesy: Emerson.)

h/D ratio (Figure 6.27), corresponding to the β ratio of an orifice plate, where h is the height of the opening below the restriction (the only critical dimension) and D is the ID of the pipe.

The slanting upstream face of the wedge element is insensitive to wear and creates a sweeping action that has a scouring effect that helps to keep it clean and free of build-up. Further, because the wedge does not restrict the bottom of the pipe, it can be used for a variety of corrosive, erosive, and highly viscous fluids and slurries.

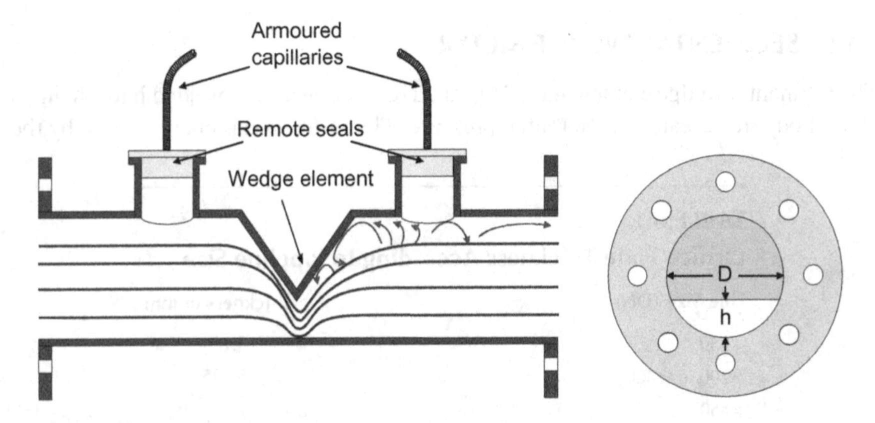

FIGURE 6.27 The segmental wedge element has a V-shaped restriction cast or welded into a flanged meter body that creates a differential pressure.

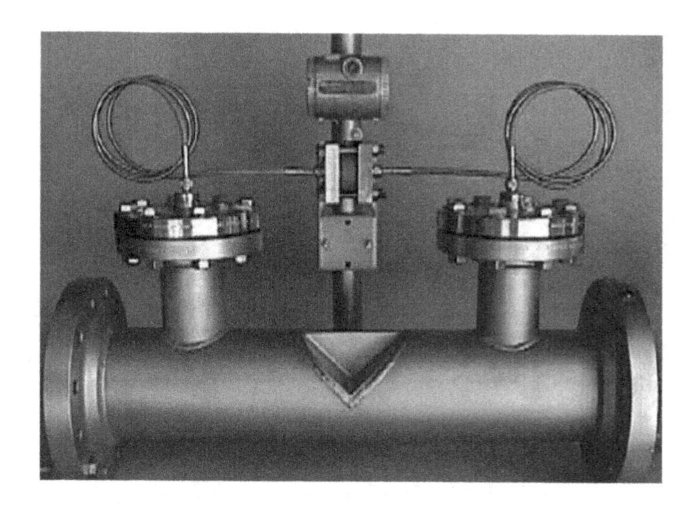

FIGURE 6.28 A typical segmental wedge meter is provided as a complete assembly combining the wedge element and the pressure taps into a one-piece unit. (Courtesy: PFS Inc.)

The discharge coefficient (C_d) is stable for Reynolds numbers of less than 500 – allowing it to be used down to laminar flow regimes. Further, because the discharge coefficient is highly insensitive to velocity profile distortion and swirl, only 5 pipe diameters of relaxation piping is required upstream of the meter for most common combinations of fittings and valves. An uncalibrated element has a C_d uncertainty of 2–5% whilst for a calibrated system it is 0.5%. The largest source of measurement error is generally due to the variations in the density which, if not measured, is taken as an assumed 'normal' value.

A typical segmental wedge meter is provided as a complete assembly combining the wedge element and the pressure taps into a one-piece unit (Figure 6.28). The upstream and downstream pressure taps are usually in the form of remote diaphragm seals – eliminating the need for lead lines.

Because of its symmetrical design the primary segmental wedge can be used to measure bidirectional flow but would require two differential pressure transmitters.

6.13 V-CONE METER

The V-Cone Flowmeter from McCrometer is a patented technology that features a centrally located cone inside the flow tube that interacts with the fluid and creates a region of lower pressure immediately downstream of the cone.

The pressure difference is measured between the upstream static line pressure tap, placed slightly upstream of the cone, and the downstream low pressure tap located in the downstream face of the cone (Figure 6.29).

Because the cone is suspended in the centre of the pipe, it interacts directly with the high velocity core of the flow – forcing it to mix with the lower velocity flow closer to the pipe walls. As a consequence, the flow profile is flattened towards the shape of a well-developed profile – even under extreme conditions, such as single or double elbows out-of-plane positioned closely upstream of the meter.

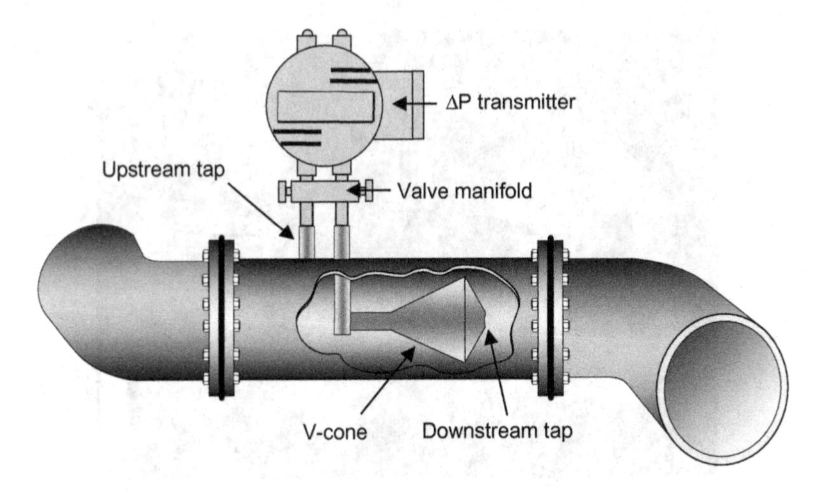

FIGURE 6.29 The pressure difference is measured between the upstream static line pressure tap, placed slightly upstream of the cone, and the downstream low pressure tap located in the downstream face of the cone. (Courtesy: McCrometer.)

Other major features of the V-cone flowmeter include:

- 0–3 D straight run piping upstream and 0 to 1 D downstream
- Primary element accuracy of ±0.5% of reading with a repeatability of ±0.1% or better
- Turndown ratio 10:1 with Reynolds numbers as low as 8,000
- Suitable for use with dirty fluids

6.14 VENTURI TUBE METER

The Venturi tube (Figure 6.30) has tapered inlet and outlet sections with a central parallel section, called the throat, where the low pressure tapping is located.

Generally, the inlet section, which provides a smooth approach to the throat, has a steeper angle than the downstream section. The shallower angle of the downstream section reduces the overall permanent pressure loss by decelerating the flow smoothly and thus minimising turbulence.

Consequently, its permanent pressure loss is only about 10% of the differential pressure – as shown previously in Figure 6.5. At the same time, its relatively streamlined form allows it to handle about 60% more flow than for example an orifice plate.

The Venturi tube also has relatively high accuracy: better than ±0.75% over the orifice ratios (d/D) of 0.3–0.75. This order of accuracy, however, can only be obtained as long as the dimensional accuracy is maintained. Consequently, although the Venturi tube can also be used with fluids carrying a relatively high percentage of entrained solids, it is not well suited for abrasive media.

Although generally regarded as the best choice of a differential type meter for bores over 1,000 mm, the major disadvantage of the Venturi type meter is its high

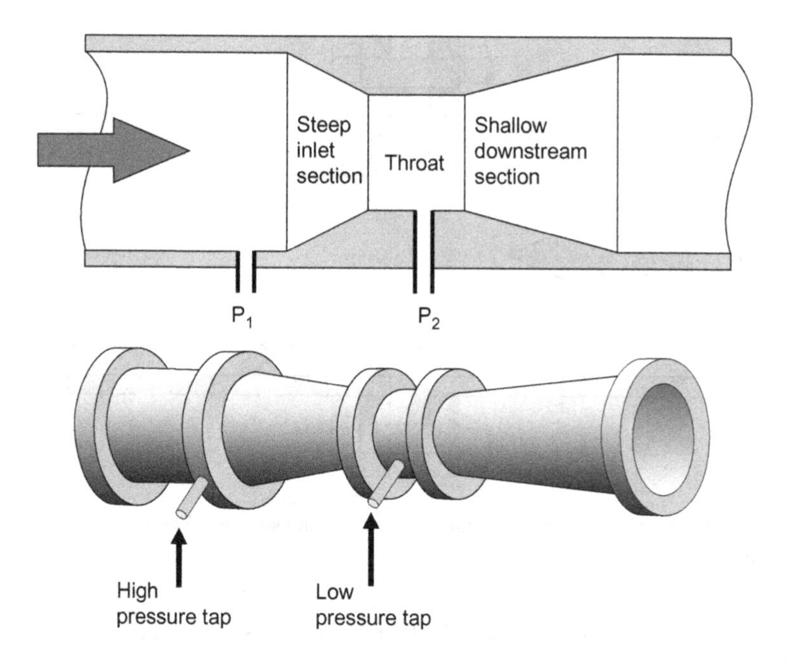

FIGURE 6.30 The Venturi tube has tapered inlet and outlet sections with a central parallel section. (Courtesy: Emerson.)

cost – about 20 times more expensive than an orifice plate. In addition, its large and awkward size makes it difficult to install since a 1 m bore Venturi is 4–5 m in length. Although it is possible to shorten the length of the divergent outlet section by up to 35%, thus reducing the high manufacturing cost without greatly affecting the characteristics, this is at the expense of an increased pressure loss.

6.14.1 Advantages of the Venturi Tube

- Less significant pressure drop across restriction
- Less unrecoverable pressure loss
- Requires less straight pipe up- and downstream

6.14.2 Disadvantages of the Venturi Tube

- More expensive
- Bulky – requires large section for installation

6.15 VENTURI NOZZLE METERS

The Venturi nozzle is an adaptation of the standard Venturi that makes use of a 'nozzle'-shaped inlet (Figure 6.31), a short throat, and a flared downstream expansion section. Whilst increasing the permanent pressure loss to around 25% of the measured differential pressure of the standard Venturi, the Venturi nozzle is cheaper, requires

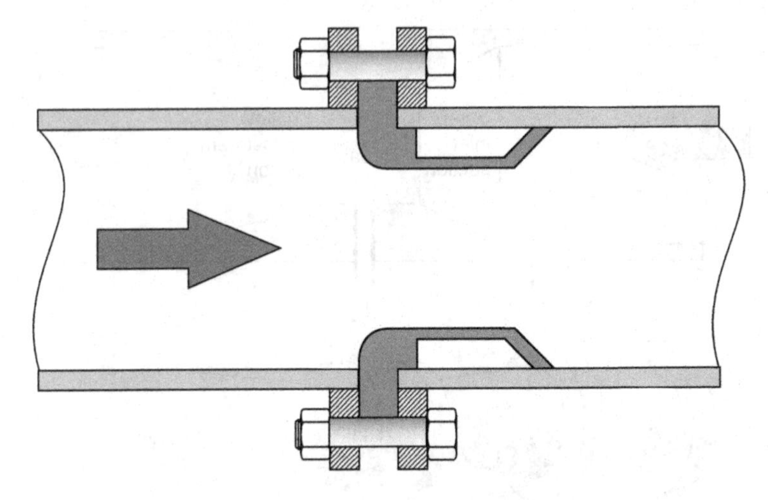

FIGURE 6.31 The Venturi nozzle is an adaptation of the standard Venturi using a 'nozzle'-shaped inlet.

less space for installation, and yet still retains the benefits of high accuracy (±0.75%) and high velocity flow.

6.16 FLOW NOZZLE METERS

The flow nozzle (Figure 6.32) is used mainly in high velocity applications or where fluids are being discharged into the atmosphere. It differs from the Venturi nozzle in that it retains the nozzle inlet but has no exit section.

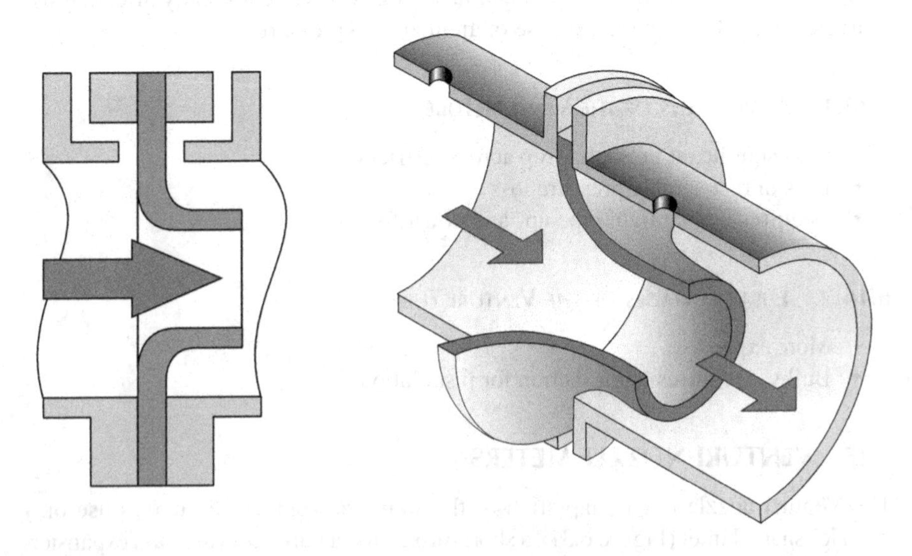

FIGURE 6.32 The flow nozzle is used mainly in high velocity applications. (Courtesy: Emerson.)

The main disadvantage of the flow nozzle is that the permanent pressure loss is increased to between 30% and 80% of the measured differential pressure – depending on its design.

Offsetting this disadvantage, however, accuracy is only slightly less than for the Venturi tube (±1–1.5%) and it is usually only half the cost of the standard Venturi. In addition it requires far less space for installation and, because the nozzle can be mounted between flanges or in a carrier, installation, and maintenance are much easier than for the Venturi.

6.17 DALL TUBE

Although many variations of low-loss meters have appeared on the market, the best-known and most commercially successful is the Dall tube (Figure 6.33).

The Dall tube essentially comprises two cones separated by an annular gap. The inlet cone has a short steep converging section that starts at a stepped buttress or inlet shoulder whose diameter is somewhat less than the pipe diameter. Following the annular gap at the 'throat', there is a diverging cone that again finishes at a step.

High pressure is measured at the upstream edge of the inlet shoulder whilst low pressure is measured at the slotted throat – the area between the two cones – providing an average 'throat' pressure.

Because of the annular gap, no breakaway of the liquid from the wall occurs at the throat and the flow leaves the throat as a diverging jet. Since this jet follows the walls of the diverging cone, eddy losses are practically eliminated, while friction losses are small because of the short length of the inlet and outlet sections.

The main disadvantages are high sensitivity to both Reynolds number and manufacturing complexity.

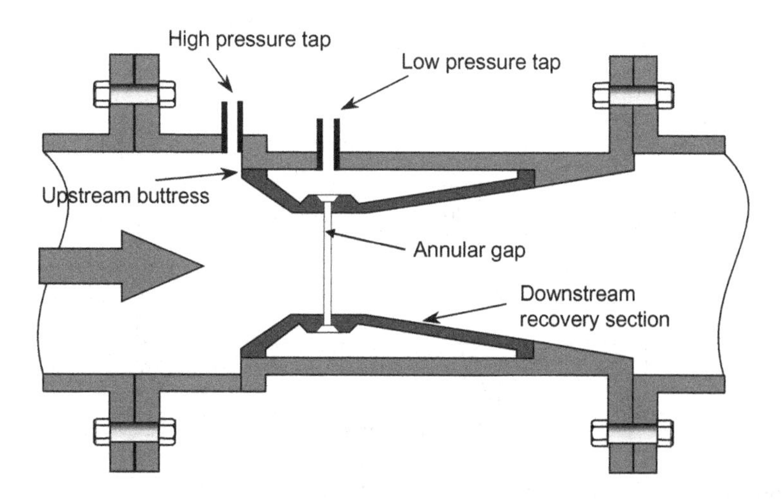

FIGURE 6.33 Dall tube low-loss meter.

6.18 PITOT TUBE

The Pitot tube is one of the oldest devices for measuring the velocity and is frequently used to determine the velocity profile in a pipe by measuring the velocity at various points.

In its simplest form the Pitot tube (Figure 6.34) comprises a small tube inserted into a pipe with the head bent so that the mouth of the tube faces into the flow. As a result, a small sample of the flowing medium impinges on the open end of the tube and is brought to rest. Thus, the kinetic energy of the fluid is transformed into potential energy in the form of a head pressure (also called stagnation pressure).

▶ Mathematically this can be expressed by applying Bernoulli's equation to a point in the small tube and a point in the free flow region. From Bernoulli's general equation:

$$P_1 + \frac{1}{2} \cdot \rho \cdot v_1^2 + \rho \cdot g \cdot h_1 = P_2 + \frac{1}{2} \cdot \rho \cdot v_2^2 + \rho \cdot g \cdot h_2 \tag{6.25}$$

since the velocity (v_1) at the impact hole is zero and dividing through by the density (ρ) we can now write:

$$\frac{P_h}{\rho} + 0 + g \cdot h_1 = \frac{P_s}{\rho} + \frac{\nu^2}{2} + g \cdot h_2 \tag{6.26}$$

where

P_s = static pressure

P_h = stagnation pressure

v = liquid velocity

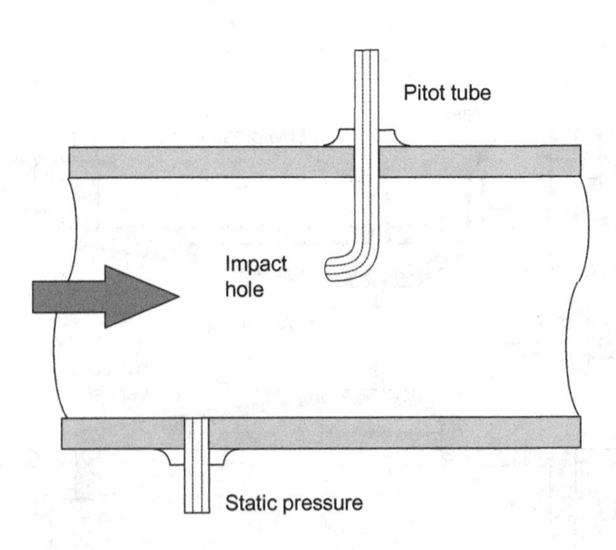

FIGURE 6.34 Basic Pitot tube illustrating the principle of operation.

g = acceleration due to gravity
h_1 and h_2 = heads of the liquid at the static and stagnation pressure measuring points, respectively

If $h_1 = h_2$ then:

$$v = \sqrt{\frac{2 \cdot (P_h - P_s)}{\rho}} \tag{6.27}$$

Because the Pitot tube is an intrusive device and some of the flow is deflected around the mouth, a compensatory flow coefficient K_p is required. Thus:

$$v = K_p \cdot \sqrt{\frac{2 \cdot \Delta P}{\rho}} \tag{6.28}$$

For compressible fluids at high velocities (e.g. >100 m/s in air) a modified equation should be used. ◀

By measuring the static pressure with a convenient tapping, the flow velocity can be determined from the difference between the head pressure and the static pressure. This difference, measured by a differential pressure cell, provides a measurement of flow that, like a conventional differential pressure measurement, obeys a square root relationship to pressure. Low flow measurement at the bottom end of the scale is thus difficult to achieve accurately.

A problem with this basic configuration is that the flow coefficient K_p depends on the tube design and the location of the static tap.

One means of overcoming this problem is to use a system as shown in Figure 6.35 that makes use of a pair of concentric tubes – the inner tube measuring the full head pressure and the outer tube using static holes to measure the static pressure.

Both these designs of Pitot tube measure the point velocity. However, provided a fully developed turbulent profile exists, a rough indication of the average velocity

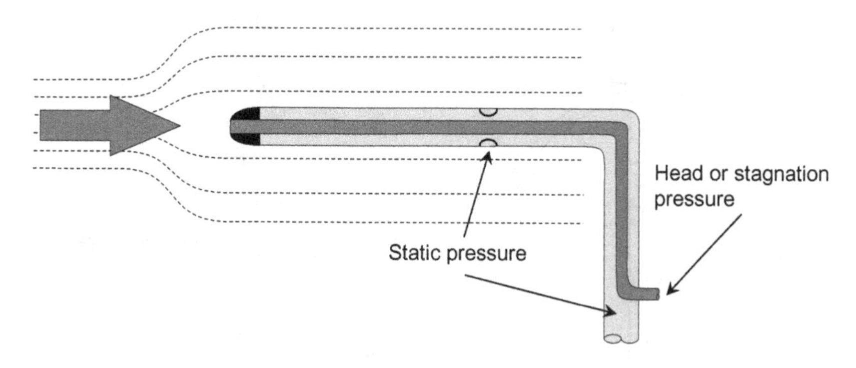

FIGURE 6.35 Integrated Pitot tube system in which the inner tube measures the head pressure and the outer tube uses static holes to measure the static pressure.

can be obtained by positioning the tube at a point three-quarters of the way between the centreline and the pipe wall.

6.19 POINT AVERAGING PITOT TUBE

Another method of determining the average velocity is with a point averaging Pitot tube system (Figure 6.36).

Essentially, this instrument comprises two back-to-back sensing bars that span the pipe, in which the up- and downstream pressures are sensed by a number of critically located holes. The holes in the upstream detection bar are arranged so that the average pressure is equal to the value corresponding to the average of the flow profile.

Because the point at which the fluid separates from the sensor varies according to the flow rate (Figure 6.37) extreme care must be taken in positioning the static pressure sensing holes. One solution is to locate the static pressure point just before the changing separation point.

Alternatively, a 'shaped' sensor, such as that typified by Rosemount's Annubar, (Figure 6.38) can be used to establish a fixed point where the fluid separates from the sensor.

As illustrated in Figure 6.39, this comprises a T-shaped bar comprising a single high-pressure plenum and three common low-pressure plenums. The impact (high (P_H)) pressure is measured with frontal slots whilst the stagnation (low (P_L)) pressure is measured with sensing ports located in the stagnation zones at the rear of the sensor.

When properly installed 'Annubar' type instruments have a turndown ratio of 14:1 and an accuracy of 0.8% of the flow rate. However, accurate alignment can be

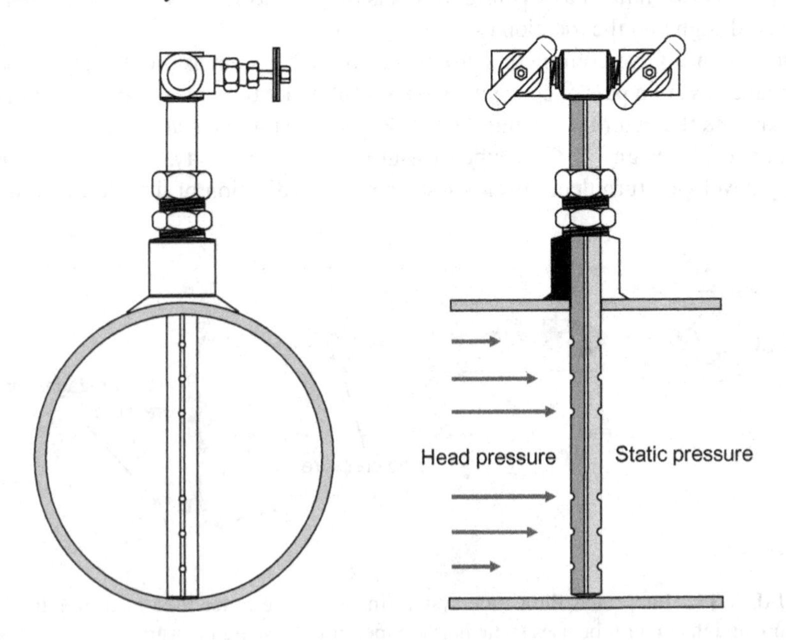

FIGURE 6.36 Multiport Pitot averaging system.

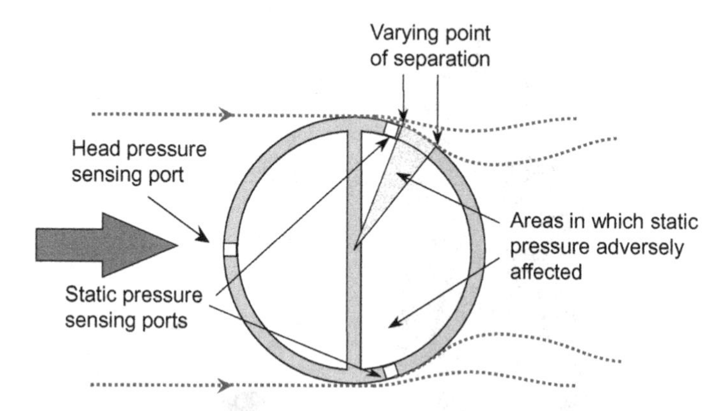

FIGURE 6.37 Variation in flow velocity can affect the point of separation and the downstream static pressure measurement.

critical. A port facing slightly upstream is subject to part of the stagnation pressure whilst a static port facing slightly downstream is subjected to a slightly reduced impact pressure.

Used mainly to meter flows in large bore pipes, multiport-type instruments offer a low pressure drop and application on a wide range of fluids – particularly water and steam. Although intrusive, they average the flow profile across the diameter of the pipe bore and are thus less sensitive to the flow profile than for example an orifice plate, and can be used as little as 2½ pipe diameters downstream of a discontinuity.

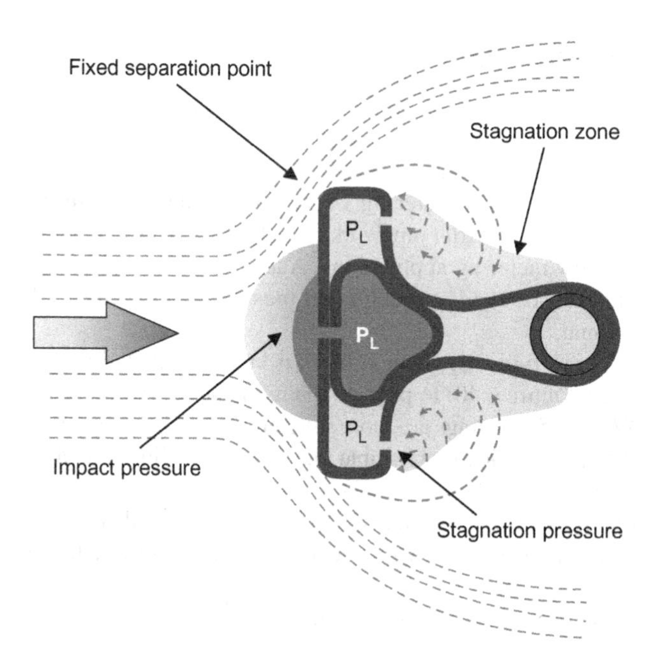

FIGURE 6.38 'Shaped' bluff body establishes a fixed separation point. (Courtesy: Emerson.)

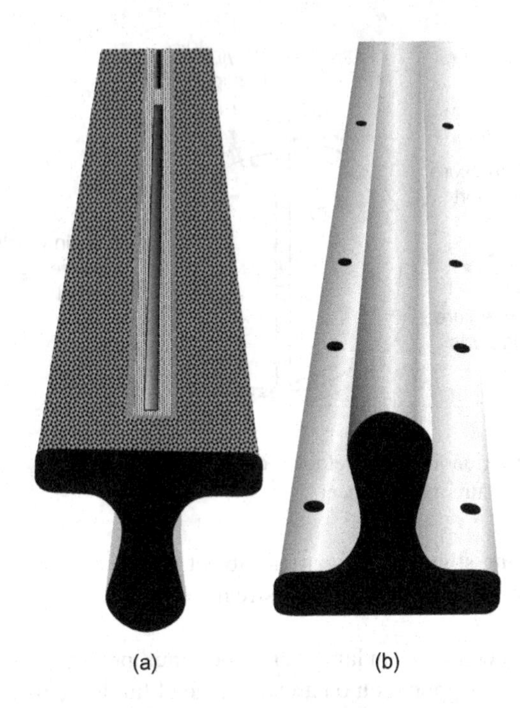

FIGURE 6.39 (a) The impact (high (P_H)) pressure is measured with frontal slots whilst (b) the stagnation (low (P_L)) pressure is measured with sensing ports located in the stagnation zones at the rear of the sensor. (Courtesy: Emerson.)

6.20 ELBOW METERING

Although not commonly used, in applications where cost is a factor and additional pressure loss from an orifice plate is not permitted, a pipe elbow can be used as a differential pressure primary device.

The application of elbow metering is certainly underrated since its low cost, together with its application after pipework completion, can be a major benefit for low accuracy liquid flow metering applications. And since most piping systems already have elbows that can be used, no additional pressure drop occurs and the expense involved is minimal.

Elbow meters are typically formed by two tappings drilled at an angle of 45° through the bend (Figure 6.40) to provide the high and low pressure tapping points, respectively. Whilst 45° tappings are more suited for bi-directional flow measurement, tappings at 22.5° can provide more stable and reliable readings and are less affected by upstream piping.

Because a number of factors contribute to producing the differential pressure, it is difficult to predict the exact flow rate accurately. Some of these factors include:

- Force of the flow onto the outer tapping
- Turbulence generated due to cross-axial flow at the bend
- Differing velocities between outer and inner radii of flow

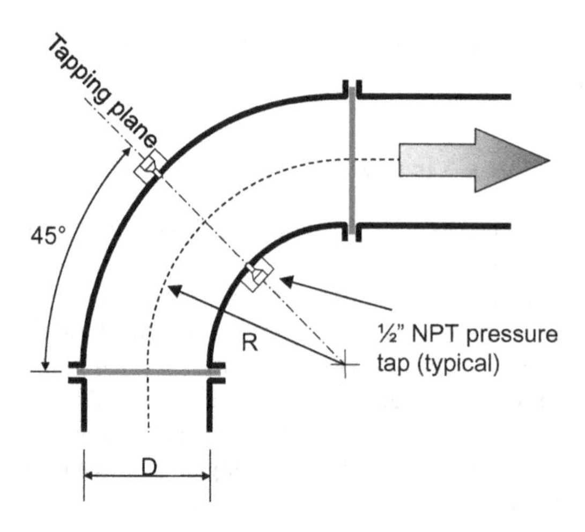

FIGURE 6.40 Elbow meter geometry.

- Pipe texture
- Relationship between elbow radius and pipe diameter

Generally, the elbow meter is only suitable for higher velocity liquid flow and cannot produce an accuracy of better than ±4–5%. However, on-site calibration can produce more accurate results, with the added advantage of high repeatability.

A general rule of thumb is that the elbow should have 25 pipe diameters of straight pipe upstream and at least 10 pipe diameters of straight pipe downstream.

6.20.1 Advantages of Elbow Meter

- No limitations on the line size
- Uncalibrated accuracy of approximately ±4–5%
- Repeatability of the order of ±0.20%
- No additional head loss attributable to the meter
- Relatively low acquisition cost
- Suitable for bi-directional flow measurement
- Minimum Reynolds number of 50,000 with no maximum limit

6.20.2 Disadvantages of Elbow Meter

- Discharge coefficient is non-linear and varies with Reynolds number
- Differential pressure produced by the meter is significantly lower than other primary head loss elements
- Limited to liquid (incompressible) fluids

Although the elbow flow meter is essentially confined to liquid (incompressible) applications it is possible to meter gas flow but requires significant additional calibration work to empirically establish the adiabatic expansion factor specific to a given meter and application.

6.21 VARIABLE AREA METER

The variable area flowmeter is a reverse differential pressure meter used to measure the flow rate of liquids and gases.

The instrument generally comprises a vertical, tapered glass tube and a weighted float whose diameter is approximately the same as the tube base (Figure 6.41).

In operation, the fluid or gas flows through the inverted conical tube from the bottom to the top, carrying the float upwards. Since the diameter of the tube increases in the upward direction the float rises to a point where the upward force on the float created by differential pressure across the annular gap, between the float and the tube, equals the weight of the float.

As shown in Figure 6.41, the three forces acting on the float are:

- Constant gravitational force W
- Buoyancy A that, according to Archimedes' principle, is constant if the fluid density is constant
- Force S, the upward force of the fluid flowing past the float

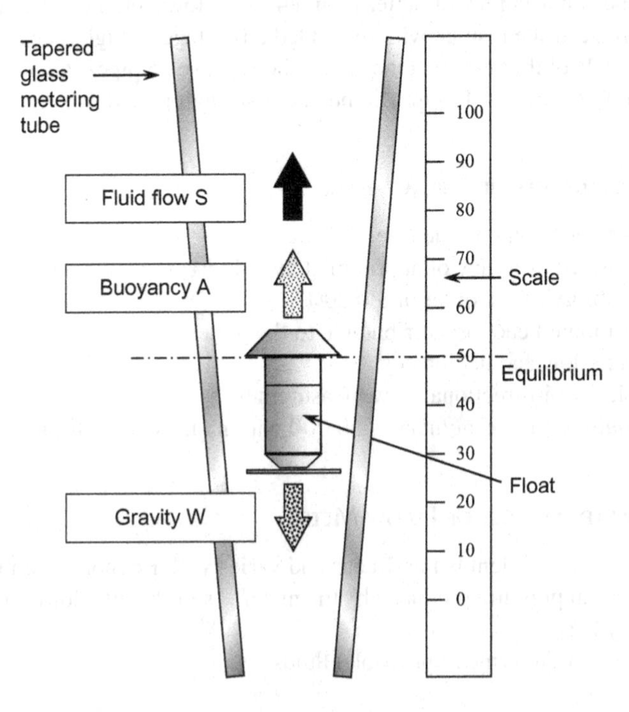

FIGURE 6.41 Basic configuration of a variable area flowmeter. (Courtesy: Brooks Instrument.)

For a given instrument, when the float is stationary, W and A are constant and S must also be constant. In a position of equilibrium (floating state) the sum of forces S + A is opposite and equal to W and the float position corresponds to a particular flow rate that can be read off a scale. A major advantage of the variable area flowmeter is that the flow rate is directly proportional to the orifice area that, in turn, can be made to be linearly proportional to the vertical displacement of the float. Thus, unlike most differential pressure systems, it is unnecessary to carry out square root extraction.

The taper can be ground to give special desirable characteristics such as an offset of higher resolution at low flows.

▶ In a typical variable area flowmeter, the flow q can be shown to be approximately given by:

$$q = C \cdot A \cdot \sqrt{\rho} \tag{6.29}$$

where

q = flow
C = constant that depends mainly on the float
A = cross-sectional area for fluid flow past the float
ρ = density of the fluid

Indicated flow, therefore, depends on the density of the fluid which, in the case of gases, varies strongly with the temperature, pressure, and composition of the gas.

It is possible to extend the range of variable area flowmeters by combining an orifice plate in parallel with the flow meter. ◀

6.21.1 Floats

The float material is largely determined by the medium and the flow range and includes stainless steel, titanium, aluminium, black glass, synthetic sapphire, polypropylene, Teflon, PVC, hard rubber, Monel, nickel, and Hastelloy C.

An important requirement for accurate metering is that the float is exactly centred in the metering tube. One of three methods is usually applied:

Slots in the float head cause the float to rotate and centre itself and prevent it sticking to the walls of the tube (Figure 6.42). This arrangement led to the term 'Rotameter' being applied to all variable area flow meters.* Slots cannot be applied to all float shapes and, further, can cause the indicated flow to become slightly viscosity dependent.

Three *moulded ribs* within the metering tube cone (Figure 6.43), parallel to the tube axis, guide the float and keep it centred. This principle allows a variety of float shapes to be used and the metering edge remains visible even when metering opaque fluids.

---

* The term 'Rotameter' is currently a registered brand name of the British company GEC Elliot automation, Rotameter Co. In many other countries, the brand name Rotameter is registered by Rota Yokogawa GmbH & Co.

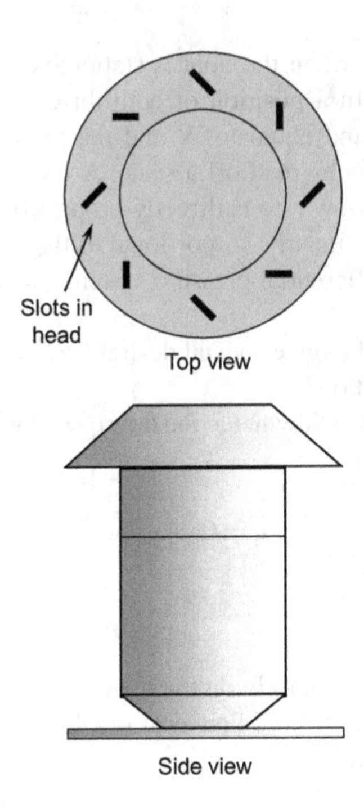

FIGURE 6.42 Float centring in which a slotted float head rotates and automatically centres itself.

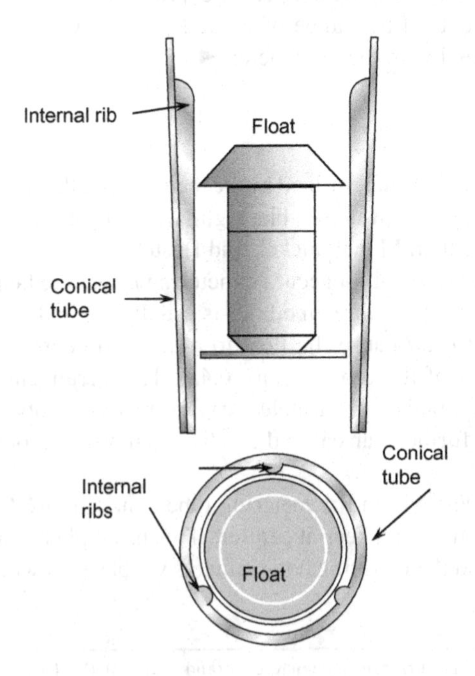

FIGURE 6.43 Float centring in which the float is centred by three moulded ribs parallel to the tube axis.

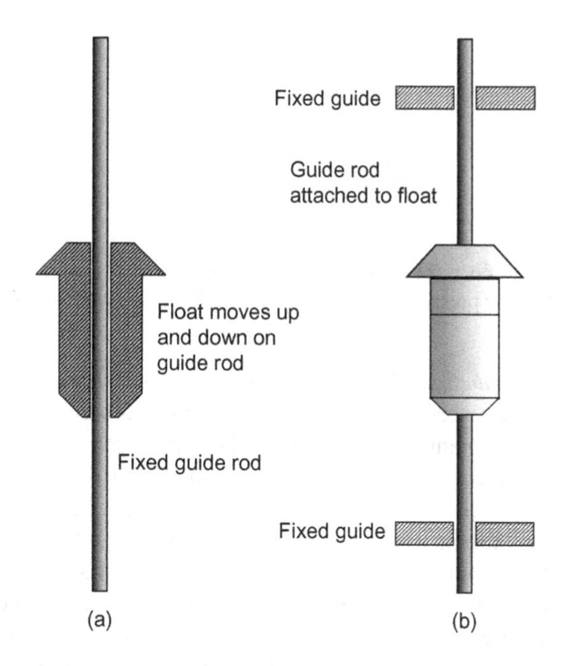

FIGURE 6.44 Float is centred by (a) fixed centre guide rod; or (b) guide rod attached to the float. (Courtesy: ABB Fischer & Porter.)

A *fixed centre guide rod* within the metering tube (Figure 6.44a) is used to guide the float and keep it centred. Alternatively, the rod may be attached to the float and moved within fixed guides (Figure 6.44b). The use of guide rods is confined mainly to applications where the fluid stream is subject to pulsations likely to cause the float to 'chatter' and possibly, in extreme cases, break the tube. It is also used extensively in metal metering tubes.

6.21.2 Float Shapes

The design of the floats is confined to four basic shapes (Figure 6.45):

- Ball float
- Rotating (viscosity non-immune) float
- Viscosity immune float
- Float for low pressure losses

The *ball float* (Figure 6.45a) is mainly used as a metering element for small flowmeters – with its weight determined by selecting from a variety of materials.

Figure 6.46 shows the effect of viscosity on the flow rate indication. Since its shape cannot be changed, the flow coefficient is clearly defined (1) and, as shown, exhibits virtually no linear region. Thus, any change in viscosity, due often to small changes in temperature, results in changes in indication.

Rotating floats (Figure 6.45b) are used in larger sized meters and are characterised by a relatively narrow linear (viscosity-immune) region as shown in Figure 6.46 (2).

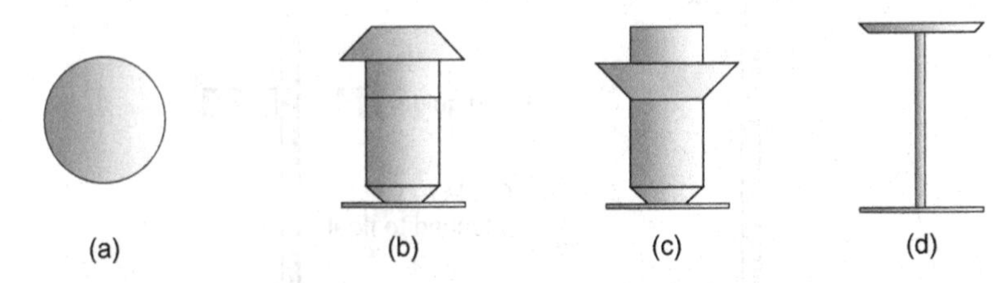

FIGURE 6.45 (a) Ball float, (b) rotating (viscosity non-immune) float, (c) viscosity immune float, and (d) float for low pressure losses. (Courtesy: ABB Fischer & Porter.)

The *viscosity immune float* (Figure 6.45c) is appreciably less sensitive to changes in viscosity and is characterised by a wider linear region as shown in Figure 6.46 (3). Although such an instrument is unaffected by relatively large changes in viscosity, the same size meter has a span 25% smaller than the previously described rotating float.

For gas flow rate metering, low pressure loss floats (Figure 6.45d) can be used. The pressure drop across the instrument is due, primarily, to the float since the energy required to produce the metering effect is derived from the pressure drop of the flowing fluid. This pressure drop is independent of the float height and is constant.

Further the pressure drop is due to the meter fittings (connection and mounting devices) and increases with the square of the flow rate. For this reason, the design requires a minimum upstream pressure.

6.21.3 Metering Tube

The meter tube is normally manufactured from borosilicate glass that is suitable for metering process medium temperatures up to 200°C (392°F) and pressures up to about 20–30 bar.

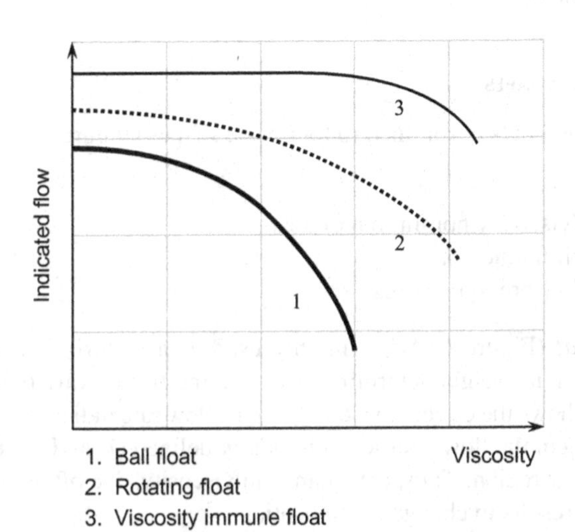

FIGURE 6.46 Viscosity effect for various float shapes. (Courtesy: ABB Fischer & Porter.)

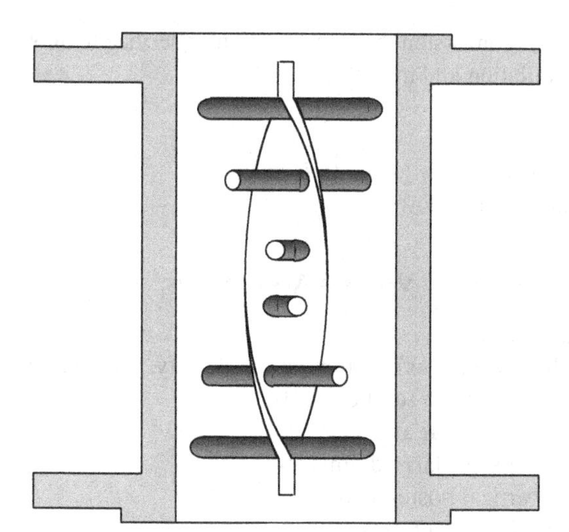

FIGURE 6.47 Typical magnetic filter. (Courtesy: Krohne.)

Because the glass tube is vulnerable to damage from thermal shocks and pressure hammering, it is often necessary to provide a protective shield around the tube.

Variable area meters are inherently self-cleaning since the fluid flow between the tube wall and the float provides a scouring action that discourages the build-up of foreign matter. Nonetheless, if the fluid is dirty, the tube can become coated – affecting calibration and preventing the scale from being read. This effect can be minimised through the use of an in-line filter.

In some applications use can be made of an opaque tube used in conjunction with a float follower. Such tubes can be made from steel, stainless steel, or plastic.

By using a float with a built-in permanent magnet, externally mounted reed-relays can be used to detect upper and lower flow limits and initiate the appropriate action.

The temperature and pressure range may be considerably extended (e.g. up to 400°C (752°F) and 7 bar) through the use of a stainless steel metering tube. Again, the float can incorporate a built-in permanent magnet that is coupled to an external field sensor that provides a flow reading on a meter.

In cases where the fluid might contain ferromagnetic particles that could adhere to the magnetic float, a magnetic filter should be installed upstream of the flowmeter. Typically (Figure 6.47) such a filter contains bar magnets, coated with polytetrafluoroethylene (PTFE) as protection against corrosion, arranged in a helical fashion.

Generally, variable area flowmeters have uncertainties ranging from 1% to 3% of full scale. Precision instruments are, however, available with uncertainties down to 0.4% of full scale.

6.21.4 Advantages of Variable Area Meters

- Linear float response to flow rate change
- 10 to 1 flow range or turndown ratio

- Easy sizing or conversion from one particular service to another
- Ease of installation and maintenance
- Simplicity
- Low cost
- High low-flow accuracy (down to 5 cm^3/min)
- Easy visualisation of flow

6.21.5 Disadvantages of Variable Area Meters

- Limited accuracy
- Susceptibility to changes in temperature, density, and viscosity
- Fluid must be clean, no solid contents
- Erosion of device (wear and tear)
- Can be expensive for large diameters
- Operates in vertical position only
- Accessories required for data transmission

6.22 DIFFERENTIAL PRESSURE TRANSMITTERS

Measurement of the differential pressure is carried out with a differential pressure transmitter whose role is to measure the differential pressure and convert it to an electrical signal (Figure 6.48).

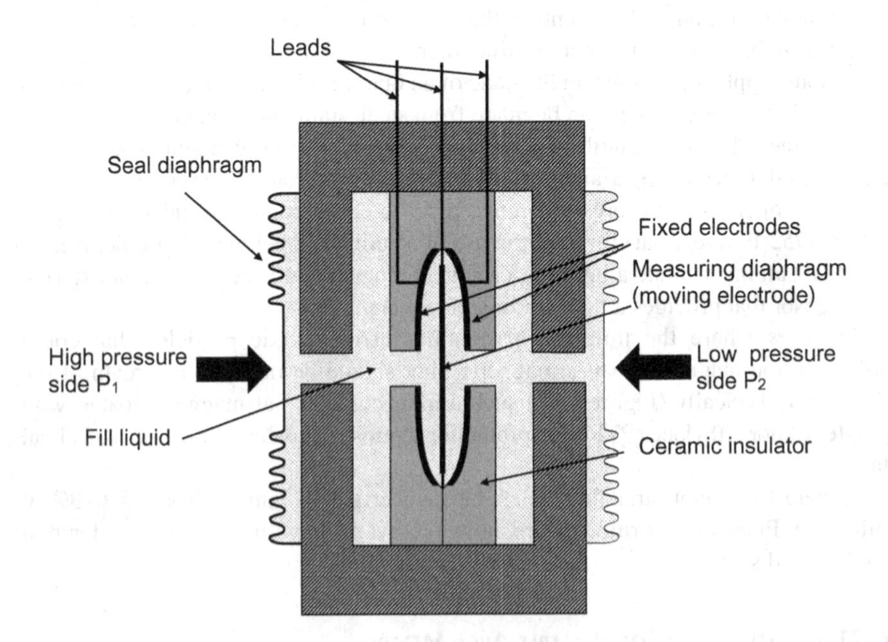

FIGURE 6.48 Basic construction of a capacitive differential pressure sensor in which the movement of the isolation diaphragms is transmitted via the isolating fluid to the measuring diaphragm whose deflection is a measure of the differential pressure.

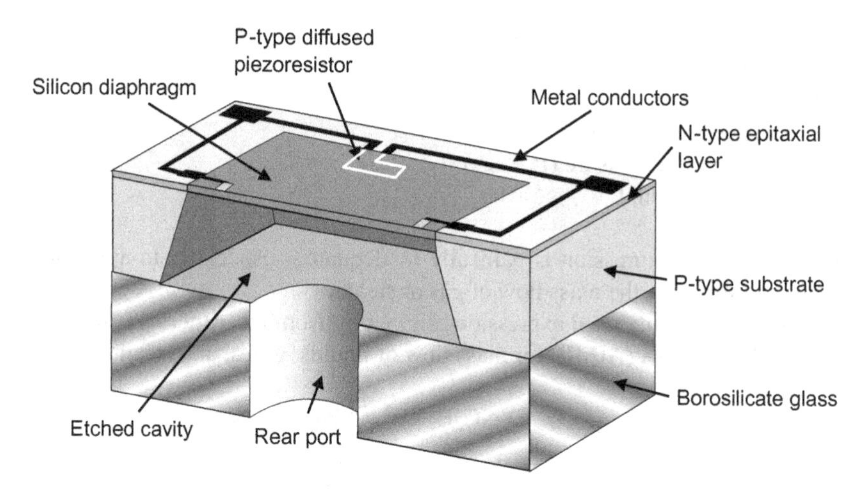

FIGURE 6.49 Piezoresistive element in which piezoresistors are diffused into the surface of an n-type silicon wafer.

Isolation diaphragms isolate the measuring cell from the process media. Variations in pressure, applied to the isolation diaphragms, are transmitted via the isolating fluid (e.g. silicon fluid) to the measuring diaphragm whose deflection is a measure of the differential pressure.

Measurement of the deflection of the measuring diaphragm may be carried out by a number of methods including inductance, strain gauge, and piezoelectric. However, the most popular method of measuring differential pressure, adopted by a large number of manufacturers, is the variable capacitance transmitter.

As illustrated (Figure 6.48), the sensing diaphragm is in the form of a movable electrode. As the electrode changes its distance from the fixed plate electrodes this produces a change in capacitance that, in turn, produces a varying electrical signal.

Capacitance-based transmitters are simple, reliable, accurate (typically 0.1% or better), small in size and weight, and remain stable over a wide temperature range. The main advantage of the capacitive transmitter is that it is extremely sensitive to small changes in pressure – down to 0.001 bar.

Several manufacturers make extensive use of a piezoresistive element, in which piezoresistors are diffused into the surface of an n-type silicon wafer (Figure 6.49). Such silicon-on-insulator devices are now capable of providing continuous operation at temperatures up to 220°C (428°F) at pressures of up to 70 bar.

6.23 MULTIVARIABLE TRANSMITTERS

► At the beginning of this chapter it was shown that the differential pressure can be related to flow by the expression:

$$Q = kC_d\sqrt{\frac{\Delta P}{\rho}} \tag{6.30}$$

where

Q = flow rate
k = constant
C_d = discharge coefficient
ΔP = differential pressure $(P_1 - P_2)$
ρ = density of fluid

In practice this expression is painfully inadequate – especially in applications involving for example the mass flow of gas or steam.

The most commonly used expression originating from the American Institute of Mechanical Engineers (AIME) for mass flow of liquids, gases, and steam is:

$$Q_m = NC_dE_vY_1d^2\sqrt{\Delta P\rho} \qquad (6.31)$$

where

Q_m = mass flow rate
N = units factor
C_d = discharge coefficient
E_v = velocity of approach factor
Y_1 = gas expansion factor (= 1 for liquid)
d = bore diameter
DP = differential pressure
ρ = density of fluid

Using this equation, the traditional approach has been to make use of three separate transmitters to measure differential pressure, static pressure, and temperature to infer the mass flow. As shown in Figure 6.50, the density of a gas may be deduced from the *measurement* of static pressure and temperature combined with the entry of certain known constants: that is the compression factor, gas constant, molecular weight, and fluid constant. ◀

Several companies have now developed a single transmitter solution (multivariable transmitter) that makes simultaneous measurement of differential pressure, static pressure, and temperature and provides the on-board computation.

Apart from providing tremendous cost savings in purchase price as well as installation, such multivariable transmitters provide accurate mass flow measurements of process gases (combustion air and fuel gases) and steam, whether saturated or superheated. Other applications include DP measurement across filters and in distillation columns where the user is concerned with the static pressure and temperature measurements to infer composition and in liquid flow rate applications where density and viscosity compensation is required due to large temperature changes.

6.24 IMPULSE LINES

Impulse lines (also referred to as sensing lines) are small-bore pipes that connect the upstream and downstream primary element taps (e.g. from the orifice plate) to the differential pressure transmitter.

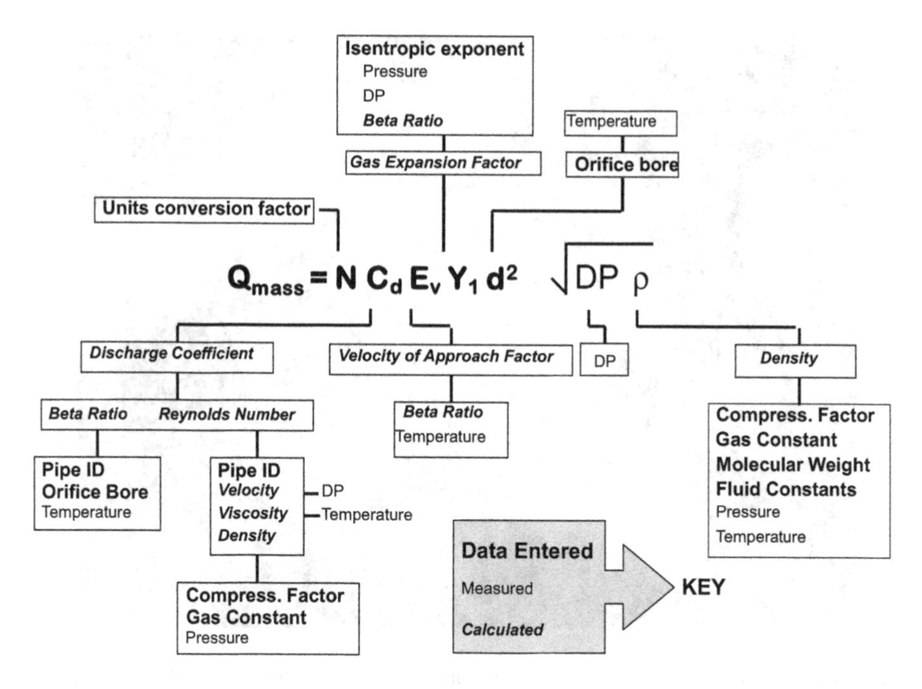

FIGURE 6.50 Computation of fully compensated mass flow requires the measurement of DP, static pressure, and temperature. (Courtesy: Emerson.)

Impulse lines also require additional ancillary equipment (Figure 6.51) including isolation valves (sometimes called root valves) and a valve manifold. A manifold valve consolidates multiple single valves into a single valve block – allowing users to perform multiple functions without removing the pressure transmitter from its installed position. The single block also minimises the number of connections and potential leak points.

The three-valve manifold (Figure 6.52), shown inside the dotted box, comprises an equalising valve and two block valves. An extra bleed valve is used to vent the trapped fluid pressure to atmosphere. In its normal working mode, the block valves are open and the equalising valve is closed. When taking the transmitter out of commission, the high-pressure (HP) block valve is first closed, the equalising valve is opened, and finally the low-pressure (LP) valve is closed. The reverse action is followed when putting the transmitter back into commission.*

The presence of a built-in bleed valve in the five-valve manifold (Figure 6.53) makes it easier for users to purge the transmitter. Checking the transmitter zero is accomplished by closing the block valves and opening the equalising valve.

Typically, use is made of 316 stainless steel – although offshore applications normally make use of Duplex steel to avoid salt spray corrosion.

* Note. The equalising valve(s) should never be opened while both block valves are also open since this would allow the process fluid to flow through the equalising valve(s) from the high-pressure to the low-pressure side of the process.

FIGURE 6.51 Impulse lines are used with a range of ancillary equipment including isolation valves and a valve manifold.

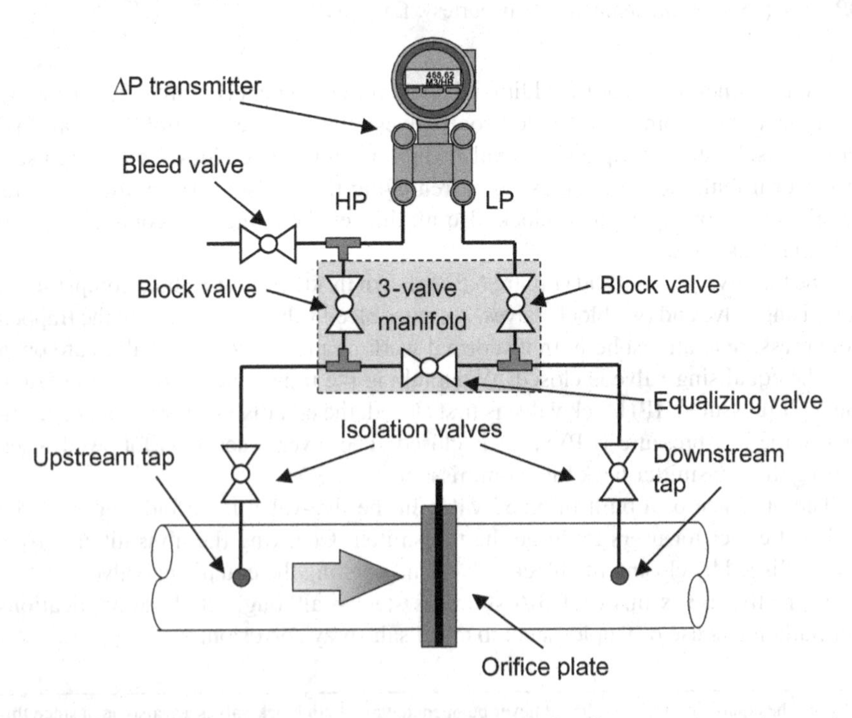

FIGURE 6.52 Typical impulse line and valve connection incorporating a three-valve manifold.

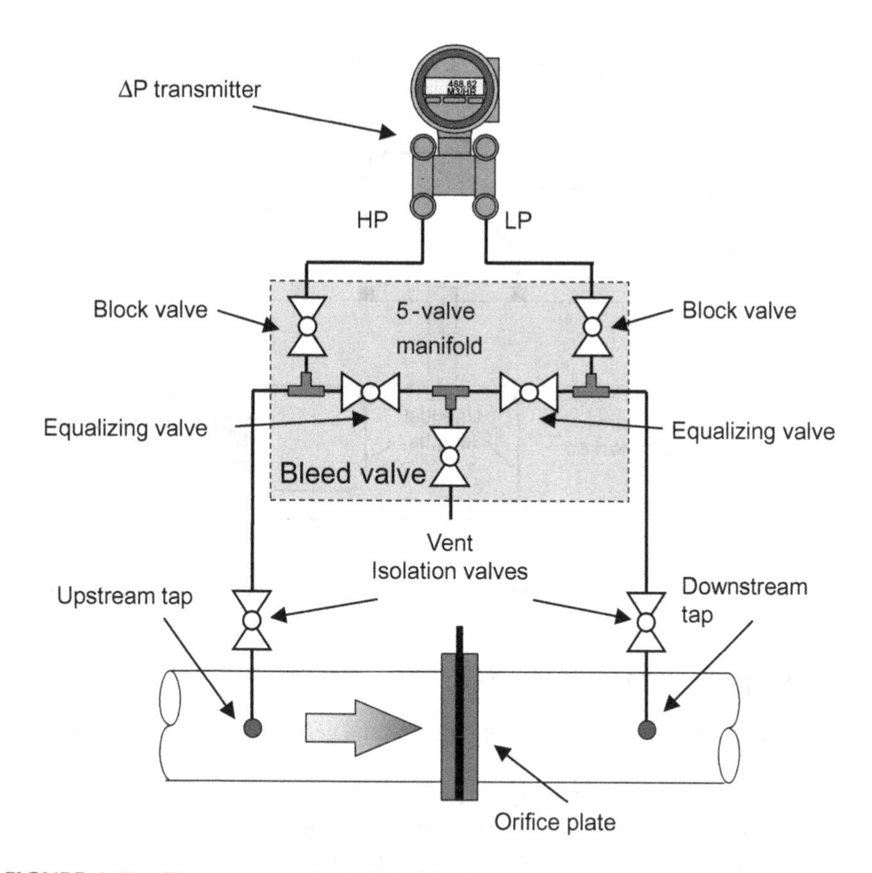

FIGURE 6.53 The presence of a built-in bleed valve in the five-valve manifold makes it easier for users to purge the transmitter.

Pipe bore diameter recommendations vary considerably. For example, ISO 2186 recommends pipe ID from as little as 7 up to 25 mm depending on the impulse tube length and application. Because of the risk of plugging or blocking of the tubes most process control applications recommend a minimum ID of 16 to 18 mm (5/8″ to 3/4″). And for high temperatures in condensing vapour service, 25 mm (1″) is specified to allow for unimpeded flow of condensate.

Despite all the foregoing recommendations it should be remembered that most transmitters have standardised on 12.7 mm (½″) ports. Consequently, it is rather pointless installing 25 mm (1″) tubing and still have a ½″ bottleneck at the transmitter.

The general recommendation for maximum impulse tube length is: '… keep it as short as possible'. Nonetheless, ISO 2186 provides for lengths of up to 45 m – although a previous reference providing for lengths up to 90 m was discarded. A common rule of thumb is that the maximum length should be kept to about 15 m if some of the problems illustrated in Figure 6.54 are to be avoided.

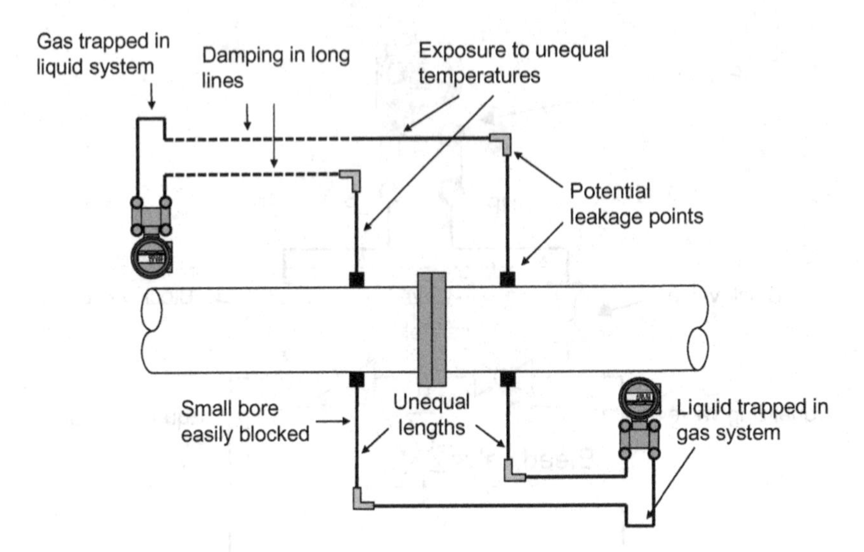

FIGURE 6.54 Potential problems and bad practice encountered with the installation of impulse lines.

6.25 PIPING ARRANGEMENTS

To avoid the problems illustrated in Figure 6.54 the general recommendations are:

- Use short impulse lines of equal length
- Provide effective venting of trapped gas in a liquid-filled line
- Provide draining of any liquid in a gas-filled line
- Use recommended configurations – determined by the process media
- If necessary, use heat tracing and/or insulation to ensure equal temperatures in both lines
- For pulsating flow:
    - Avoid changes in the diameter of the impulse lines diameter
    - Ensure pulsation and impulse line frequencies do not coincide

To avoid gas or vapour bubbles in liquid-filled lines and liquid in gas-filled lines, liquid-filled lines should slope downwards from the pressure tapping to the transmitter and gas-filled lines should flow upwards.

6.25.1 Liquid in Horizontal Pipes

As indicated, liquid-filled lines should generally slope downwards from the pressure tapping to the transmitter (Figure 6.55a). Consequently, the primary element taps may lie somewhere between the horizontal centreline and 60° below the centreline (3 to 5 o'clock) (Figure 6.55b). Taps at the bottom dead-centre should be avoided since they may accumulate solids and taps above the centreline may accumulate air or non-condensing gases.

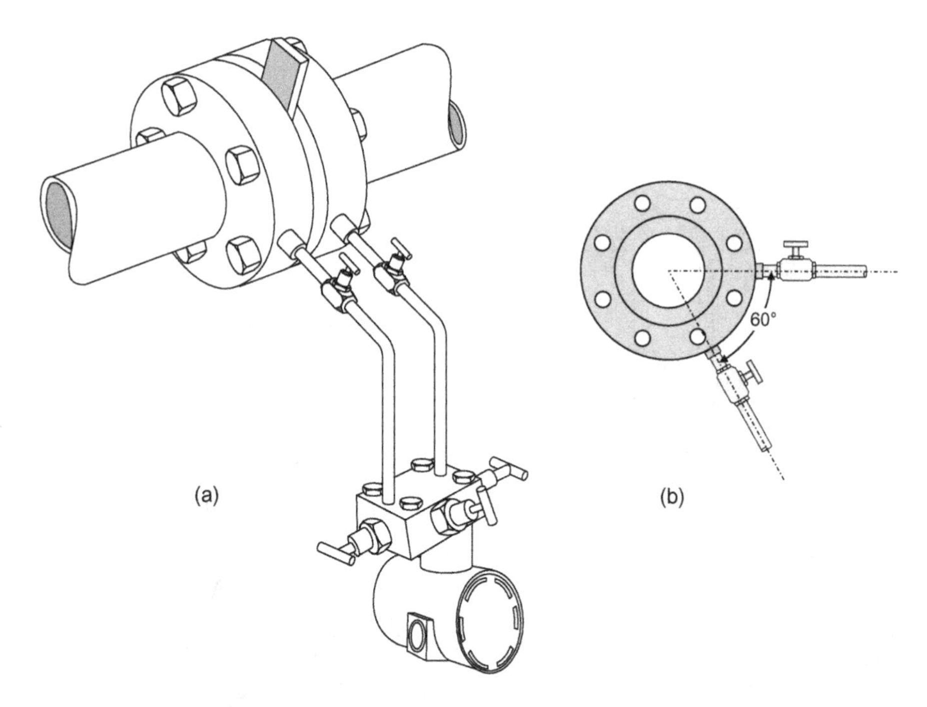

FIGURE 6.55 (a) Liquid-filled lines should generally slope downwards from the pressure tapping to the transmitter. (b) The taps may lie somewhere between the horizontal centreline and 60° below the centreline (3 to 5 o'clock).

6.25.2 Gas in Horizontal Pipes

To avoid liquid collecting in gas-filled lines the lines should flow upwards – with the transmitter mounted above the pipe (Figure 6.56a). Here, the taps may lie somewhere between the vertical centreline and 60° below it (12 to 2 o'clock) (Figure 6.56b).

6.25.3 Steam in Horizontal Pipes

Since standard ΔP transmitters typically have a maximum temperature limit of around 100°C (212°F), due consideration must be observed when measuring steam where temperatures can run as high as 800°C (1,500°F). Furthermore, depending on the temperature and pressure it can be in the liquid or gaseous phase. Consequently the pipework must cater for the presence of the gas or liquid.

As illustrated in Figure 6.57, the pressure taps should lie on the horizontal centreline (3 o'clock) – with the vertical lines filled with condensate.

During start-up the transmitter could be exposed to the steam temperature before the lines fill with condensate. Consequently, plugged T-fittings are normally provided to enable the vertical impulse lines and transmitter to be filled with water prior to start-up.

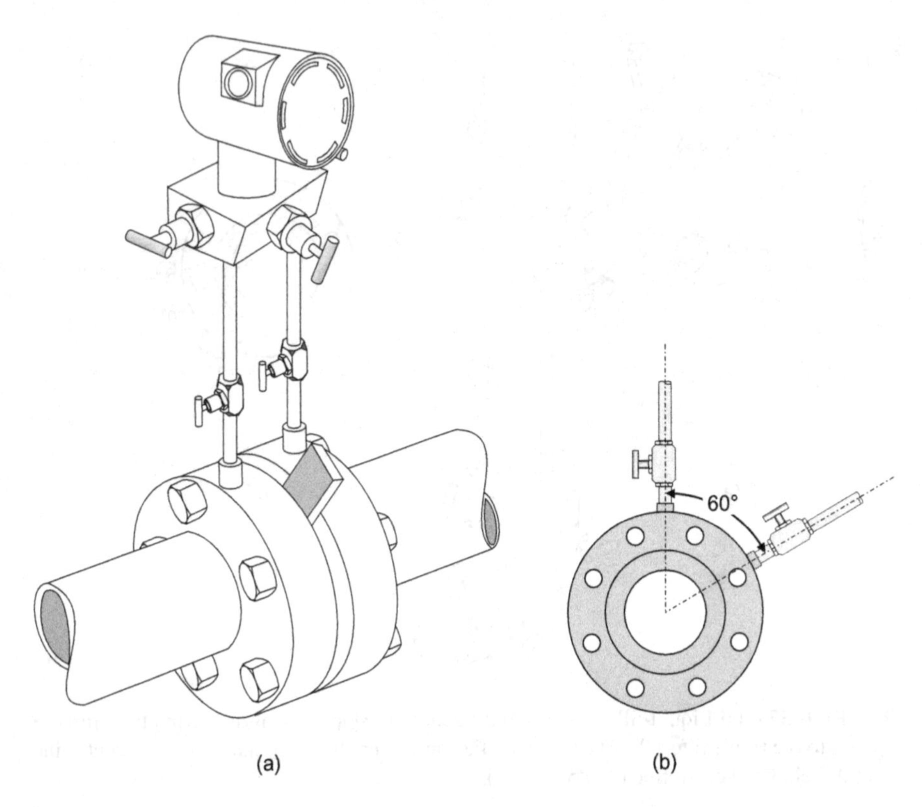

FIGURE 6.56 Gas-filled lines the lines should flow upwards – with the transmitter mounted above the pipe. (b) The taps may lie somewhere between the vertical centerline and 60° below it (12 to 2 o'clock).

6.25.4 Liquid in Vertical Pipes

In vertical flow, the slight difference in the height of the tappings (Figure 6.58) may require a calibration offset on the transmitter – depending on the density of the liquid. The piping should be brought out horizontally for as short a distance as possible and then down to the transmitter below both taps.

6.25.5 Gas in Vertical Pipes

As in horizontal pipes the lines should flow upwards in order to avoid liquid collecting in the gas-filled lines – with the transmitter mounted above the pipe (Figure 6.59).

6.25.6 Steam in Vertical Pipes

In horizontal pipes where the vertical impulse lines and transmitter are filled with liquid, the liquid heads are identical. However, in a vertical pipe there is a difference

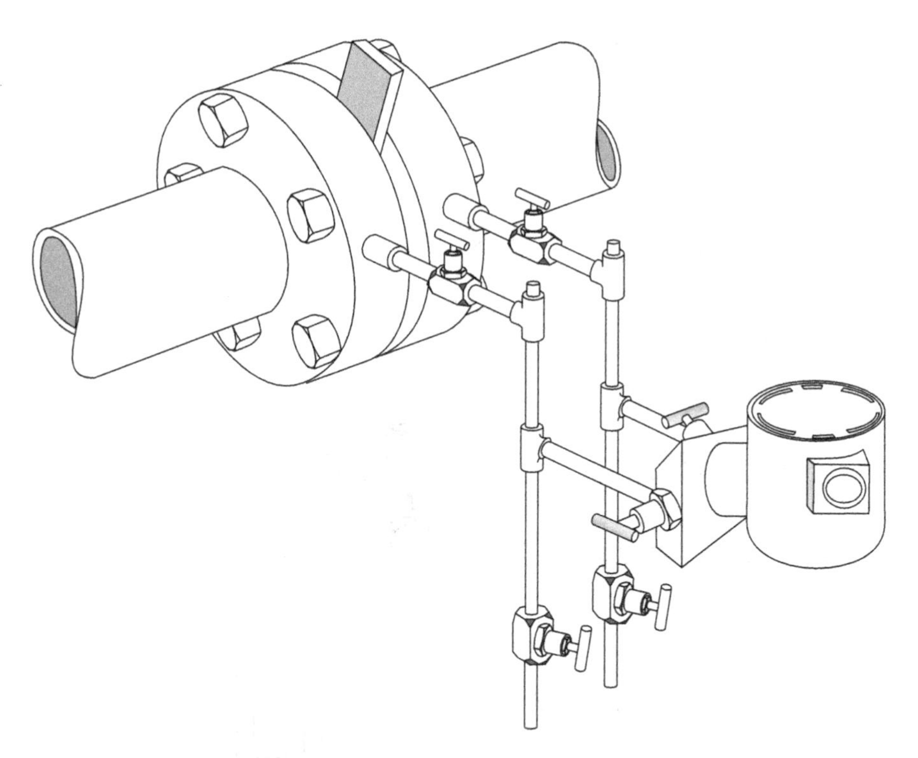

FIGURE 6.57 For steam service, the pressure taps should lie on the horizontal centreline – with the vertical lines filled with condensate.

in the liquid head between the tappings. Consequently, there is a pressure difference between the tappings even under static conditions. Compensation for this pressure difference can be made in one of two ways.

In one method (Figure 6.60), the lower impulse line is brought up to the same height as the upper impulse line before dropping down to the transmitter. This arrangement equalises the liquid head above the transmitter in the two impulse lines and avoids the need for offset correction to be made during calibration.

Alternatively, both impulse lines leave the tappings horizontally and then turn down to the transmitter. Offset correction to the transmitter is then performed during calibration to compensate for the difference in the liquid head.

6.26 PLUGGED LINE DETECTION

Because of their small diameter impulse tubes are prone to plugging – often due to freezing or to products solidifying in the line. Blocking may also occur as a result of poor installation in which isolation valves are improperly aligned or seated or the impulse tubes are actually crimped. Unfortunately, despite blocking of one or

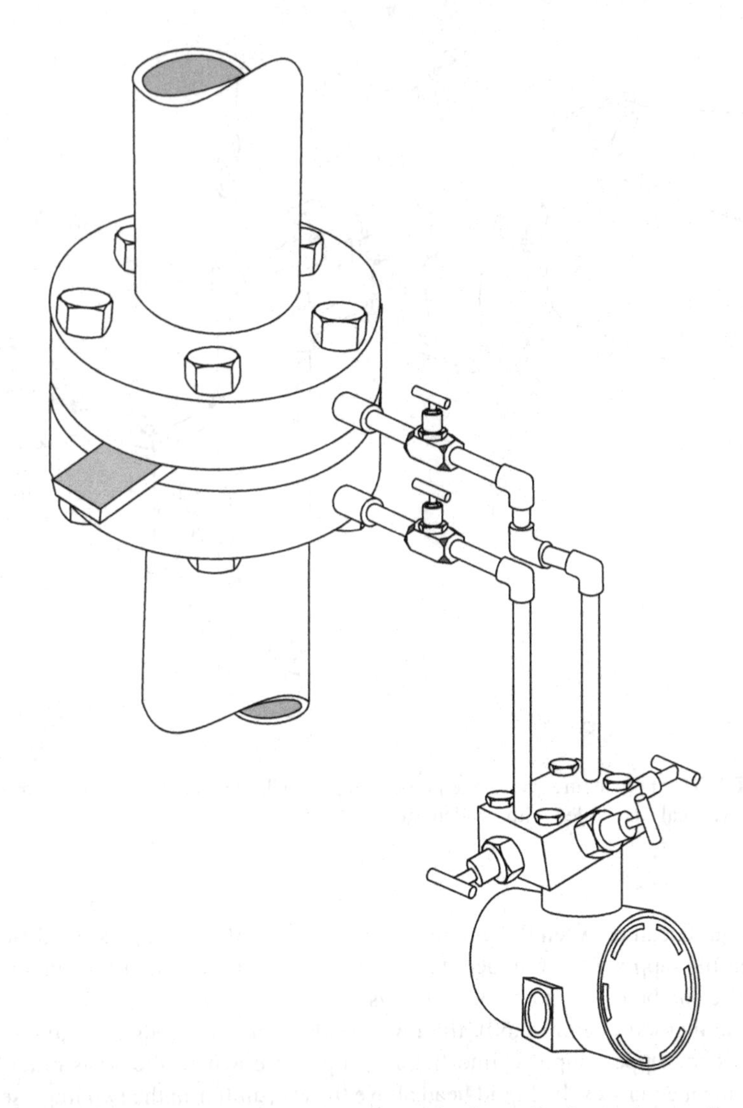

FIGURE 6.58 The piping should be brought out horizontally for as short a distance as possible and then down to the transmitter below both taps.

even both lines, the measurements continue to look acceptable. Consequently, such problems are often extremely difficult to detect.

To help operators identify the existence of plugged impulse lines, several manufacturers have developed a technology that uses the background noise of the process by comparing the current performance of the transmitter to a background 'signature' determined during commissioning.

Flow measurement is essentially 'noisy' due to any number of factors that include: variations in vortex shedding from upstream/downstream pipe discontinuities,

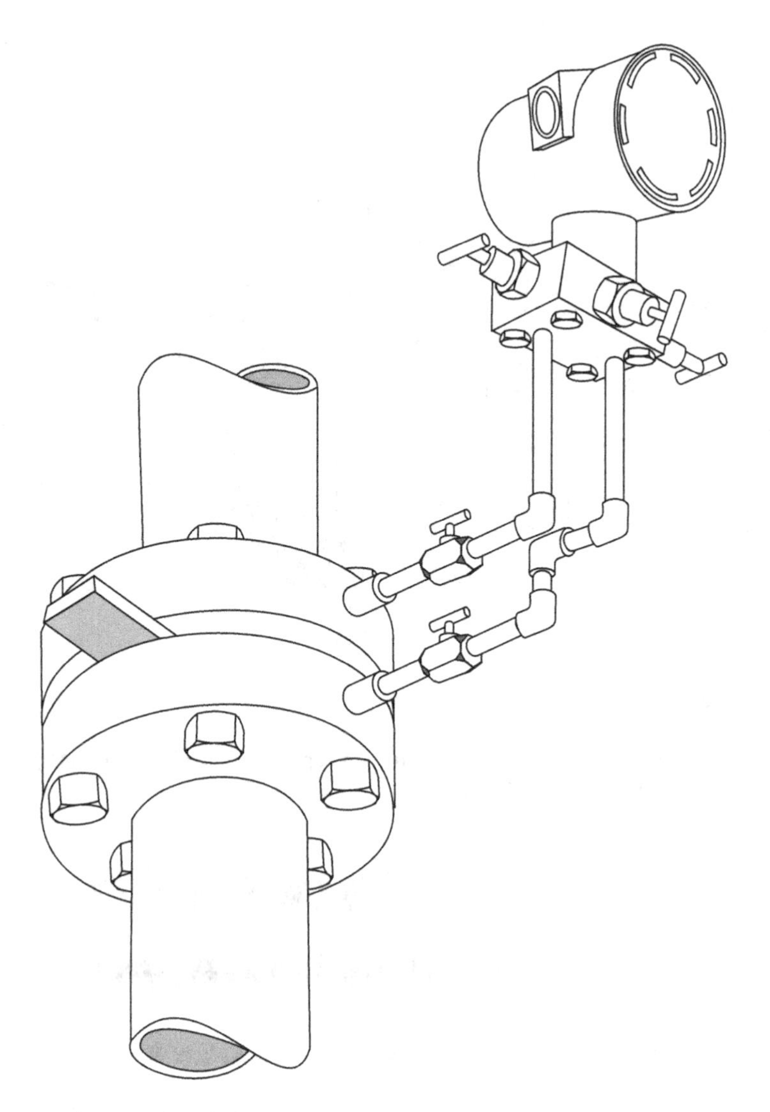

FIGURE 6.59 In vertical pipes the lines should flow upwards in order to avoid liquid collecting in the gas-filled lines – with the transmitter mounted above the pipe.

pipe roughness, pulsations, etc. Under normal operating conditions, when applied to the high- (H) and low-pressure (L) inputs of the transmitter (Figure 6.61) this process noise is effectively cancelled out since the pressure transmitter measures the differential pressure.

In the event that one or other of the impulse lines becomes blocked, cancellation does not occur and the differential pressure signal rises to that of the unblocked line (Figure 6.62). However, when both impulse lines are plugged, the noise level of the measured differential pressure falls virtually to zero (Figure 6.63).

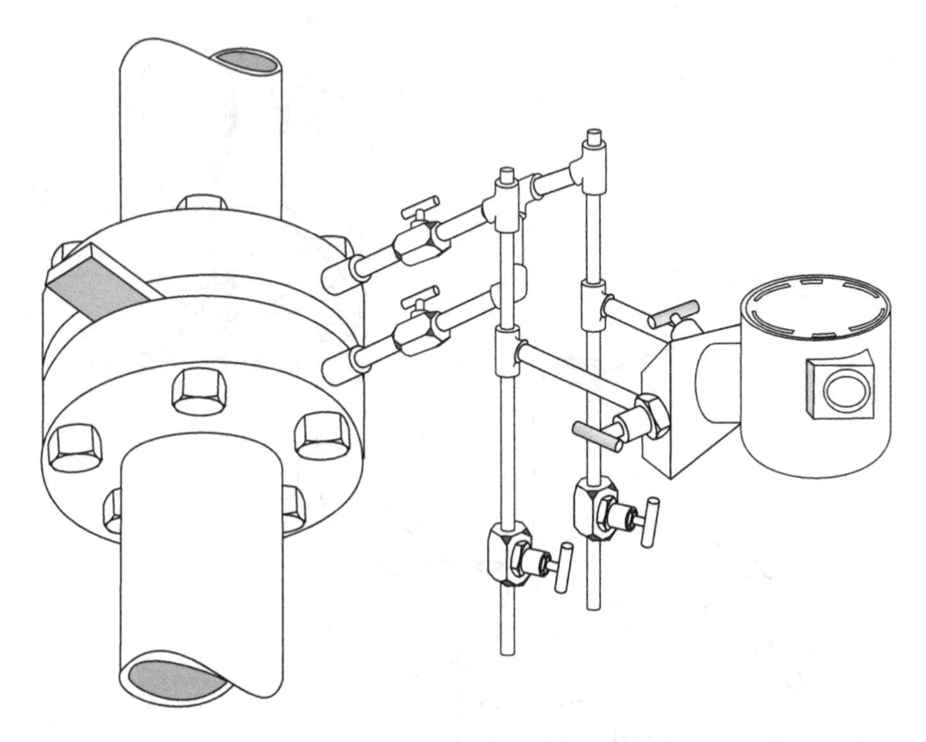

FIGURE 6.60 By bringing the lower impulse line up to the same height as the upper line, the liquid head in the two lines is equalised.

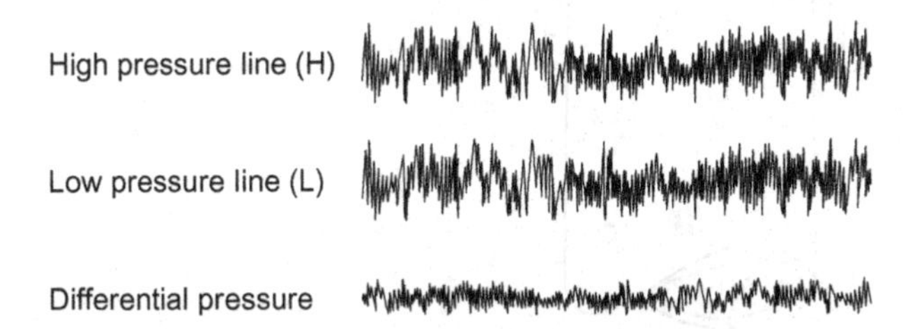

FIGURE 6.61 With neither line blocked the process noise is effectively cancelled out since the pressure transmitter measures the differential pressure.

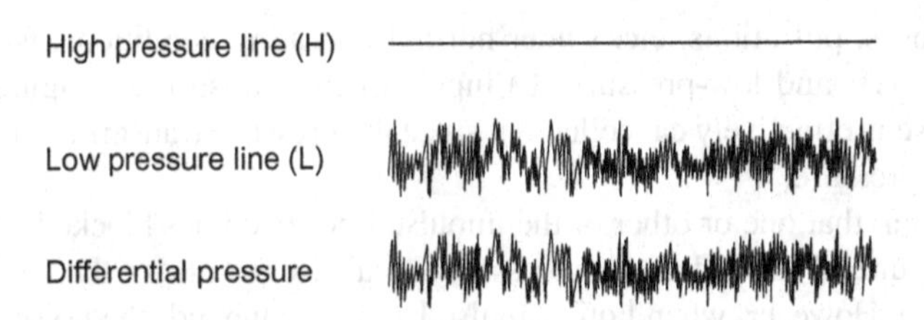

FIGURE 6.62 If one or other of the impulse lines becomes blocked, cancellation does not occur and the differential pressure signal rises to that of the unblocked line.

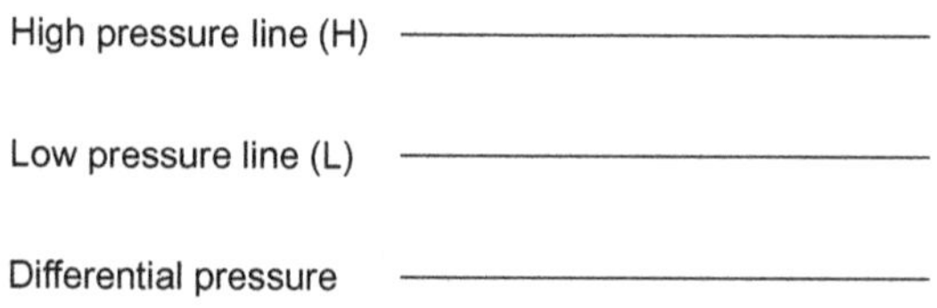

FIGURE 6.63 When both impulse lines are plugged, the noise level of the measured differential pressure falls to zero.

7 Inferential Volumetric Meters

7.1 INTRODUCTION

Inferential volumetric meters are indirect volumetric totalisers, in which packages of the flowing media are separated from the flow stream and moved from the input to the output. However, unlike the positive displacement meter, the enclosed volume is not geometrically defined but 'inferred'.

Inferential volumetric meters have rotor-mounted blades in the form of a vaned rotor or turbine that is driven by the medium at a speed proportional to the flow rate. The number of rotor revolutions is proportional to the total flow and is monitored by either a gear train or by a magnetic or optical pick-up.

Competing with the positive displacement meter for both accuracy and repeatability, inferential volumetric meters are used extensively in custody transfer applications in the oil and gas industries.

7.2 TURBINE METERS

Available in sizes from 5 to 600 mm, the turbine meter usually comprises an axially mounted bladed rotor assembly (the turbine) running on bearings and mounted concentrically within the flow stream by means of upstream and downstream support struts (Figure 7.1). The support assembly also often incorporates upstream- and downstream-straightening sections to condition the flow stream. The rotor is driven by the medium (gas or liquid) impinging on the blades.

Turbine rotation may be detected by several methods. Commonly, permanent magnets are embedded in the tips of the rotor blades that induce electrical pulses in an externally mounted pick-up coil. As each blade passes the coil, a voltage is generated in the coil – with each pulse representing a discrete volume of liquid.

In an improved design the externally mounted pick-up coil is integrated with a permanent magnet (Figure 7.2) and the rotor blades are made of a magnetically permeable ferrous material. As before, as each blade passes the pick-up coil, it cuts the magnetic field produced by the magnet and induces a voltage pulse in the coil. As illustrated, a second sensor, mounted at a lagging angle, provides automatic direction sensing for bi-directional turbine meters and also redundancy. Comparison of the two signals also provides a check on signal transmission security.

Rotors may be open bladed (rimless) or rimmed type. Open bladed (rimless) rotors (as illustrated in Figure 7.1) are most commonly used in smaller meters with line sizes of less than 150 mm (6″) and are typically constructed of 400 series stainless steel or another paramagnetic material that can be easily detected by reluctance-type pickup coils.

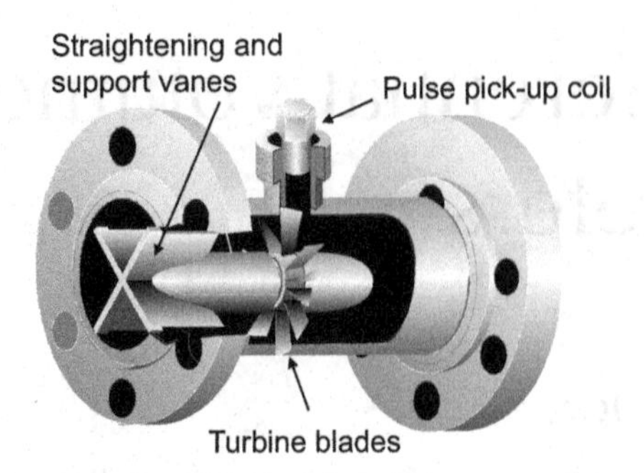

FIGURE 7.1 Turbine meter consists of a bladed rotor suspended in the flow stream. Upper and lower-straightening vanes are normally included. (Courtesy: Emerson.)

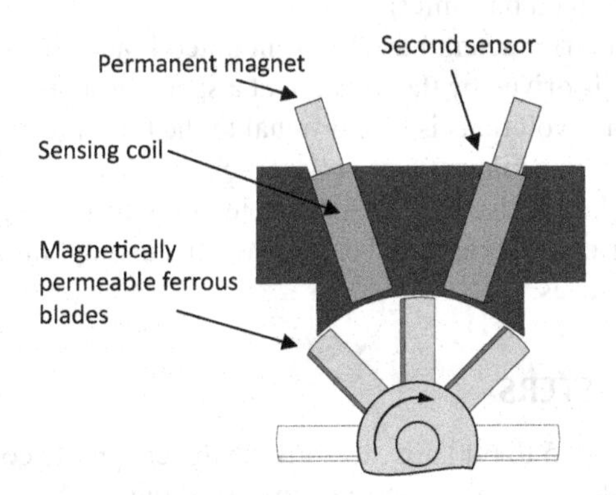

FIGURE 7.2 Externally mounted pick-up coil is integrated with a permanent magnet and the rotor blades are made of a magnetically permeable ferrous material. A second sensor may be used to detect the rotational direction in bi-directional measurement devices.

Rimmed rotors are typically used in larger meters – 150 mm (6″) and above. As shown in Figure 7.3, a lightweight stainless steel rim or shroud carries small magnetic or paramagnetic buttons that provide greater flow resolution by generating more pulses per unit volume.

7.2.1 K-FACTOR

The number of pulses produced per unit volume is termed the K-factor.

Ideally, the meter would exhibit a linear relationship between the meter output and the flow rate – a constant K-factor. In reality, however, the driving torque of the fluid on the rotor is balanced by the influence of viscous, frictional, and magnetic drag effects.

Since these vary with the flow rate, the shape of the K-factor curve (Figure 7.4) depends on viscosity, flow rate, bearing design, blade edge sharpness, blade roughness, and the nature of the flow profile at the rotor leading edge. In practice, all these influences have differing effects on the meter linearity and thus all turbine meters, even from the same manufacturing batch, should be individually calibrated.

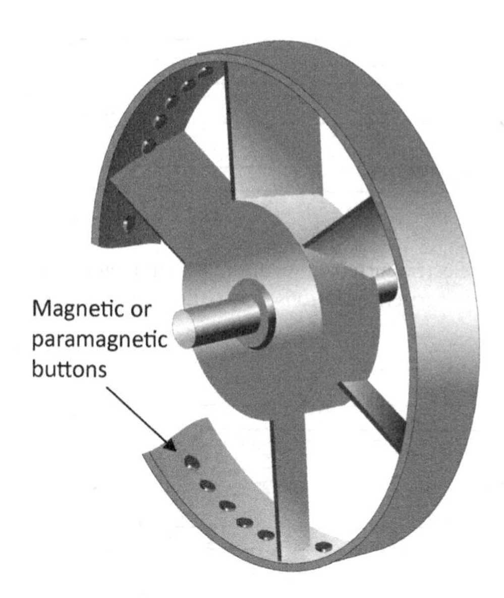

FIGURE 7.3 In rimmed rotors, a lightweight stainless steel rim or shroud carries small paramagnetic buttons that provide greater flow resolution by generating more pulses per unit volume.

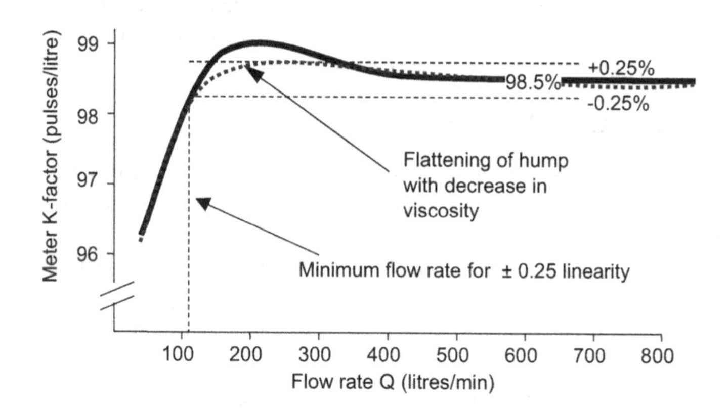

FIGURE 7.4 K-factor (the meter 'constant') should, ideally, be flat. The actual plot exhibits a drop off at low flow rates and a viscosity hump.

The linear relationship of the K-factor is confined to a flow range of about 10:1 – sometimes extending up to 20:1.

At low flows, the poor response of the meter is due to bearing friction, the effect of fluid viscosity and magnetic drag on the rotor due to the use of a magnetic pick-off.

It is possible to extend the lower limit of the meter's response by using for example a Hall-effect device coupled with the use of high-quality rotor bearings. The humping section of the curve flattens as the viscosity decreases – with a resultant increase in accuracy.

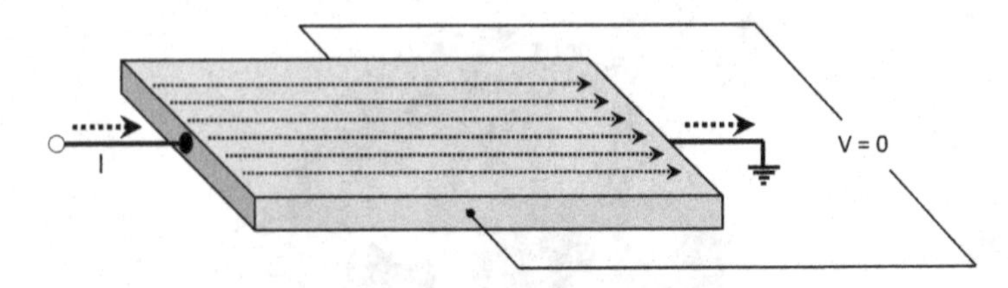

FIGURE 7.5 When an electric current flows through a semiconductor material its distribution is uniform and the charge carriers (electrons) move through it in pretty much a straight line.

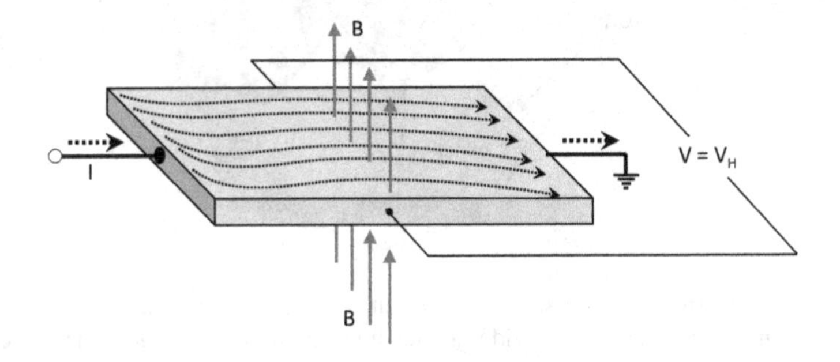

FIGURE 7.6 When the conductor is placed within a magnetic field a transverse force is exerted on the moving negatively charged electrons and deflects them to one side of the semiconductor material – creating an excess of electrons on one side and a depletion on the other to create a potential difference – the Hall voltage (V_H).

7.3 HALL-EFFECT SENSORS

The Hall effect refers to the phenomenon in which a current-carrying conductor, when placed within a magnetic field, generates a voltage that is perpendicular to both the current and the field.

The basic principle is illustrated in Figure 7.5 which shows that when an electric current flows through a semiconductor material its distribution is uniform and the charge carriers (electrons) move through it in pretty much a straight line. Consequently, the voltage measured across the output, perpendicular to the current flow, is zero.

If now the conductor is placed within a magnetic field (Figure 7.6), perpendicular to both the current flow and the output, a transverse force is exerted on the moving negatively charged electrons and deflects them to one side of the semiconductor material. This force (the Lorentz force) produces an excess of electrons on one side and a depletion on the other to create a potential difference – the Hall voltage (V_H).

Because Hall-effect sensors have no magnetic drag, they can operate at lower flow velocities (0.06 m/s) than magnetic pick-up designs (0.15–0.3 m/s).

7.4 HELICAL TURBINE METERS

As illustrated in Figure 7.7, the rotor of the helical turbine meter has only two blades that are much wider and are helical-shaped. In addition the internals are mounted in a measurement tube that is removable from the housing.

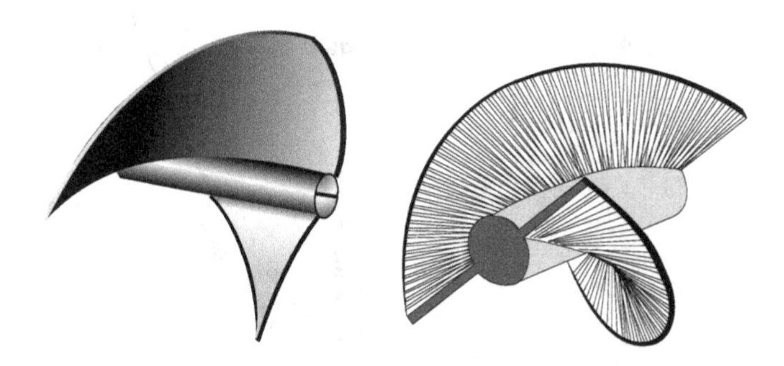

FIGURE 7.7 The rotor of the helical turbine meter has only two blades that are much wider and are helical-shaped.

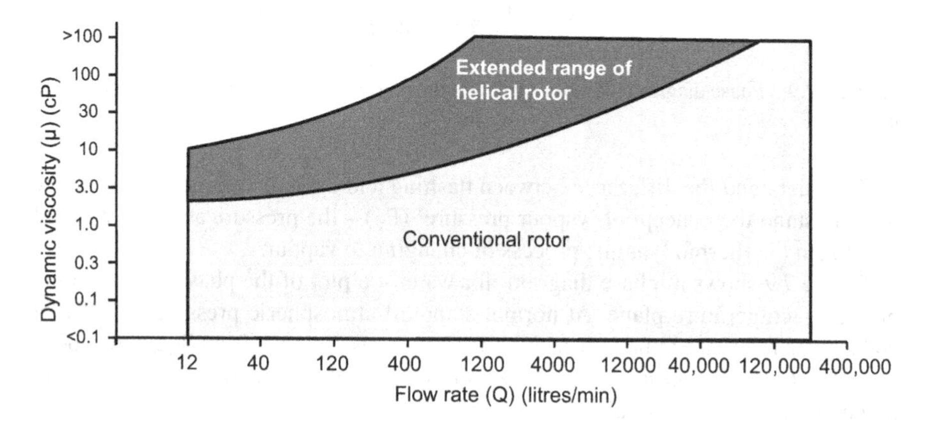

FIGURE 7.8 The two-bladed rotor gives the meter the ability to accurately measure higher viscosity liquids. (Courtesy: Faure Herman.)

Since the flow velocity in a helical rotor is parallel to the blade, whatever the position, this improves the flow profile. In addition, the vortices at the end of the blade are minimised.

Most importantly the two-bladed rotor gives the meter the ability to accurately measure higher viscosity liquids (Figure 7.8). This is as a result of the reduced effect of the stagnant boundary layer that builds up on the rotor surfaces when higher viscosity oils are being measured.

Although this type of meter has been around for many years it did not find wide application because of the difficulty in proving – posed by its inherently low pulse resolution. However, with the acceptance of pulse interpolation techniques for proving low-resolution meters, the helical turbine meter is now finding a wider audience.

7.5 FLASHING AND CAVITATION

Both flashing and cavitation should be avoided. Flashing causes the meter to read high and cavitation results in rotor damage.

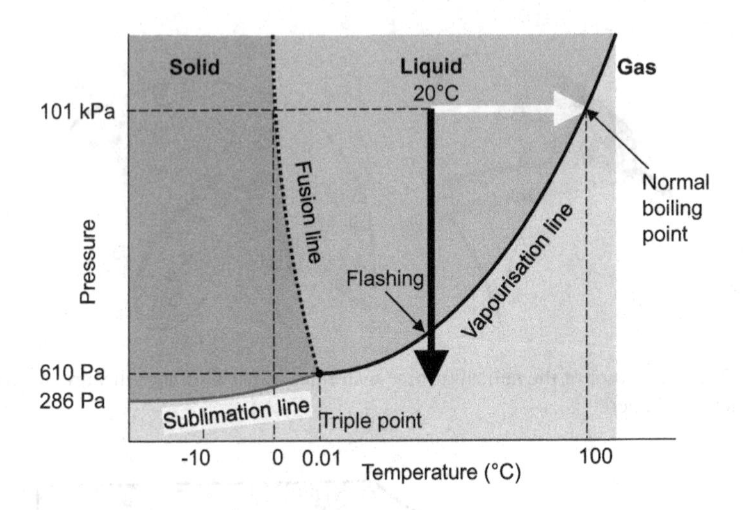

FIGURE 7.9 Phase diagram for water – a plot of the phase of water on a pressure–temperature plane.

To understand the difference between flashing and cavitation, it is first necessary to understand the concept of 'vapour pressure' (P_V) – the pressure at which a liquid will begin the thermodynamic process of changing to vapour.

Figure 7.9 shows a 'phase diagram' for water – a plot of the phase of water on a pressure–temperature plane. At normal standard atmospheric pressure of 101 kPa and a given reference temperature (e.g. 20°C) water will transit from liquid to vapour at 100°C (normal boiling point). However, keeping the temperature constant and reducing the pressure will also lead to a transition from a liquid to a vapour state – in other words, it will boil but at a much reduced temperature. This is termed 'flashing'.

Any restriction in the flow line gives rise to a pressure drop. This is illustrated in Figure 7.10 in which the pressure falls below the vapour pressure (P_V) and remains below this level. The resultant flashing produces vapour bubbles that remain intact

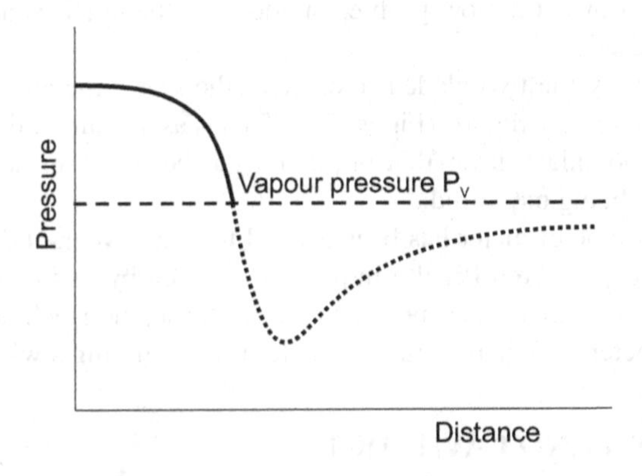

FIGURE 7.10 Flashing occurs when the pressure falls below the vapour pressure (P_V) and remains below this level.

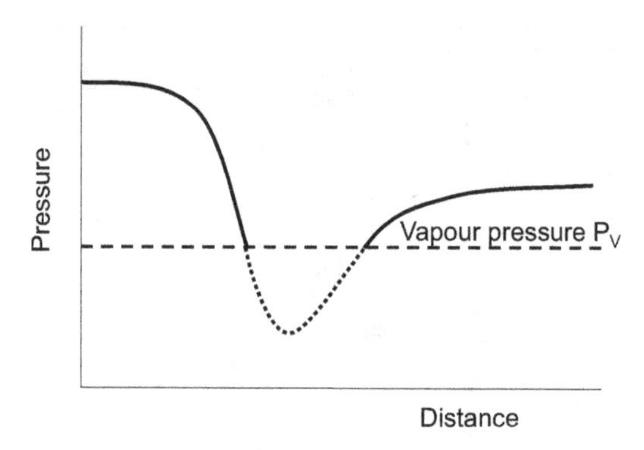

FIGURE 7.11 Cavitation is caused by the pressure dropping to the vapour pressure of the fluid and rising to a higher pressure further downstream.

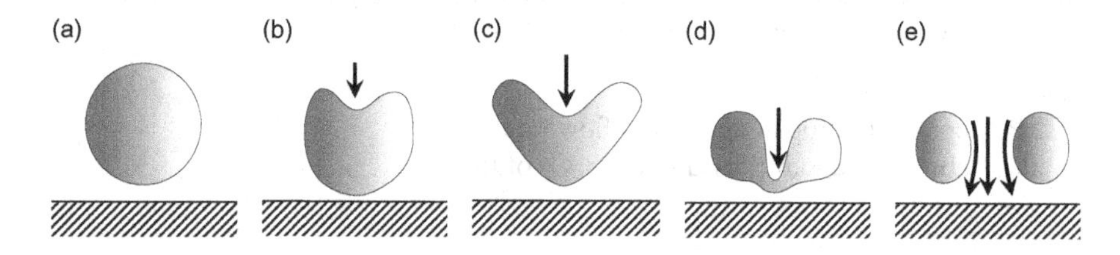

FIGURE 7.12 Schematic of a single vapour bubble collapsing as the surrounding fluid pressure recovers to above the vapour pressure.

and proceed further downstream. The vaporisation of the liquid causes a large increase in volume and therefore higher overall fluid velocity.

If, however, the pressure recovers to a point above the vapour pressure of the fluid the vapour bubbles will collapse (Figure 7.11). This two-stage phenomenon, vapour bubble formation and their subsequent collapse, is known as cavitation.

Figure 7.12 shows a schematic of a single vapour bubble collapsing as the surrounding fluid pressure recovers to above the vapour pressure. (a) Initial formation of spherical bubble adjacent to metering body; (b) the collapsing bubble starts to experience shape instability; (c) upper fluid starts to penetrate the upper side; (d) increasing asymmetry forms a rapidly accelerating jet of fluid, entering the bubble from the boundary most distant from the proximate solid surface which receives damage; and (e) final micro-jet achieves high speeds and shock waves of up to 7,000 bar or more.

If the micro-jet bubble implosions collapse on or near solid surfaces, the material is chipped away (Figure 7.13).

7.6 FLASHING AND CAVITATION PREVENTION

Flashing and cavitation may be eliminated by increasing the system back pressure on the meter. The API recommendation is:

$$\text{Minimum back pressure} = 2 \cdot \Delta P + 1.25 \cdot V_P \tag{7.1}$$

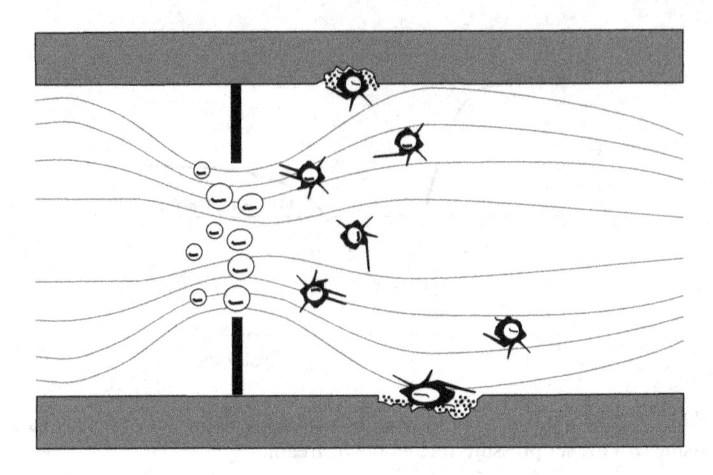

FIGURE 7.13 Cavitation damage occurs when the bubbles collapse on or near solid surfaces within the valve or piping and the material is chipped away.

where

ΔP = cpressure drop at maximum flow rate
P_V = absolute vapour pressure (at 38°C) of the fluid

7.6.1 Selection and Sizing

Although turbine meters are sized by the volumetric flow rate, the main factor that affects the meter is viscosity. Typically, larger meters are less affected by viscosity than smaller meters. This may indicate that larger meters would be preferred; in fact the opposite is true. By using a smaller meter, operation is more likely to occur towards the maximum permitted flow rate and away from the non-linear 'hump' response at low flows.

Generally turbine meters are suitable for flows between 0.3 and 3 m/s – with maximum velocities of 10 m/s.

For good rangeability, it is recommended that the meter be sized such that the maximum flow rate of the application is about 70–80% of that of the meter.

If the pipe is oversized (with flow velocity under 0.3 m/s), select a Hall-effect pick-up or use a meter smaller than the line size. However, bear in mind that meters should be sized for a differential pressure drop of between 0.2 and 0.35 bar at maximum flow. Consequently, because the pressure drop increases with the square of flow rate, reducing the meter to the next smaller size increases the pressure drop considerably.

Care should also be taken to avoid any form of build-up on the meter blades. For example, a 25 μm build-up on all surfaces of a 100 DN rotor will decrease the flow area through the rotor, and thus the K-factor, by about 0.5%.

Straight runs of 10–15D upstream and 5D downstream are normally required. In addition, the following upstream requirements are required for:

- 90° elbow, T, filter, strainer, or thermowell –20D
- Partially open valve –25D
- Two elbows in different planes –50D or more

7.6.2 Advantages of Turbine Meters

- Suitable for pressures of up to 64 MPa
- High accuracy (up to ±0.2% of flow)
- Excellent repeatability (±0.05%)
- Wide rangeability up to 20:1
- Wide range of temperature applications from –220°C to 600°C
- Measurement of non-conductive liquids
- Capability of heating measuring device

7.6.3 Disadvantages of Turbine Meters

- Not suitable for high viscous fluids
- Viscosity must be known – with each meter requiring recalibration for changes in product
- 10 diameter upstream and 5 diameter downstream of straight pipe are required
- Not effective with swirling fluids
- Subject to wear
- Only suitable for clean liquids and gases
- Pipe system must not vibrate
- Specifications critical for measuring range and viscosity
- Subject to erosion and damage
- Relatively expensive
- A turbine for liquids should never be subjected to gas flow (danger of over-speeding) – never clean with compressed air

7.7 WOLTMAN METER

The Woltman meter, used primarily as a water meter, is very similar in basic design to the turbine meter. The essential difference is that the measurement of rotation is carried out mechanically using a low friction gear train connecting the axle to the totaliser.

The Woltman meter is available in two basic designs – one with a horizontal turbine (Figure 7.14) and one with a vertical turbine (Figure 7.15). The vertical design offers the advantage of minimal bearing friction and therefore a higher sensitivity resulting in a larger flow range. Whilst the pressure drop of the vertical turbine meter is appreciably higher, because of the shape of the flow passage, it is widely used as a domestic water consumption meter.

In many designs, an adjustable regulating vane is used to control the amount of deflection and thus adjust the meter linearity.

7.8 PROPELLER TYPE METER

Used mainly in the field of irrigation, the propeller-type flowmeter (Figure 7.16) makes use of a three-bladed propeller with the body of the meter positioned in the centre of the flow-stream. With large clearances between each blade and only

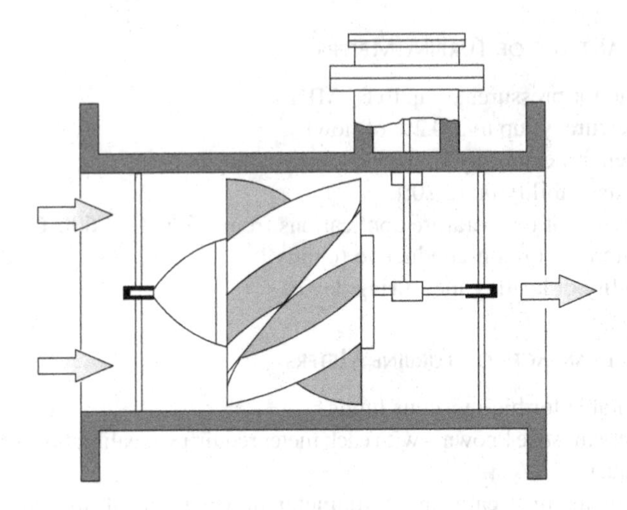

FIGURE 7.14 Horizontal turbine Woltman meter.

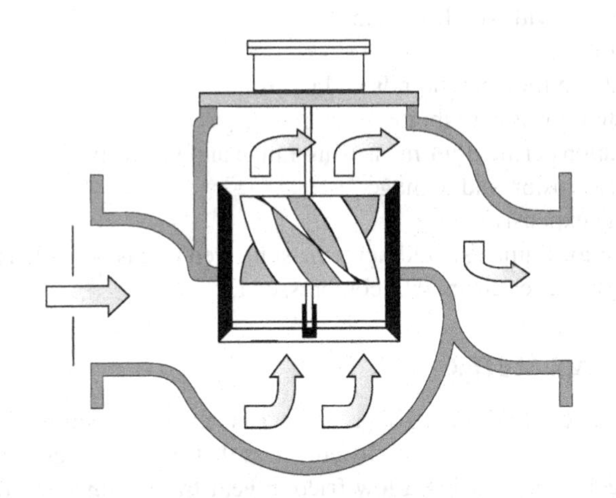

FIGURE 7.15 Vertical turbine Woltman meter.

the propeller in the flow line, particles in suspension are able to pass with ease. In addition, the transmitter and all working parts can be removed and replaced in a few minutes, without breaking the pipeline.

The saddle-type (as illustrated) can be welded or clamped.

Typically available for flow rates up to 280 kL/min; turndown ratios of 15:1, and accuracies of ±2% with repeatability ±0.25%.

7.9 IMPELLER METERS

As opposed to the vane-axial blades of turbine-type models, the rotating blades of impeller-type sensors are perpendicular to the flow – making them inherently less accurate than turbine sensors. However, their typical 1% accuracy and excellent repeatability makes them ideal for many applications.

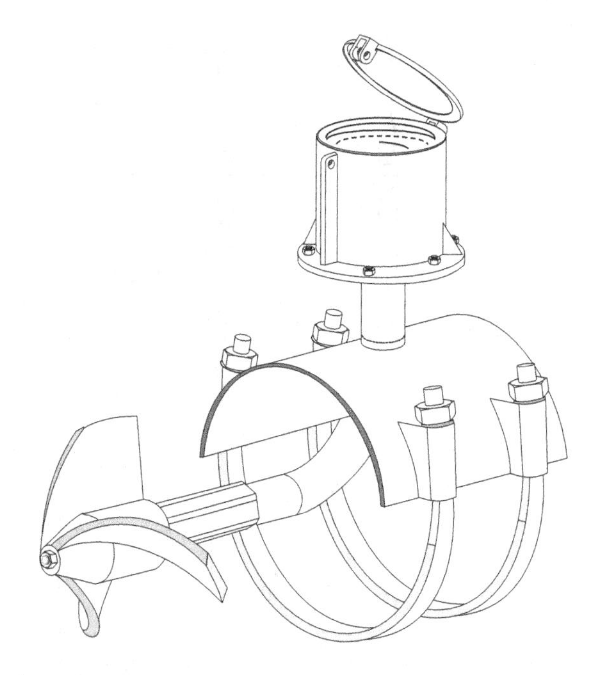

FIGURE 7.16 Propeller-type flowmeter with only the propeller in the flow line. (Courtesy: McCrometer.)

Impeller sensors are especially suitable for measuring flow rates of low-viscosity liquids that are low in suspended solids over line velocities of between 0.15 and 10 m/s.

At lower flow rates, the fluid cannot maintain the force needed to overcome bearing friction, impeller mass inertia, and fluid drag. And at flow rates above 10 m/s, cavitation can occur and cause readings to increase more than the increase in flow velocity. As velocity continues to increase under cavitation conditions, the reading eventually decreases with respect to true velocity.

The most common form of impeller-type meter is the in-line insertion format in which the main bearing is located out of the main flow stream and thus provides only a minimal pressure drop. Figure 7.17 illustrates a T-mount flow sensor suitable for pipe sizes ranging from 10 to 100 mm.

Other versions are available for use with welded-on pipe threads that allow the same meter to be used on pipe sizes ranging from 75 mm to 2.5 m diameter. This technique also allows its use in a 'hot tap' mode whereby it may be removed and replaced on high pressure lines without the need for a shutdown.

Another form of the impeller type meter, the Pelton wheel turbine (Figure 7.18), is able to measure extremely low flow rates down to 0.02 L/min, coupled with a turndown ratio of up to 50:1.

The incoming low velocity fluid is concentrated into a jet that is directed onto a lightweight rotor suspended on jewel bearings. The rotational speed is linear to the flow rate and is detected by means of ferrite magnets, located in the rotor tips, which induce voltage pulses in a sensing coil. One drawback is that the nozzle can cause a rather large pressure drop.

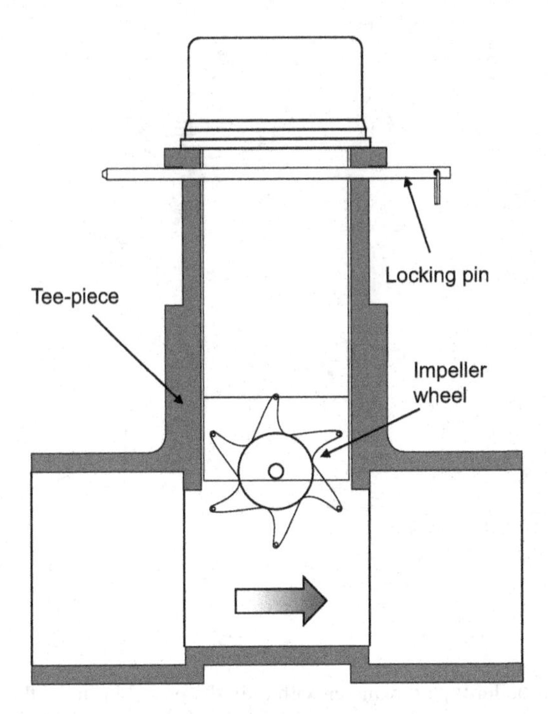

FIGURE 7.17 A T-mount flow sensor suitable for pipe sizes ranging from 10 to 100 mm. (Courtesy: GLI International.)

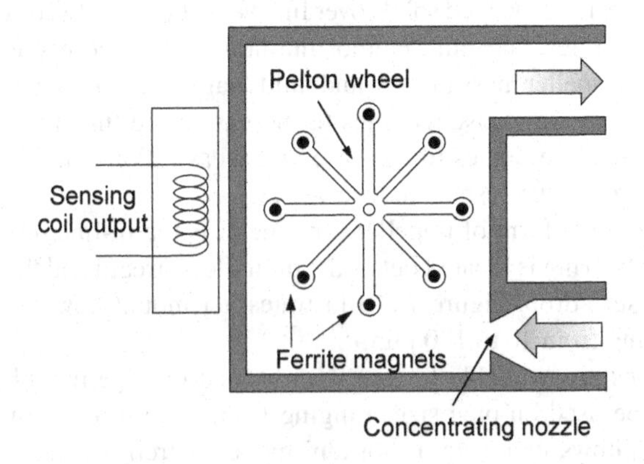

FIGURE 7.18 Cross-section of the Pelton wheel system.

7.9.1 Application Limitations

As with turbine meters, most such sensors employ multiple blades with a permanent magnet embedded in each blade. A pick-up coil in the sensor acts as a generator stator – generating an electrical pulse each time the blade passes near it.

The use of such a magnetic pick-up, however, has some serious drawbacks. First, the signal is susceptible to interference by extraneous magnetic fields in the vicinity of

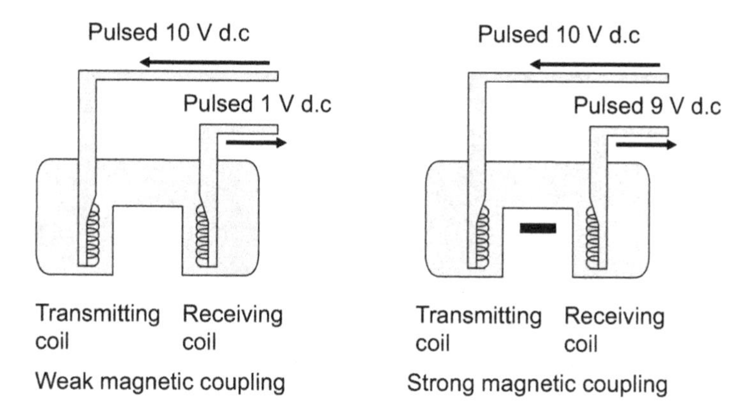

FIGURE 7.19 In the absence of a ferrite rod, the magnetic coupling is loose and the signal produced by the receiving coil is small. When a ferrite rod is present, the magnetic coupling is strong resulting in a much larger output signal. (Courtesy: GLI International.)

the coil. In addition, ferrous contamination, present in many industrial applications, causes particles to be attracted to the magnets in each blade. This not only affects the sensor accuracy, but can impede or stop the impeller from rotating. Further, at low flows, the magnetic attraction between each rotating blade and the pick-up coil increases the force required to turn the impeller – resulting in poor linearity.

One method of overcoming this problem is to make use of a Hall-effect transducer.

Alternatively use may be made of non-magnetic ferrite rods embedded in the impeller blades that form a low permeable path for a magnetic field.

As shown in Figure 7.19 the pickup comprises a composite transmitting and sensing coil. In the absence of a ferrite rod the magnetic coupling is loose and the signal produced by the receiving coil is small. However, in the presence of a ferrite rod, the magnetic coupling is strong – resulting in a much larger output signal.

Because permanent magnets are not used, there is no magnetic drag and no accumulation of magnetic particles to degrade the accuracy or cause clogging.

8 Oscillatory Flow Meters

8.1 INTRODUCTION

Oscillatory flow measurement systems involve three primary metering principles: vortex, vortex swirl (precession), and Coanda effect. In all three, the primary device generates an oscillatory motion of the fluid whose frequency is detected by a secondary measuring device to produce an output signal that is proportional to fluid velocity.

8.2 VORTEX FLOWMETERS

Vortex flowmeters for industrial flow measurement were first introduced in the mid-1970s but the technology was poorly applied by several suppliers. As a result, the vortex flow metering developed a bad reputation and several manufacturers dropped the technology. However, since the mid-1980s, many of the original limitations have been overcome and vortex flow metering has become a fast growing flow technology – especially in the field of steam metering where other technologies such as ultrasonic and Coriolis fall short due to the high temperatures encountered.

Vortex meters are based on the phenomenon known as vortex shedding that takes place when a fluid (gas, steam, or liquid) meets a non-streamlined obstacle – termed a bluff body. Because the flow is unable to follow the defined contours of the obstacle, the peripheral layers of the fluid separate from its surfaces to form vortices in the low pressure area behind the body (Figure 8.1). These vortices are swept downstream to form a so-called Karman vortex street. Vortices are shed alternately from either side of the bluff body at a frequency that, within a given Reynolds number range, is proportional to the mean flow velocity in the pipe.

In vortex meters, the differential pressure changes that occur as the vortices are formed and shed are used to actuate the sealed sensor at a frequency proportional to the vortex shedding.

8.2.1 Formation of Vortices

At very low velocities – the laminar flow region (Figure 8.2a) – the fluid flows evenly around the body without producing turbulence. As the fluid velocity increases, the fluid tends to shoot past the body, leaving a low pressure region behind it (Figure 8.2b). As the fluid velocity increases even further, this low pressure region begins to create a flow pattern as shown in Figure 8.2c – the beginning of the turbulent flow region. This action momentarily relieves the pressure void on one side of the low pressure region and the fluid forms into a vortex. The interaction of the vortex with the main

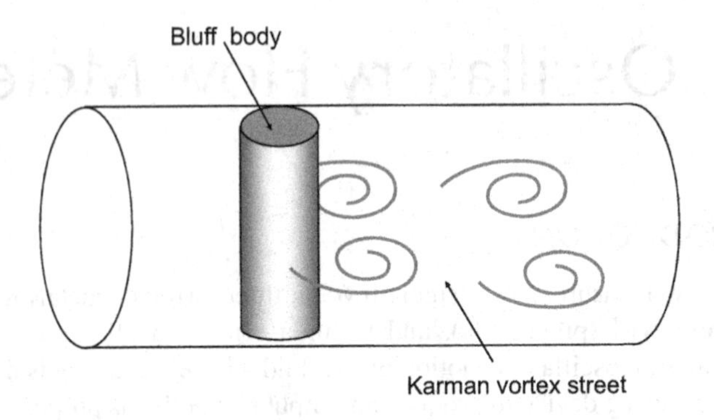

FIGURE 8.1 The Karman vortex street – with vortices formed on alternate sides in the low pressure area of the bluff body.

stream fluid releases it from the surface of the body and it travels downstream. Once released, the low pressure region shifts towards the other rear side of the body to form another vortex. This process is repeated, resulting in the release of vortices from alternate sides of the bluff body as illustrated in Figure 8.1.

Vortex shedding occurs naturally throughout nature and can be observed in the whistling tone that the wind produces through telephone wires or in a flag waving from a flagpole. Because the flagpole acts as a bluff body, vortex shedding occurs. As the wind speed increases, the rate of vortex shedding increases and causes the flag to wave faster.

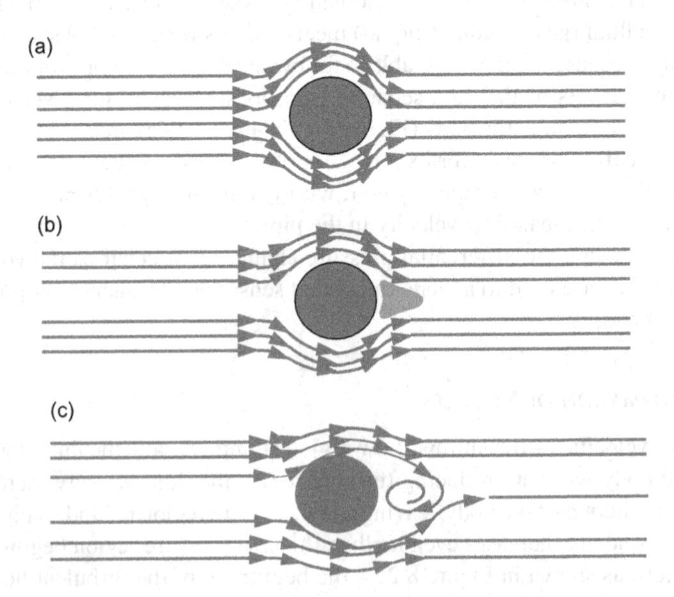

FIGURE 8.2 Formation of vortices: (a) laminar flow region with fluid flowing evenly around the body; (b) at higher velocities a low pressure region starts to form behind the bluff body; and (c) beginning of the turbulent flow region and formation of vortex.

8.2.2 Strouhal Factor

► In 1878, Strouhal observed that the frequency of oscillation of a wire, set in motion by a stream of air, is proportional to the flow velocity. He showed that:

$$f = \frac{St \cdot v}{d} \tag{8.1}$$

where

f = vortex frequency (Hz)
d = diameter of the bluff body (m)
v = velocity of liquid (m/s)
St = Strouhal factor (dimensionless)

Unlike other flow-sensing systems, because the vortex shedding frequency is directly proportional to flow velocity, drift is not a problem as long as the system does not leave its operating range. Furthermore, the frequency is unaffected by the medium's density, viscosity, temperature, pressure, and conductivity, as long as the Reynolds number (Re) stays within defined limits. Consequently, irrespective of whether the meter is used for measuring steam, gas, or liquids, it will have virtually the same calibration characteristics and the same meter factor – although not necessarily over the same volumetric flow velocity ranges.

In reality, the Strouhal factor is not a constant but, as illustrated in Figure 8.3, varies with the shape of the bluff body and the Reynolds number. The ideal vortex flowmeter would, therefore, have a bluff body shape that features a constant Strouhal number over as wide a measuring range as possible. ◄

Meters based on this relationship are shown to have a linearity of better than ±0.5% over a wide flow range of as high as 50:1 for liquids and 100:1 for gases. The

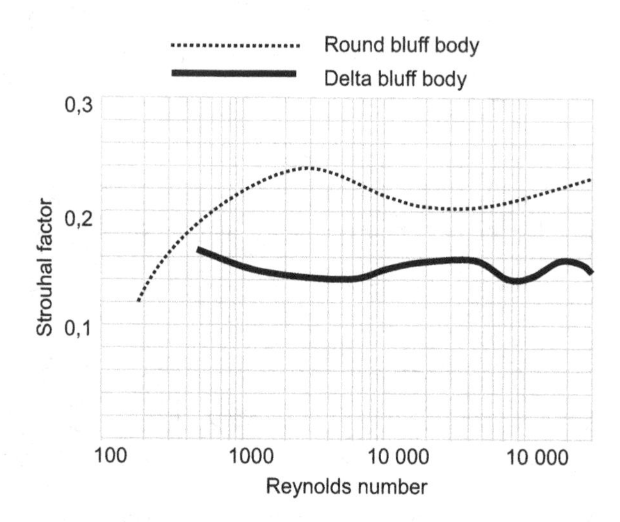

FIGURE 8.3 Relationship between Strouhal factor and Reynolds number for both a round and a delta bluff body. (Courtesy: Endress+Hauser.)

limits are determined at the low-end by viscosity effects and at the upper end by cavitation or compressibility.

Another major advantage of the vortex meter is that it has a constant, long-term calibration that does not involve any in-service adjustment or tuning. For a given size and shape of the bluff body, the vortex shedding frequency is directly proportional to the flow rate.

8.2.3 Shedder Design

Meters differ only in the shape of the bluff body and in the sensing methods used – with each manufacturer claiming specific advantages. Some of the bluff body shapes are shown in Figure 8.4.

Tests have shown that changes in the dimensions of the bluff body have a negligible effect on calibration. For example, tests with a rectangular bluff body indicate that with a body-to-meter bore ratio of 0.3, the body width can vary as much as ±10% to produce a change in the meter factor of <0.4%. Similarly, radiussing the edges of the bluff body by as much as 4 mm will not cause the calibration to deviate outside the standard accuracy band.

(Compare this with an orifice plate where radiussing the sharp edge of the orifice by as little as 0.4 mm produces a reading inaccuracy of approximately 4%.)

The major benefit of this insensitivity to dimensional changes of the bluff body is that the vortex meter is virtually unaffected by erosion or deposits.

8.2.3.1 Cylindrical Bodies

Early bluff bodies were cylindrical. However, as the boundary layer changes from laminar to turbulent, the vortex release point fluctuates backwards and forwards,

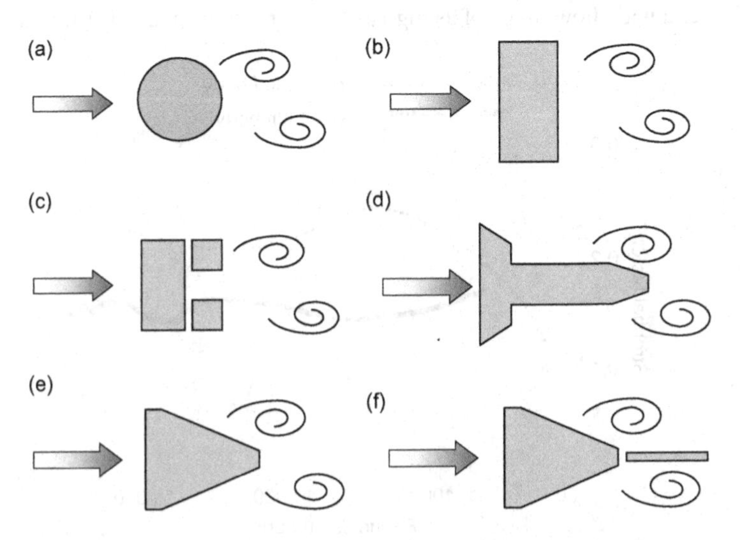

FIGURE 8.4 Various bluff body shapes: (a) round; (b) rectangular; and (c) two-part rectangular; (d) T-shaped bar; (e) delta-shaped; (f) delta-shaped twopart bodies. (Courtesy: Endress+Hauser.)

depending on the flow velocity, and the frequency is, subsequently, not exactly proportional to velocity. As a result, use is made of bluff bodies having a sharp edge that defines the vortex shedding point.

8.2.3.2 Rectangular Bodies

Following the cylindrical body, the rectangular body was used for many years. However, current research indicates that this body shape produces considerable fluctuation in linearity in varying process densities.

8.2.3.3 Rectangular Two-Part Bodies

In this configuration, the first body is used to generate the vortices and the second body to measure them.

The two-part body generates a strong vortex (hydraulic amplification) that requires the use of less complicated sensors and amplifiers. On the negative side, the pressure loss is almost doubled.

8.2.3.4 Delta-Shaped Bodies

The delta-shaped shedder has a clearly defined vortex shedding edge and tests (including those carried out by NASA) indicate that the delta shape provides excellent linearity. Accuracy is not affected by pressure, viscosity, or other fluid conditions. Many variations of the delta shape exist and are in operation.

8.2.3.5 Delta-Shaped Two-Part Bodies

Claimed to combine the best features of modern technology, here, the delta-shaped bluff body generates the vortices and the second body is used to measure them.

8.2.3.6 T-Shaped Bar

Also claimed to combine the best features of the delta-shaped body with a high hydraulic amplification.

8.2.4 Sensors

► Since the shedder bar is excited by kinetic energy, the amplitude of the vortex signal depends on the dynamic pressure of the fluid:

$$p_d = \frac{1}{2} \cdot \rho \cdot v^2 \tag{8.2}$$

where

p_d = dynamic pressure
ρ = fluid density
v = velocity

As shown, the sensor amplitude is thus proportional to the fluid density and to the square of the velocity (Figure 8.5).

Consequently, the dynamic sensitivity range of the vortex sensor needs to be quite large. For a turndown ratio of 1:50 in flow velocity, the magnitude of the vortex signal

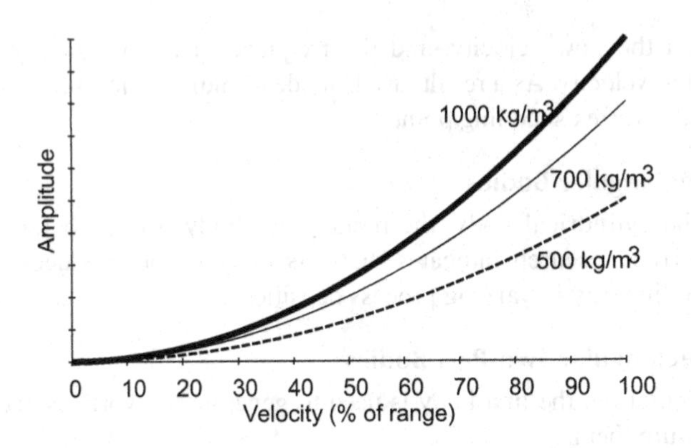

FIGURE 8.5. Amplitude as a function of velocity and process density.

would vary by 1:2,500. This leads to very small signal levels at the low end of the measuring range. ◄

While the vortex shedding frequency decreases as the size of the bluff body or meter increases, the signal strength falls off as the size decreases – thus, generally, limiting the meter size to within the range of 15–200 mm bore. While there are several methods available for measuring the vortex frequency, there is no sensor currently available that will suit all operating conditions.

Many vortex meters use non-wetted, external sensors connected to internal parts that move or twist due to vortex shedding. Formerly, this technology was plagued by sensitivity to pipeline vibration that produces a similar motion to vortex shedding when there is no flow in the pipe and can cause an erroneous output at zero flow. However, modern instruments have largely overcome this problem and systems are generally insensitive to vibrations in each axis up to at least 1 g covering the frequency range up to 500 Hz.

8.2.4.1 Capacitive Sensors

In this example (Figure 8.6), a sensor bar, placed behind the shedding bar, projects into the pipe and is deflected by the vortices. As shown in Figure 8.7, this changes the distance between the central electrode and the outer electrodes and varies the capacitance of the detection system.

8.2.4.2 Piezoelectric Sensors

Like the capacitive sensor, the alternating vortices, shed on each side of the shedder, act on two diaphragms mounted on each side of the sensor. In this case (Figure 8.8), the flexing motion is coupled to a piezoelectric sensor, outside the flow line, which senses the alternating forces and converts them to an alternating signal.

The piezo elements produce a voltage output that is proportional to the applied pressure. Whilst piezoceramic materials produce a high output for a given pressure (a high 'coupling factor') they have a limited operating temperature range (about 250°C).

The piezoelectric material lithium niobate ($LiNbO_3$) offers only medium coupling factors but can be operated at temperatures above 300°C.

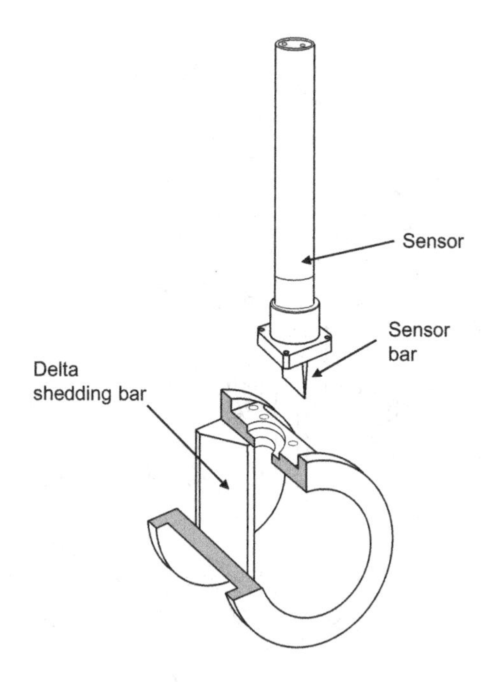

FIGURE 8.6 Use of a separate mechanically balanced sensor positioned behind the bluff body. (Courtesy: Endress+Hauser.)

Generally, piezoelectric materials are unsuitable for temperatures below −40°C since below this point, the piezoelectric effect becomes too small.

Because the piezoelectric element produces an output that is affected by movement or acceleration, it is also sensitive to external pipe vibration. This problem can be overcome by using a second piezoelectric element to measure the vibration and use

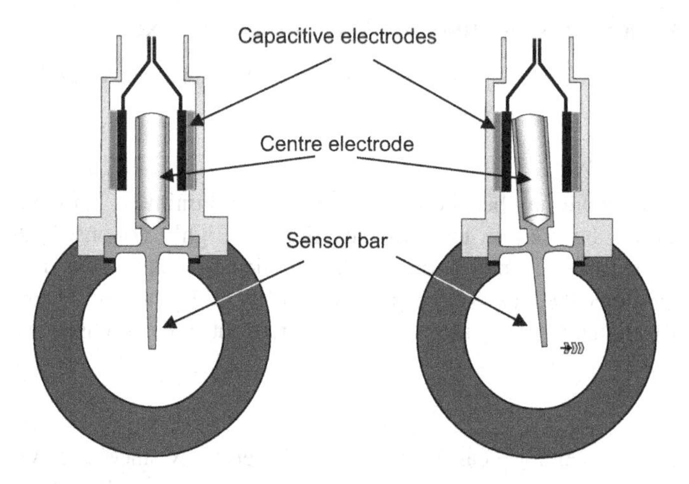

FIGURE 8.7 Deflection of the sensor bar changes the distance between the central electrode and the outer electrodes and varies the capacitance of the detection system. (Courtesy: Endress+Hauser.)

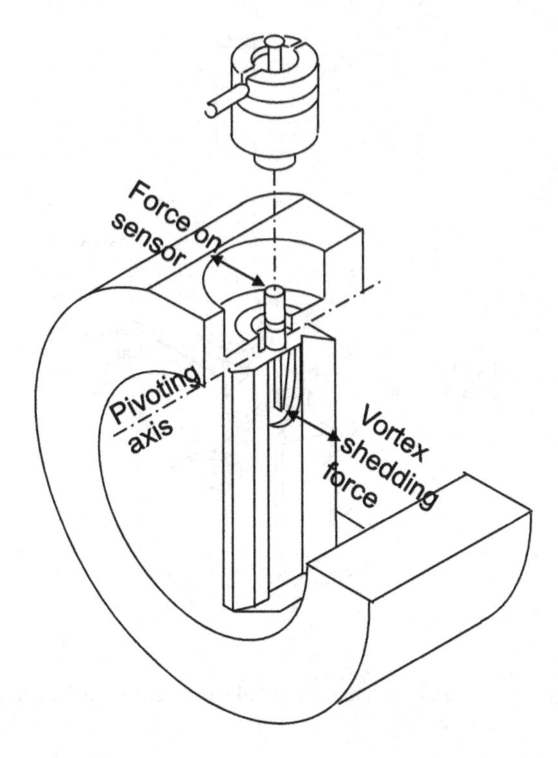

FIGURE 8.8 Use of piezoelectric sensor positioned outside the flow line. (Courtesy: Emerson.)

it in a compensating circuit to ensure that only the clean vortex shedding frequency is obtained.

8.2.5 Application Guidelines for Vortex Flowmetering*

In general, a vortex shedding flowmeter works well on relatively clean low viscosity liquids, gases, and steam to obtain specified accuracy.

8.2.5.1 Viscosity

The system Reynolds number should be above 30,000 minimum. This means vortex meters can only be used on low viscosity liquids. Highly viscous fluids (>3 Pa s (30 cP)) and slurries are not recommended applications. As a rule of thumb, the viscosity should be 0.8 Pa s (8 cP) or less (a viscosity of 0.8 Pa s would correspond to cooking oil). Higher viscosity fluids can be metered, but at the expense of rangeability and head loss.

8.2.5.2 Low Flow

The vortex meter cannot measure flow down to zero flow since, at low flow rates, vortex shedding becomes highly irregular and the meter is totally inaccurate. This

* These application guidelines have been compiled from a series of notes supplied by Krohne.

generally corresponds to a Reynolds number between 5,000 and 10,000 and therefore depends on the pipe diameter and the fluid viscosity. For water, typical minimum velocity flow rate values would vary from about 2.4 m/s for a 15 DN pipe to 0.5 m/s for a 300 DN pipe.

Whilst the minimum Reynolds number requirement imposes a limitation on the usability of the vortex meter, this is not a serious limitation for many applications. For example, water flow in line sizes 25 DN and higher generally corresponds to Reynolds numbers in the tens of thousands to hundreds of thousands. Gas and steam applications generally correspond to Reynolds numbers in the low hundreds of thousands to the millions.

Most vortex meters include a low flow cut-in point, below which the meter output is automatically clamped at zero (e.g. 4 mA for analogue output).

For many applications, the low flow cut-off point does not pose a problem. However, it can be a serious drawback for applications that see low flows during start-up and shut down operations (i.e. flows much lower than normal conditions, often by a factor of 10 or more). While users may not want to measure the flow accurately during such times, they may want to get some indication of the flow. The vortex meter is not a good choice for such an application.

8.2.5.3 Batching Operations

Vortex meters may or may not be suitable for typical batching applications involving intermittent (on/off) flow – especially if the pipe does not remain full at zero flow. The vortex meter will not register the flow as the fluid accelerates from zero to the low flow cut-in value, and again when the flow decelerates from the low flow cut-in value to zero. This lost flow may or may not create a significant error depending on the dynamics of the system and the size of the batch being measured. In addition, the vortex meter can only measure the flow in one direction. Any back flow through the meter (e.g. the result of turning a pump off) will not be measured and will not be deducted from the registered batch total. One way to minimise errors on intermittent flows is to install check valves with the vortex meter on horizontal lines to keep the line full during zero flow conditions.

8.2.5.4 Measuring Range

Note that in vortex meters, the measuring range is fixed for a given application and meter size. Although it depends on the specific application, it is generally >20:1 on gases and steam and >10:1 on liquids.

A 50 mm vortex meter has, typically, a flow range over the range of 1–15 L/s on water (15:1 rangeability). If we need to measure over the range 0.5–3 L/s there is nothing that can be done to the 50 DN meter to allow it to measure a lower range and it would be necessary to use a 25 DN meter. For this reason, vortex meters are sized to the desired flow range, rather than to the nominal pipe diameter. To get the proper measuring range (Figure 8.9), it is often necessary to use a smaller diameter meter than the nominal diameter of the pipe.

When buying a flow meter, the instrument engineer often does not know the exact flow range and has to make an educated guess. Since the vortex meter rangeability

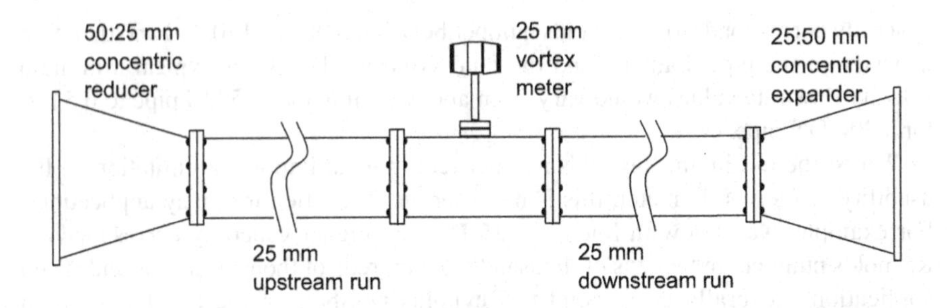

FIGURE 8.9 Use of reducer and expander to obtain the correct measuring range. (Courtesy: Krohne.)

is fixed for a given line size by the process conditions, a meter sized on an educated guess may not meet the process conditions.

Consequently if the user does not have a good 'ball park' figure with regards to rangeability it is often better to opt for a more forgiving technology such a magnetic flowmeter.

8.2.5.5 Process Noise

Process noise from pumps, compressors, steam traps, valves, etc., may cause the meter to read high, by triggering a higher than expected frequency output from the sensor, or by indicating a false flow rate when the system is at zero flow. Process noise is generally not a problem on liquids because the sensor's signal-to-noise ratio is at a maximum. However, gases and steam produce a much weaker sensor signal, which may not be as easily discernible from process noise at low flow.

Process noise cannot be quantified before the meter is installed and, therefore, it should always be assumed that some process noise exists. It can be eliminated using built-in noise filtering circuitry. However, this raises the threshold value of the low flow cut-off. Thus, the more filtering used to eliminate process noise, the less the net rangeability of the meter. To avoid this, vortex flowmeters need to be sized properly to ensure the desired rangeability. There are two general sizing guidelines that should be followed:

1. The user URV must not be less than 20% of the meter's URL

Note: URL is the highest flow rate that a meter can be adjusted to measure whilst the URV is the highest flow rate that a meter is adjusted to measure. The URV will always be equal to or lower than the URL.

2. The minimum desired flow rate must be >2 times the value of the meter's low flow cut-in rate

8.2.5.6 Accuracy

Vortex meter accuracy is based on the known value of the meter factor (K-factor), determined from a water calibration at the factory. The accuracy for liquids is typically ±0.75% of flow rate for Reynolds numbers above 30,000.

Water calibration data cannot precisely predict K-factor values for gases and steam, which can flow at Reynolds numbers well outside the test data range. For this reason, gas and steam accuracy is typically stated as ±1.0% of flow rate for Reynolds numbers above 30,000.

Long-term accuracy depends on the stability of the internal dimensions of the flow-tube and shedder body. Only significant changes in these dimensions (due to corrosion, erosion, coatings, etc.) can affect accuracy with time. Whilst vortex meter K-factors can only be determined by wet calibration, the dimensions of the flow-tube inside diameter and bluff body thickness can be used as a 'flag' to determine if recalibration is necessary. Prior to installation, inspect the flow-tube and carefully measure and record the two reference dimensions. After a period of time in service, the meter can be removed, cleaned, and re-measured. The meter does not require recalibration if there has been no significant change in the two reference dimensions.

8.2.5.7 Effect of Erosion

Although vortex shedding flowmeters are primarily designed for measuring the flow of clean liquids and gases, they can still be used if small amounts of foreign matter are present. Since there are no moving parts, or ports with active flow, there is little concern for erosion, physical damage, or clogging. The effect of erosion on the salient edges of the bluff body is small and often poses no significant accuracy degradation.

8.2.5.8 Low Density Gases

Measuring gas flows can be a problem when the process pressure is low (i.e. low-density gases) because a vortex produced under such conditions does not have a strong enough pressure pulse to enable a sensor to distinguish it from flow noise. For such applications, a minimum measurable flow becomes a function of the strength of the pressure pulse (a function of the product of fluid density and the square of fluid velocity) rather than Reynolds number. Low-density gases can be measured with a vortex meter; however, a minimum measurable flow may correspond to a high-fluid velocity and rangeability may be significantly less than 20:1.

8.2.5.9 Orientation

Vortex meters can be installed vertically, horizontally, or at an angle. However, for liquid measurements the meter must be full at all times. The meter should also be installed to avoid formation of secondary phases (liquid, gas, or solid) in the internal sensor chambers.

8.2.5.10 Pressure Drop

If the inside meter diameter is the same as the nominal diameter of the process piping (i.e. a 50 DN meter is used in a 50 DN line), then the pressure drop will normally be less than 40 kPa on liquid flow at the URL (usually in the 14–20 kPa range at the user's URV). However, when downsizing the vortex meter to achieve a desired rangeability, the unrecoverable pressure loss through the meter is increased.

It must be ensured that this increased pressure loss is not enough to cause a liquid to flash or cavitate within the pipe. Flashing and cavitation have an adverse effect on the meter accuracy and can cause damage to the meter itself.

8.2.5.11 Multiphase Flow

Measurement of two- or three-phase flow (e.g. water with sand and air, or 'wet' steam with vapour and liquid) is difficult and if multi-phase flow is present the vortex meter will not be as accurate.

Because the vortex meter is a volumetric device, it cannot distinguish which portions of the flow are liquid and which portions of the flow are gas or vapour. Consequently, the meter will report all the flow as gas, or all the flow as liquid, depending on the original configuration of the device. Thus, for example, if the meter is configured to measure water in litres, and the actual water has some entrained air and sand mixed in, a litre registered by the meter will include the water, air, and sand that is present. Therefore, if the area of interest was the amount of water, the reading from the meter would be consistently high, based on the proportions of air and sand present. A user would, consequently, need to separate the phases prior to metering or live with this inherent error.

8.2.5.12 Material Build-up

Fluids that tend to form coatings are bad applications for vortex meters. Coating build-up on the bluff body will change its dimensions, and therefore, the value of the K-factor.

8.2.5.13 Piping Effects

The specification for the vortex meter accuracy is based on a well-developed and symmetrical fluid velocity profile, free from distortion or swirl, existing in the pipe. The most common way to prevent errors is to provide sufficient lengths of straight, unobstructed pipe, upstream, and downstream of the meter, to create a stable profile at the meter site.

Generally, vortex meters require similar amounts of upstream and downstream pipe runs to orifice plates, turbine meters, and ultrasonic meters. Vortex meters are not usually recommended for 'tight' piping situations, with limited runs of straight pipe, unless repeatability is more important than accuracy. Typical manufacturers' recommendations are shown in Figure 8.10, when flow conditioners are not being used.

Most performance specifications are based on using schedule 40 process piping. This pipe should have an internal surface free from mill scale, pits, holes, reaming scores, bumps, or other irregularities for a distance of 4 diameters upstream and 2 diameters downstream of the vortex meter. The bores of the adjacent piping, the meter and the mating gaskets must be carefully aligned to prevent measurement errors.

For liquid control applications, it is recommended that the vortex meter be located upstream of the control valve for a minimum of 5 diameters. For gas or steam control applications, it is recommended that the vortex meter be located a minimum of 30 diameters downstream of the valve. The only exception to this rule is for butterfly

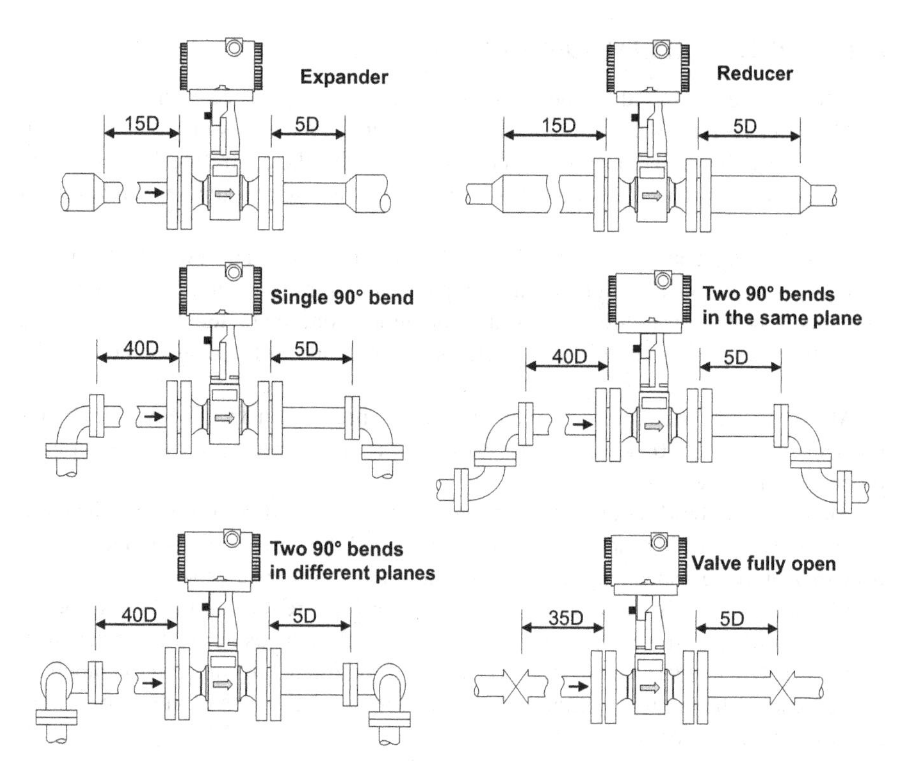

FIGURE 8.10 Typical manufacturers' recommendations for straight pipe lengths. (Courtesy: Emerson.)

valves. In this instance the recommended distances are increased to 10 diameters for liquids and 40–60 diameters for gases and steam.

8.2.6 Avoiding Problems

The following guidelines will help prevent application and measurement problems with a vortex meter and ensure premium performance:

- Improper configuration
- Improper sizing
- Insufficient upstream/downstream relaxation piping
- Improper meter orientation
- Partially full piping
- Accumulation of secondary phase (gas, liquid, or solid) inside the meter
- Improper temperature/pressure taps
- Flows below Reynolds numbers of 30,000
- Flows below the low flow cut-in
- Process noise (at low flows or zero flow)
- Presence of multiple phases

8.3 VORTEX PRECESSION METERS

The 'Swirlmeter' is based on the principle known as vortex precession.

The inlet (Figure 8.11) uses guide vanes, whose shape is similar to a turbine rotor, to force the fluid entering the meter to spin about the centreline. This swirling flow then passes through a Venturi, where it is accelerated and then expanded in an expansion chamber.

The expansion changes the direction of the axis about which the swirl is spinning – moving the axis from a straight to a helical path. This spiralling vortex is called vortex precession. A flow straightener is used at the outlet from the meter. This isolates the meter from any downstream piping effects that may affect the development of the vortex.

Above a given Reynolds number, the vortex precession frequency, which lies between 10 and 1,500 Hz and is measured with a piezoelectric sensor, is directly proportional to the flow rate.

Although the Swirlmeter can be used with both gases and liquids, it finds its main application as a gas flowmeter. A major advantage of the vortex precession technique over that of vortex shedding is that it has a much lower susceptibility to the flow profile and hence only three diameters of straight line are required upstream of the meter. In addition, the Swirlmeter features linear flow measurement, rangeability between 1:10 and 1:30, no moving parts, and installation at any angle in the pipeline.

Because of the higher tolerance in manufacturing of this type of meter, it is more expensive than comparative meters.

8.4 FLUIDIC FLOWMETER

The fluidic flowmeter is based on the wall attachment or *Coanda* effect. Wall attachment occurs when a boundary wall is placed in proximity to a fluid jet – causing the jet to bend and adhere to the wall.

This effect is caused by the differential pressure across the jet, deflecting it towards the boundary (Figure 8.12). Here it forms a stable attachment to the wall, which is little affected by any downstream disturbances.

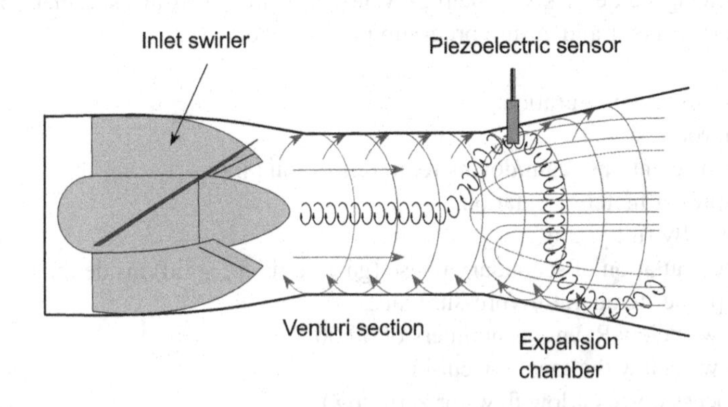

FIGURE 8.11 Basic principle of a vortex precession Swirlmeter. (Courtesy: ABB.)

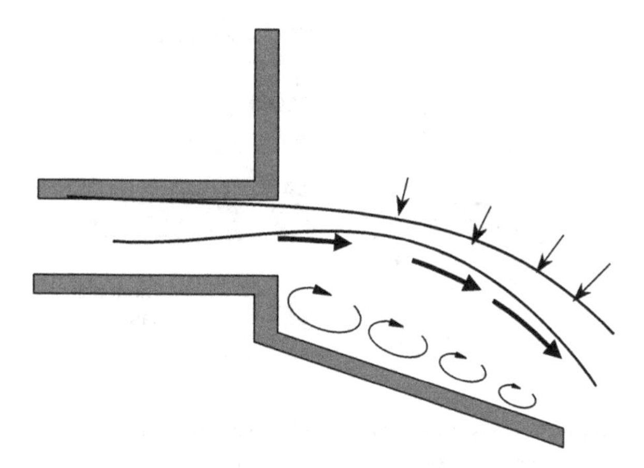

FIGURE 8.12 Explanation of the Coanda effect – resulting in stable attachment of the flow stream to the wall.

In the fluidic meter (Figure 8.13), the flow stream attaches itself to one of the walls – with a small portion of the flow fed back through a passage to a control port (Figure 8.13).

This feedback diverts the main flow to the opposite side wall, where the same feedback action is repeated (Figure 8.14).

The result is a continuous oscillation of the flow between the sidewalls of the meter body whose frequency is linearly related to the fluid velocity. The flow in the feedback passage cycles between zero and maximum which is detected by using a built-in thermistor sensor.

The main benefit offered by the fluidic meter is that feedback occurs at much lower Reynolds numbers and it may thus be used with fairly viscous media. In addition, since a fluidic oscillator has no moving parts to wear with time, there is no need for recalibration during its expected lifetime. Other benefits include:

- Rugged construction
- High immunity to shock and pipe vibration

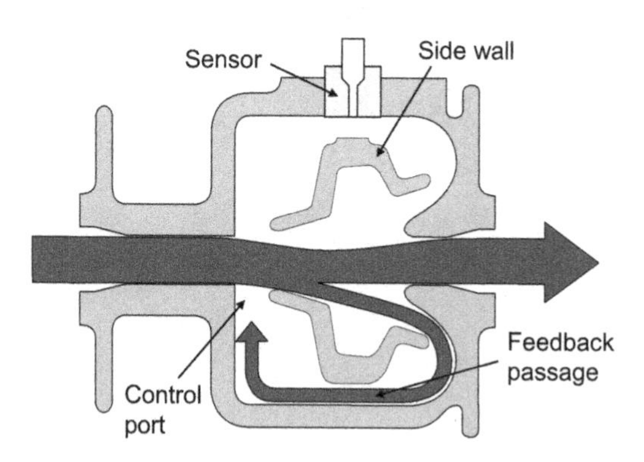

FIGURE 8.13 Once attached to one side of the wall a feedback passage diverts a portion of stream back onto the main flow. (Courtesy: Fluidic Flowmeters.)

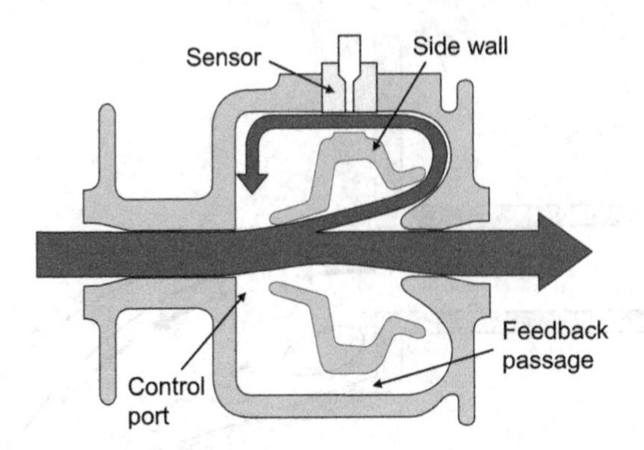

FIGURE 8.14 Main stream is diverted to the other wall by the feedback control action, and the procedure is then repeated. (Courtesy: Fluidic Flowmeters.)

- High turndown ratio (typically 30:1)
- Linear output
- Operation down to Re of 400
- Operating pressure (38 mm unit) 10 bar

The main drawback of the fluidic oscillator is its unsuitability for use with gases and its relatively high unrecoverable pressure loss – which varies with the flow rate. For example, for the largest size meter (38 mm) a flow rate of 26 L/min produces an unrecoverable pressure drop of 0.007 bar. However at a maximum flow rate of 750 L/min, the unrecoverable pressure loss rises to nearly 7 bar.

Because downstream flashing can lead to errors, a minimum backpressure is required as determined by:

$$\text{Minimum back pressure} = 1.5 \cdot \Delta P + 1.25 \cdot V_P \tag{8.3}$$

where

ΔP = pressure drop at maximum flow rate

P_V = absolute vapour pressure (at 38°C) of the fluid

9 Electromagnetic Flowmeters

9.1 INTRODUCTION

Electromagnetic (EM) flowmeters, also known as *Magflows* or *Magmeters*, have now been in widespread use throughout industry for more than 40 years and were the first of modern meters to exhibit no moving parts and zero pressure drop.

9.2 MEASURING PRINCIPLE

The principle of the EM flowmeter is based on Faraday's Law of induction that states that if a conductor is moved through a magnetic field, voltage will be induced in it that is proportional to the velocity of the conductor.

Referring to Figure 9.1, if the conductor of length (l) is moved through the magnetic field having a magnetic flux density (B) at a velocity (v), then voltage will be induced where:

$$e = B \cdot l \cdot v \tag{9.1}$$

and

e = induced voltage (V)
B = magnetic flux density (Wb/m^2)
l = length of the conductor (m)
v = velocity of conductor (m/s)

In the EM flowmeter (Figure 9.2) a magnetic field is produced across the cross-section of the pipe – with the conductive liquid forming the conductor (Figure 9.3). Two sensing electrodes, set at right angles to the magnetic field, are used to detect the voltage that is generated across the flowing liquid and which is directly proportional to the flow rate of the media.

► It can thus be seen that since v is the flow rate (the parameter to be measured), the generated voltage is limited by the length of the conductor (the diameter of the pipe) and the flux density. In turn, the flux density is given by:

$$B = \mu \cdot H \tag{9.2}$$

where

B = flux density
μ = permeability
H = magnetising field strength (ampere-turns/m)

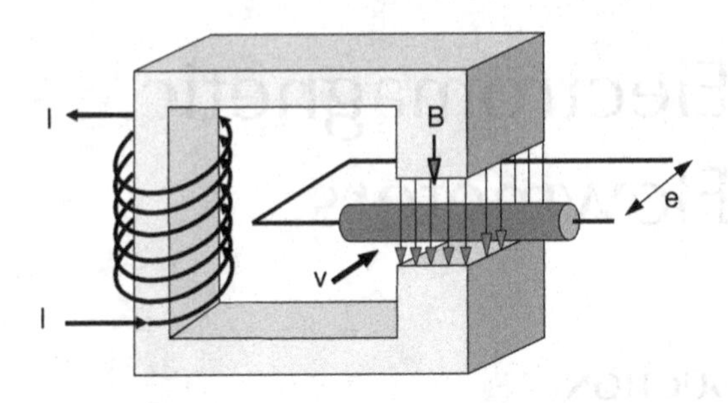

FIGURE 9.1 Illustration of Faraday's Law of EM induction.

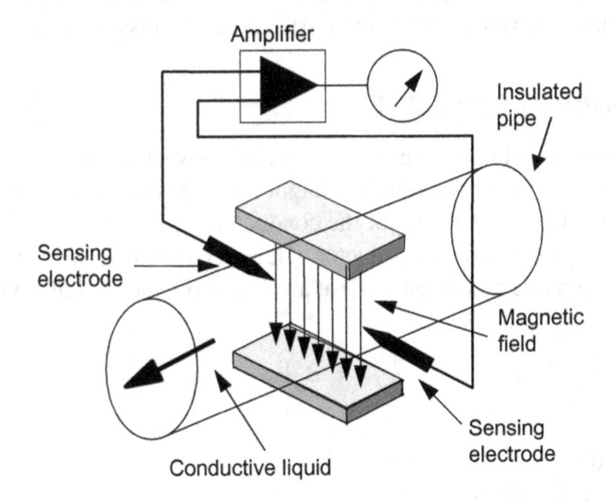

FIGURE 9.2 Basic principle of the EM flowmeter.

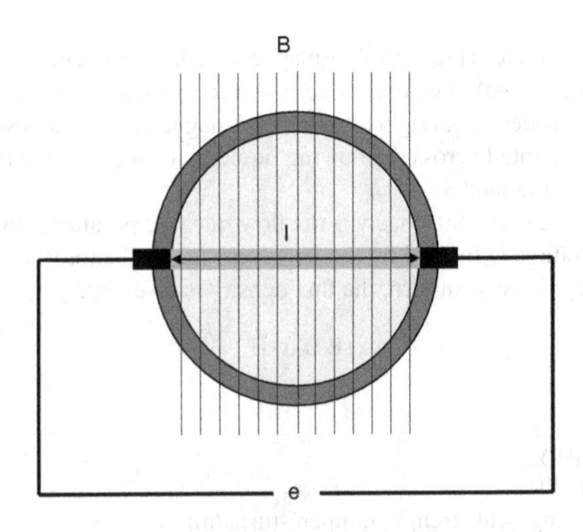

FIGURE 9.3 The conductive liquid forms the conductor in contact with the electrodes.

Because the permeability of the magnetic circuit is largely determined by the physical constraints of the pipe (the iron–liquid gap combination), the magnetic flux density B (and hence the induced voltage) can only be maximised by increasing H – a function of the coil (number of windings and its length) and the magnetising current. ◀

9.3 CONSTRUCTION

Because the working principle of the EM flowmeter is based on the movement of the conductor (the flowing liquid) through the magnetic field, it is important that the pipe carrying the medium (the metering tube) should have no influence on the field. Consequently, in order to prevent short circuiting of the magnetic field, the metering tube must be manufactured from a non-ferromagnetic material such as stainless steel or nickel–chromium.

It is equally important that the signal voltage detected by the two sensing electrodes is not *electrically* short circuited through the tube wall. Consequently, the metering tube must be lined with an insulating material. Such materials have to be selected according to the application and their resistance to chemical corrosion, abrasion, pressure, and temperature.

9.3.1 Lining Materials

9.3.1.1 PTFE (Teflon®)

PTFE is a warm deformable resin and is the most widely used liner material. It comprises an extruded tube inserted into the flowmeter without bonding – with the ends bevelled to form the flange face.

The PTFE liner may be adversely affected by exposure to vacuum conditions – increasing as the line size and temperature increase.

Note: A vacuum situation can occur during transient conditions such as draining a line.

Features:

- Wide temperature range –20°C to 180°C (–4°F to 356°F)
- Excellent anti-stick characteristics reduce build-up
- Best chemical resistance – inert to a wide range of acids and bases
- Approved in food and beverage applications
- Suitable for pressures up to 50 bar
- *Not* suitable for abrasive service for example slurries

9.3.1.2 PFA

Perfluoroalkoxy (PFA) is a melt-processable resin that is moulded directly in a reinforced stainless steel tube – resulting in extremely good mechanical performance during temperature fluctuations and under vacuum conditions. The stainless steel tube reinforcement also caters for high temperatures without deformation.

Features:

- Compared with the PTFE, PFA offers
  - Better shape accuracy
  - Better abrasion resistance since there are no bulges or deformations
- Better vacuum strength
- Wide temperature range –20°C to 150°C (–4°F to 300°F)
- Excellent anti-stick characteristics reduce build-up
- Excellent chemical resistance – inert to a wide range of acids and bases
- Suitable for pressures up to 50 bar
- Generally *not* suited for abrasive service

9.3.1.3 Polyurethane

Polyurethane is resilient and is much more resistant than PTFE/PFA to abrasive process fluids for example slurries. Consequently, polyurethane is commonly used in mining applications. However, polyurethane is limited to a maximum process temperature of 140°F.

Features:

- High resistance to abrasive process fluids
- High-pressure withstand – up to 200 bar
- Poor chemical resistance – cannot be used with strong acids or bases
- Maximum process temperature of 40°C

9.3.1.4 Neoprene

Neoprene (polychloroprene or pc-rubber) is a versatile synthetic soft rubber originally developed as an oil-resistant substitute for natural rubber. Consequently, well-suited to wastewater applications (e.g. process water) where oil is present.

Features:

- Resistant to chemical attack
- Good degree of abrasion resistance – but lower than polyurethane
- Maximum temperature of 70°C (158°F)
- Risk of swelling in water
- Suitable for pressures up to 100 bar

9.3.1.5 Natural Soft Rubber

Vulcanised raw natural rubber provides an excellent abrasion-resistant performance – especially in heavy slurry applications.

Linatex® is a 95% natural rubber that is often ranked as the premium wear-resistant rubber for sliding or wet abrasion service and is the only liner that tolerates low temperature applications down to –40°C.

Features:

- Excellent abrasion resistance particularly to sand, slurries, and particles since particles bounce off the soft rubber instead of causing damage
- Low temperature applications down to –40°C (–40°F)

- Adversely affected by oil and solvents
- Suitable for pressures up to 40 bar

9.3.1.6 Hard Rubber (Ebonite)

Formerly called vulcanite, Ebonite is obtained by vulcanising natural rubber for prolonged periods and contains about 25–80% sulphur and linseed oil.

Ebonite liners exhibit extremely low water absorption and high levels of stability – especially in high-pressure applications and temperatures in excess of 70°C.

Features:

- Inexpensive general purpose liner
- Wide corrosion resistance
- Relatively good chemical resistance and resistance to hydrocarbons
- Main application in the water and wastewater industries
- Temperatures up to 95°C (203°F)
- Extremely low water absorption
- Suitable for pressures up to 100 bar

9.3.1.7 EPDM

Ethylenepropylenediene (EPDM) rubber is an elastomer that is particularly suitable for drinking water applications.

Features:

- Main application in drinking water
- May be used in some food and beverage applications with pipe sizes greater than 100 DN (4″)
- Temperatures up to 70°C (158°F)
- Not suitable for wastewater applications where hydrocarbons are present
- Extremely low water absorption
- Suitable for pressures up to 40 bar

9.3.1.8 Nitrile Rubber

Nitrile rubber, also known as NBR, Buna-N, and acrylonitrile (ACN) butadiene rubber, is a synthetic rubber copolymer of ACN and butadiene that is highly resistant to both water and hydrocarbons.

Features:

- Lowest priced liner
- Highly resistant to both water and hydrocarbons
- Temperatures up to 70°C (158°F)
- Suitable for pressures up to 40 bar

9.3.1.9 Phenolic Resins

A variety of phenol formaldehyde (PF) resins have been developed as economic alternatives to PTFE liners in high-temperature applications. The liner is spray coated to form a smooth, hard non-porous surface and finish that is highly resistant to corrosion.

They are typically used in pulp and paper applications as well as in the chemical industry due to its excellent chemical resistance.

Features:

- Reasonably robust under vacuum conditions
- Compatible with chemicals with pH values between 3 and 13
- Temperatures up to 130°C (266°F)
- Suitable for pressures up to 40 bar
- *Not* suitable for media containing ozone

9.3.1.10 Ceramic (Al_2O_3)

Aluminium oxide liners are resistant to both acids and alkalis and are highly resistant to wear and abrasion. Being ultra-smooth, they are frequently used in the food and beverage industries since they do not provide cavities in which bacteria can accumulate. They are also frequently used in the nuclear industry since they are immune to radiation damage.

Since there is little expansion or contraction with temperature variations, the cross-sectional area, and therefore, the measurement of volumetric flow, remains virtually constant.

Electrode leakage may be totally eliminated through the use of capacitive-coupled electrodes or sintered platinum electrodes that are integral with the ceramic material.

The downside of aluminium oxide is that because of brittleness they are susceptible to tensioning or bending of the external pipework. Furthermore, cracking can occur as a result of thermal shock exceeding 30°C step change. A further limitation is that it cannot be used in oxidising acid or hot concentrated caustic applications over 50°C.

Features:

- Best suited to small diameter, high accuracy applications
- Best possible long-term accuracy
- Chemically inert in the presence of most substances, even at elevated temperatures
- High temperatures up to 200°C (390°F)
- Totally vacuum resistant
- Suitable for pressures up to 40 bar
- Sensitive to tensioning or bending of the external pipework
- Sensitive to thermal shocks

9.3.1.11 Ceramic (ZrO_2)

Zirconium oxide is an advanced ceramic material that overcomes one of the main limitations of aluminium oxide since it has no thermal shock limitations.

Table 9.1 provides a summary of these materials.

TABLE 9.1
Commonly Used Magnetic Flowmeter Liner Materials

Material	General	Shortcomings	Abrasion Resistance	Acid Resistance	Temperature Limit	Pressure Limit (bar)
PTFE	Warm deformable resin with excellent anti-stick properties and suitable for food and beverage	Adversely affected by vacuum. Not suitable for abrasive service	Poor	Excellent	−20°C to 180°C (−4°F to 356°F)	50
PFA	Melt-processable resin with better shape accuracy, abrasion resistance, and vacuum strength than PTFE		Good	Excellent	−20°C to 150°C (−4°F to 300°F)	50
Polyurethane	Extreme resistance to wear and erosion	Not suitable for strong acids or bases	Excellent	Poor	40°C (104°F)	200
Neoprene	Combines some of the resistance to chemical attack of PTFE with a good abrasion resistance	Risk of swelling in water	Good	Good	70	100
Natural soft rubber (Linatex)	Excellent abrasion resistant – good for heavy slurries Tolerates low temperatures down to −40°C	Adversely affected by oil and solvents	Excellent	Poor	Down to −40°C	40
Hard rubber (Ebonite)	Inexpensive Extremely low water absorption. Main application in potable and wastewater industries		Good	Good	95°C (203°F)	100
EPDM rubber	Elastomer particularly suitable for drinking water	Not suitable for wastewater applications where hydrocarbons are present	Excellent	Poor	70°C (158°F)	40

(Continued)

TABLE 9.1 (*Continued*)
Commonly Used Magnetic Flowmeter Liner Materials

Material	General	Shortcomings	Abrasion Resistance	Acid Resistance	Temperature Limit	Pressure Limit (bar)
Nitrile rubber	Highly resistant to both water and hydrocarbons Lowest priced liner		Good	Poor	Up to 70°C (158°F)	40
Modified phenolic	Developed for harsh environments containing acids Robust under vacuum conditions	Not suitable for media containing ozone	Good	Excellent	130°C (266°F)	Yield strength of pipe
Ceramic (Al_2O_3)	Highly recommended for very abrasive and/or corrosive applications Ultra-smooth finish No cavities Immune to radiation damage Cross-sectional area remains constant Electrode leakage may be eliminated through the use of a capacitive-coupled electrodes or sintered platinum electrodes	Susceptible to tensioning/ bending of pipework. Cracking can occur as a result of thermal shock exceeding 30°C	Excellent	Excellent	200°C (390°F)	40
Ceramic (ZrO_2)	No thermal shock limitations		Excellent	Excellent	200°C (390°F)	40

9.3.2 Electrodes

The electrodes, like the liners, are in direct contact with the process medium and again the materials of construction must be selected according to the application and their resistance to chemical corrosion, abrasion, pressure, and temperature. Because of the relatively small size of the electrode and the importance of the seal between the electrode and liner, only small corrosion rates are acceptable – typically less than 50 μm (0.002 in) per year.

Commonly used materials include 316 stainless steel, Hastelloy C, tantalum, platinum/rhodium, and titanium.

A further consideration is the material's electrode potential. The electrode potential for platinum for example is +1.2 V whilst that of titanium is −1.63 V. Consequently, depending on the materials used, the electrode potential becomes the common-mode voltage (CMV) that needs rejection at the sensor output. Stainless steel electrodes have only a couple of hundred millivolts of CMV, so the common mode can be more easily rejected.

9.3.2.1 316 Stainless Steel

This, the most commonly used electrode material provides a good combination of corrosion resistance and abrasion resistance. Although suitable for use with nitric acid, in most instances it is not suitable for use with sulphuric or hydrochloric acid.

9.3.2.2 Hastelloy C

It provides increased corrosion resistance over 316 SST in oxidising conditions, sea water, and many reducing media. Furthermore, its high strength makes it the preferred choice in slurry applications.

9.3.2.3 Tantalum

Tantalum gives improved service over both 316 SS and Hastelloy C and is very resistant to hydrochloric acid, nitric acid, and aqua regia (mixture of nitric and hydrochloric acid). However, it should not be used with hydrofluoric acid, fluorosilicic acid, or sodium hydroxide.

9.3.2.4 90% Platinum–10% Iridium

Although being the most expensive material, it provides the greatest resistance to attack all the available electrode materials. However, it should not be used with aqua regia.

9.3.2.5 Titanium

Titanium is more resistant to nitric acid than 316 SS and gives excellent performance in sea water and dilute caustic solutions. It should not be used with hydrofluoric, hydrochloric, or sulphuric acid.

9.3.3 Electrode Leakage Concerns

One of the main concerns is the need to ensure that there is no leakage of the process medium. In the construction design shown in Figure 9.4, the electrode seal is maintained through the use of five separate sealing surfaces and a coil spring.

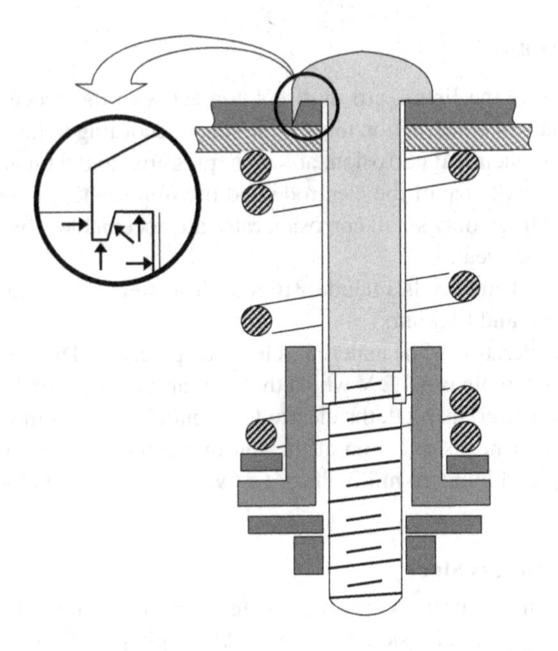

FIGURE 9.4 The electrode seal is maintained through the use of five separate sealing surfaces and a coil spring. (Courtesy: Emerson.)

However, to ensure that the overall integrity of the system is maintained, even if a process leak should occur past the liner/electrode interface, the electrode compartment can also be separately sealed. Usually rated for full line pressure, such containment ensures that in the event of a leak, no contamination of the field coils occurs.

Where heavy abrasion or contamination of the electrodes might occur, many manufacturers offer the option of field replaceable electrodes (Figure 9.5).

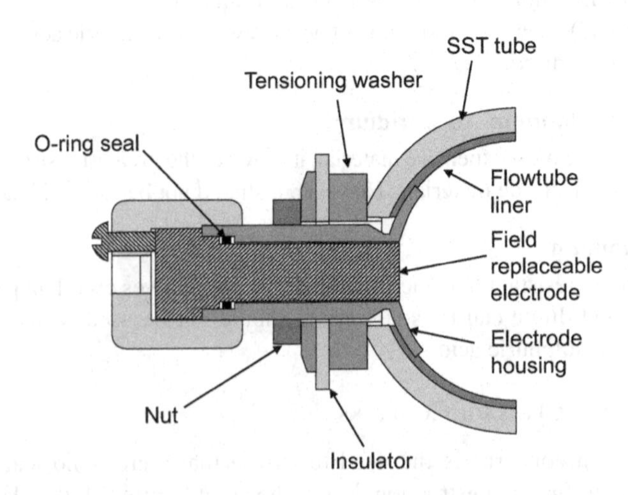

FIGURE 9.5 Field replaceable electrode. (Courtesy: Emerson.)

Fouling of the electrodes by insulating deposits can considerably increase the internal resistance of the signal circuit – changing the capacitive coupling between the field coils and signal circuitry.

9.4 CONDUCTIVITY

► The two main characteristics of the process medium that need to be considered are its conductivity and its tendency to coat the electrode with an insulating layer. As shown in Figure 9.6, to develop most of the electrode potential (e) across the input impedance (R_i) of the meter amplifier, and to minimise the effect of impedance variations due to changes in temperature, R_i needs to be at least 1,000 times higher than the maximum electrode impedance R_s.

Modern high input impedance amplifiers are available in the range of 10^{13} to 10^{14} Ω. Consequently, with an amplifier having for example an input impedance of 10^{13} Ω, the error due to impedance matching is less than 0.01% and a change in electrode impedance from 1 to 1,000 MΩ will affect the voltage by only 0.001%.

The electrode impedance depends on the fluid conductivity and varies with the size of the metering tube. In older AC driven instruments, the minimum conductivity of the fluid usually lay between 5 and 20 μS/cm. For DC field instruments, the minimum conductivity was about 1 μS/cm. However modern instruments employ a variety of technologies, including capacitively coupled meters that can be used in liquids with conductivity levels down to 0.05 μS/cm. ◄

Most refinery products, and some organic products, have insufficient conductivity to allow them to be metered using EM flowmeters (Table 9.2).

It should be noted that the conductivity of liquids can vary with temperature and care should be taken to ensure that the performance of the liquid in marginal conductivity applications is not affected by the operating temperatures. Most liquids have a positive temperature coefficient of conductivity. However negative coefficients are possible in a few liquids.

At first glance, the extended range of an EM flowmeter down to 0.05 μS/cm would appear to cover the conductivity of ultrapure water down to 0.1 μS/cm. However, in order to meet these requirements, use is made of very high input impedance amplifiers in the

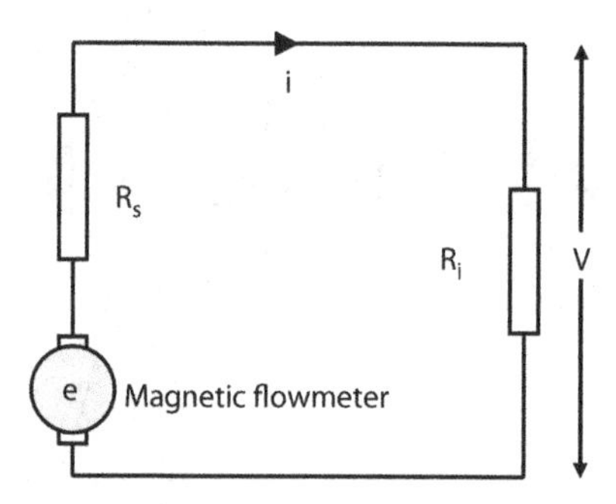

FIGURE 9.6 To develop most of the electrode potential (e) across the input impedance (R_i) of the meter amplifier the R_i needs to be at least 1,000 times higher than the maximum electrode impedance R_s.

TABLE 9.2
Conductivities of Some Typical Fluids

Liquid	Conductivity (μS/cm)
Carbon tetrachloride at 18°C	4×10^{-12}
Toluene	10^{-8}
Kerosene	0.017
Analine at 25°C	0.024
Soya bean oil	0.04
Ultrapure water	0.1
Phosphorous	0.4
Benzole alcohol at 25°C	1.8
Acetic acid (1% solution)	5.8×10^2
Acetic acid (10% solution)	16×10^2
Latex at 25°C	5×10^3
Sodium silicate	24×10^3
Sulphuric acid (90% solution)	10.75×10^4
Ammonium nitrate (10% solution)	11×10^4
Sodium hydroxide (10% solution)	31×10^4
Hydrochloric acid (10% solution)	63×10^4

range of 10^{13} to 10^{14} Ω or more, that are very susceptible to electrical noise. Unfortunately water, being a bipolar vibratory molecule, produces relatively large amplitudes of electrical noise that tends to swamp the amplifiers used to gain this sensitivity.

In some applications, coating of the electrodes is cause for concern and, over the years, a number of solutions have been offered including a mechanical scraper assembly (Figure 9.7) and ultrasonic cleaning.

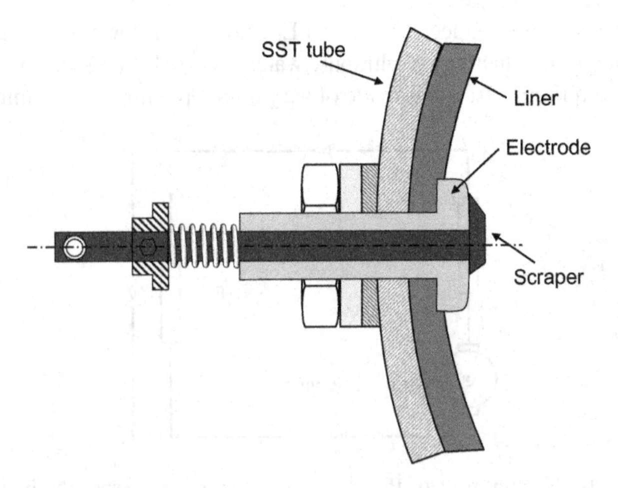

FIGURE 9.7 Following encrusted of the electrode a mechanical scraper arrangement allows the fouling to be removed.

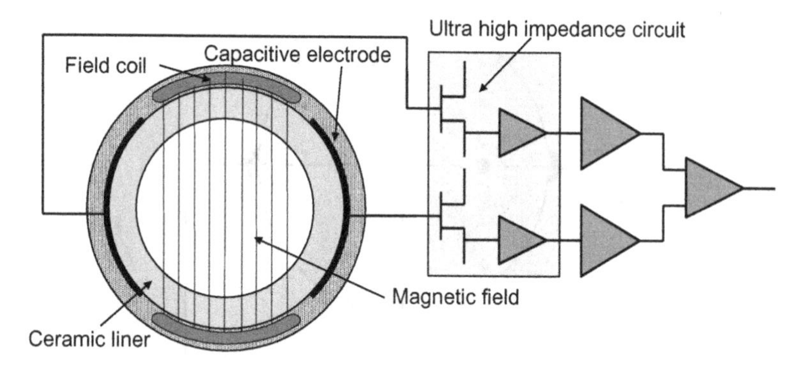

FIGURE 9.8 The traditional wetted electrodes have been replaced by capacitive electrodes bonded to the outside of the ceramic liner.

9.4.1 Capacitive-Coupled Electrodes

The foregoing solutions do not solve the problem of 'electrode coating' in which an insulating deposit effectively isolates the electrodes. These insulating deposits are often found in the paper manufacturing industry and in sewage treatment applications where grease and protein conglomerates can develop into thick insulating layers.

In the capacitive-coupled flowmeter the electrodes, which are normally wetted by the process liquid, have been replaced by two large plates bonded to the outside of a ceramic flowtube (Figure 9.8) – with an ultra-high impedance preamplifier mounted directly on the flow tube. Being capable of use with liquids with conductivity levels down to 0.05 μS/cm, the capacitive-coupled magnetic flowmeter features no gaps or crevices, no risk of electrode damage due to abrasion, no leakage, and no electrochemical effects.

9.5 FIELD CHARACTERISATION

The purpose of a flowmeter is to measure the true average velocity across the section of pipe, so that this can be related directly to the total volumetric quantity in a unit of time. The voltage generated at the electrodes is the summation of the incremental voltages generated by each elemental volume of cross-section of the flowing fluid as it crosses the electrode plane with differing relative velocities.

Initially, designs assumed the magnetic field to be homogeneous over the measured cross-section and length of the pipe in order to achieve precise flow measurement. However, early investigators showed that, for a given velocity, the medium does not generate the same voltage signal in the electrodes at all points. Thus, for a given velocity (v), the medium flowing at position A_1 (Figure 9.9) does not generate the same voltage signal as that flowing in position A_2.

Rummel and Ketelsen plotted the medium flowing at various distances away from the measuring electrodes (Figure 9.10) and showed how these contribute in different ways towards the creation of the measuring signal. This shows that a flow profile that concentrates velocity in the area of one electrode will produce eight times the output of that at the pipe centre – leading to errors that cannot be overlooked.

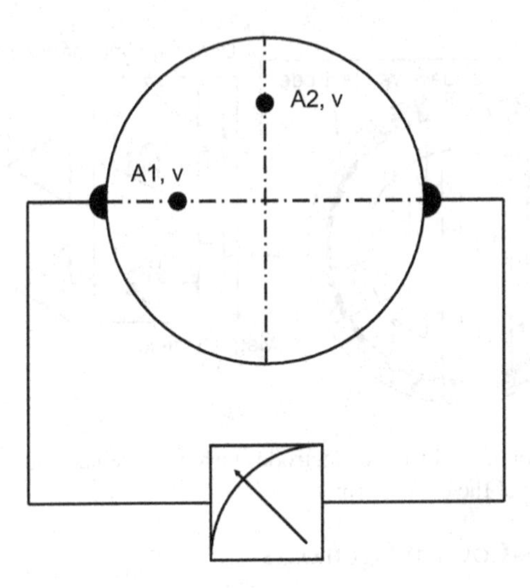

FIGURE 9.9 For a given velocity (v), the medium flowing at position A_1 does not generate the same voltage signal as that of flowing position A_2. (Courtesy: Endress+Hauser.)

One solution to this problem is to use a non-homogeneous field that compensates for these non-linear concentrations.

Subsequent to his research, Ketelsen designed a magnetic flowmeter making use of a 'characterised field'. As distinct from the homogeneous field in which the magnetic flux density (B) is constant over the entire plane (Figure 9.11a), the 'characterised field' is marked by an increase in B in the *x*-direction and a decrease in the *y*-direction (Figure 9.11b).

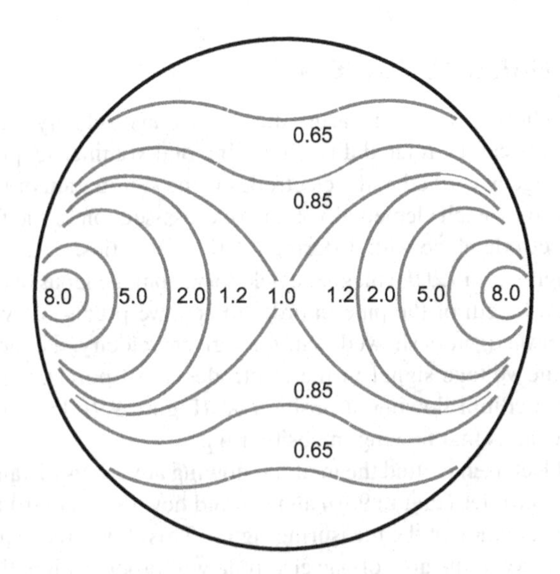

FIGURE 9.10 Weighting factor distribution in the electrode plane (Rummel and Ketelsen).

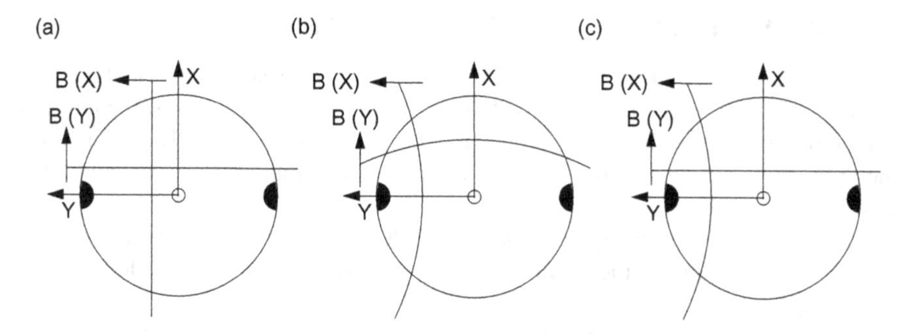

FIGURE 9.11 The three most common forms of magnetic fields: (a) the homogeneous field in which B is constant over the entire plane, (b) a 'characterised field' in which B increases in the x-direction but decreases in the y-direction, and (c) the modified field in which B increases in the x-direction but is constant in the y-direction.

Because commercial exploitation of this design is limited in terms of a patent in the name of B. Ketelsen, assigned to Fischer & Porter GmbH, a 'modified field' has been developed in which the lines of magnetic flux, at any place in the electrode plane, are characterised by an increase in B in the x-direction, from the centre to the wall, but is constant in the y-direction (Figure 9.11c). This 'modified field' is, therefore, a compromise between the 'characterised field' and the 'homogeneous field'.

9.6 FIELD DISTORTION

In view of the importance of generating a 'characterised' magnetic field, users should be aware of certain applications that can offset the efforts taken in this direction.

One such area is in the mining and mineral beneficiation fields where the slurry may contain high quantities of solid magnetic materials such as magnetite (Fe_3O_4). Magnetite is a naturally occurring ferrimagnetic rock appearing in almost all igneous and metamorphic rocks.

Because it is easily magnetised, the presence of magnetite can considerably modify the geometry of the field (Figure 9.12) – leading to significant errors.

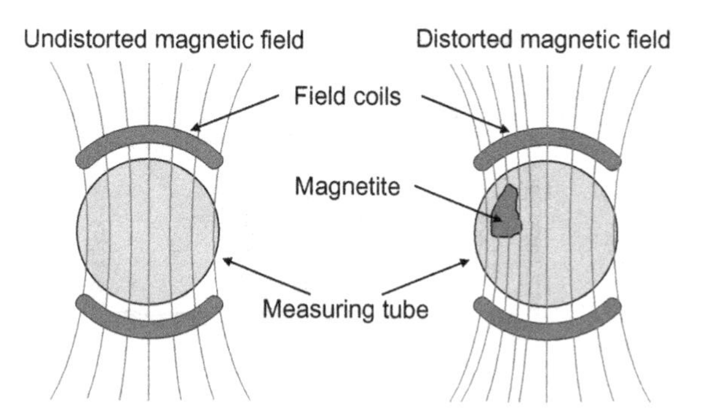

FIGURE 9.12 The presence of magnetite can considerably modify the geometry of the field and lead to significant errors.

9.7 MEASUREMENT IN PARTIALLY FILLED PIPES

A fundamental requirement for accurate *volumetric* flow measurement is that the pipe should be full. Given a constant velocity then, as the fill level decreases, the induced potential at the electrodes is still proportional to the media *velocity*. However, since the cross-sectional area of the media is unknown, it is impossible to calculate the volumetric flow rate.

In the water utility industry where large bore flowmeters are used and the hydraulic force is based on gravity, the occurrence of a partially filled pipe, due to low flow, is quite frequent.

Although installing the flowmeter at the lowest point of the pipeline in a U-section or an invert (Figure 9.13) will combat this problem, there are still many situations where even the best engineering cannot guarantee a full pipe – thus giving rise to incorrect volume readings.

One answer to this problem would be to actually determine the cross-sectional area and thus calculate the volumetric flow.

In the solution offered by ABB in their Parti-MAG, two additional electrode pairs are located in the lower half of the meter to cater for partial flowrate measurements down to 10%. In addition, the magnetic field is switched successively from a series to a reverse coil excitation. The series excitation mode (Figure 9.14) corresponds to the excitation mode for a conventional meter. As a result of this field, a voltage is induced in the electrode pairs that is related to the media velocity.

In the reverse excitation mode (Figure 9.15), the induced voltages in the upper and lower halves of the meter are of equal magnitude but opposite sign. Thus, in a full pipe, the potential would be zero at the electrode pair E_1 and some definite value

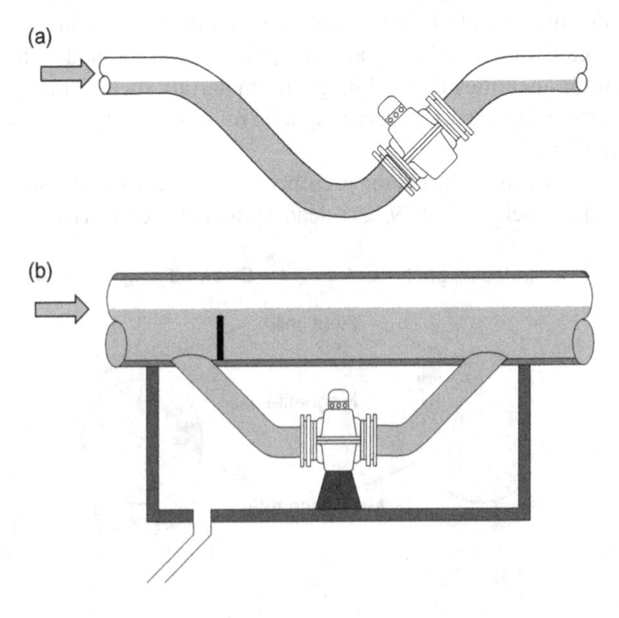

FIGURE 9.13 Flowmeter installed in (a) a U-section or (b) an invert can often ensure that the meter remains full when the media pipe is only partially full. (Courtesy: ABB.)

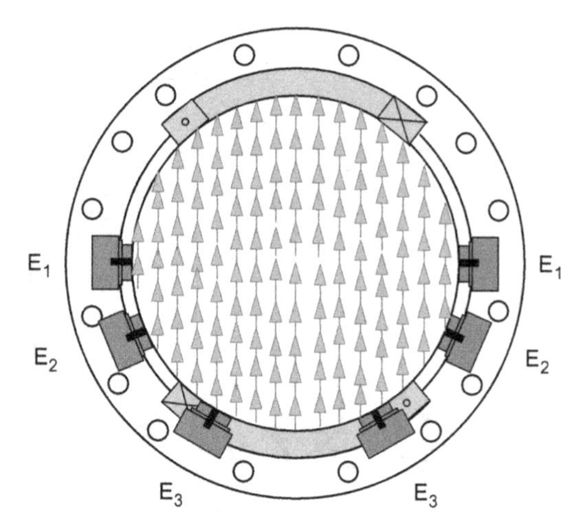

FIGURE 9.14 The series excitation mode corresponds to the excitation mode for a conventional meter. (Courtesy: ABB.)

at the electrode pairs E_2 and E_3. As the level falls, the signal contribution from the upper half decreases while that from the lower half remains the same – resulting in a change in the potential at the various electrode pairs that can be related directly to the change in media level. Microprocessor technology is then used to compute the cross-sectional area and thus the volumetric flow.

A slightly different scheme is used in Krohne's TIDALFLUX meter. This instrument combines an EM flowmeter with an independent capacitive level measuring system.

The EM flow measuring section functions like a conventional EM flowmeter using a single set of electrodes that are placed near the bottom of the pipe as shown in

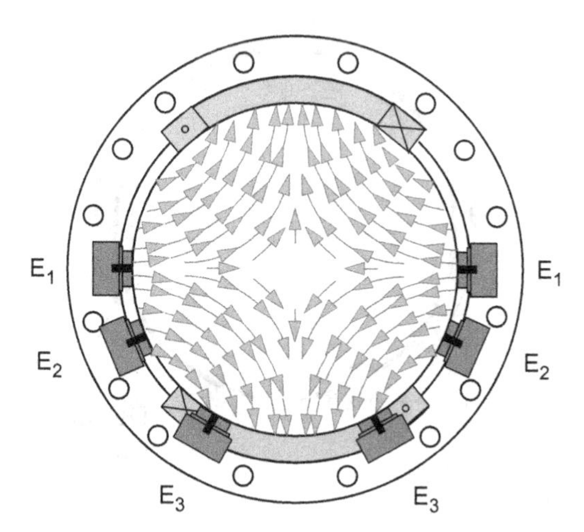

FIGURE 9.15 In the reverse excitation mode, the induced voltages in the upper and lower halves of the meter are of equal magnitude but of opposite sign. (Courtesy: ABB.)

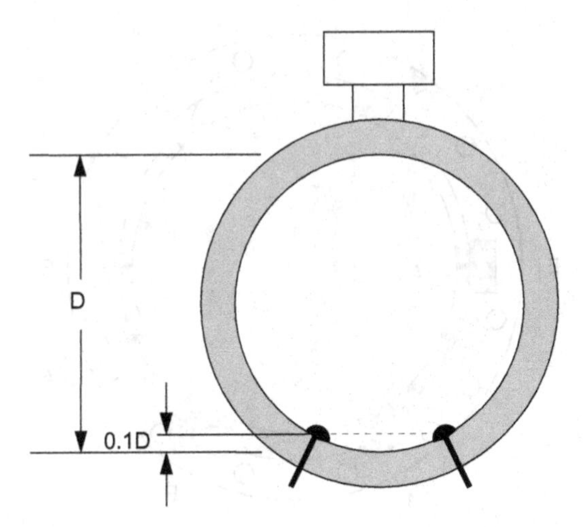

FIGURE 9.16 The two sensing electrodes are positioned so that the electrodes are still covered when the filling level falls to less than 10% of the pipe diameter. (Courtesy: Krohne.)

Figure 9.16. In this manner, even when the filling level falls to less than 10% of the pipe diameter, the electrodes are still covered and capable of providing a flow velocity related output.

The level measuring section makes use of a system of insulated transmission and detection plates embedded in the flowmeter liner (Figure 9.17) in which the change in capacitive coupling is proportional to the wetted cross-section.

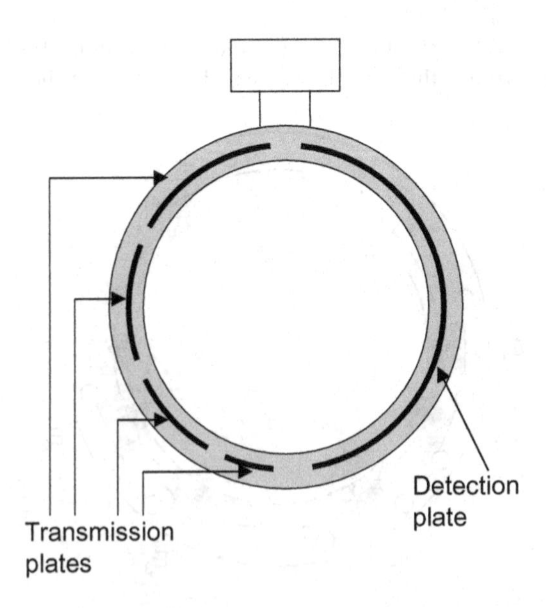

FIGURE 9.17 The level measuring section makes use of insulated transmission and detection plates embedded in the flowmeter. (Courtesy: Krohne.)

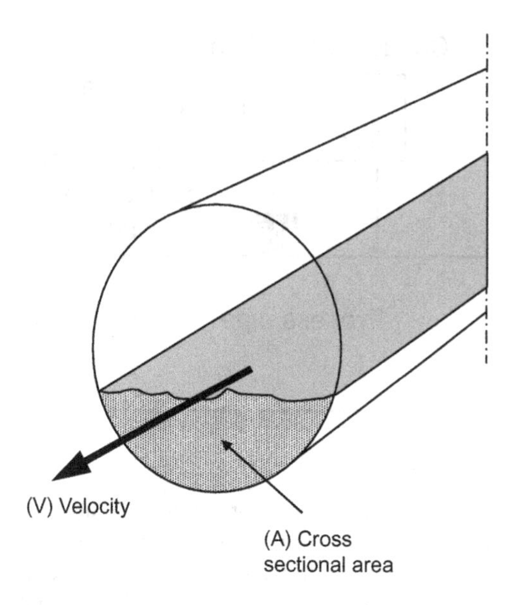

FIGURE 9.18 The volumetric flow is computed using the two measured values of velocity and cross-sectional area. (Courtesy: Krohne.)

Using these two measured values it is now possible to compute the actual volumetric flow (Figure 9.18) from:

$$Q = v \cdot A \tag{9.3}$$

where

Q = volumetric flow
v = velocity related signal
A = wetted cross-sectional area

9.8 EMPTY PIPE DETECTION

In many cases, the measurement of partially filled pipes in not required. Nonetheless, in order to draw attention to this situation, many meters incorporate an 'empty pipe detection' option.

In the most common system (Figure 9.19), a conductivity probe, mounted on top of the pipe, senses the presence of the conductive medium. If the medium clears the sensor, due to partial filling of the pipe, the conductivity falls and an alarm is generated.

An alternative scheme is to use a high frequency current generator across the flowmeter-sensing probes. Because normal flow measurement uses relatively low frequencies, the high frequency signal used to measure the conductivity is ignored by the flow signal amplifier.

'Empty pipe detection' is not only used to indicate that the volume reading is incorrect. For example, in a two-line standby system, one line handles the process and the other line is used for standby. Since the standby line does not contain any of the process medium,

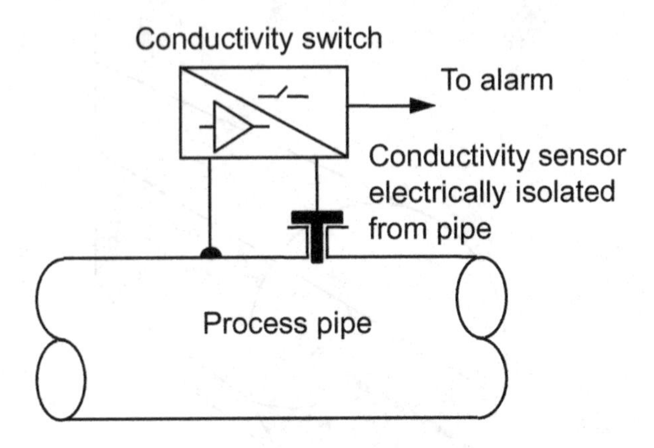

FIGURE 9.19 Conductivity probe for empty pipe detection.

the flowmeter-sensing electrodes are 'open circuit' and the amplifier output signal will be subject to random drifting. The resultant falsely generated inputs to any process controllers, recorders, etc., connected to the system will give rise to false status alarms. Here, the 'empty pipe detection' system is used to 'freeze' the signal to reference zero.

Another application for 'empty pipe detection' is to prevent damage to the field coils. Magnetic flowmeters based on a 'pulsed-DC' magnetic field generate relatively low power to the field coils – typically between 14 and 20 VA. This is usually of little concern regarding heat generation in the field coils. However, flow sensors based on an 'AC generated' magnetic fields, consume power in excess of a few hundred VA. To absorb the heat generated in the field coils, a medium is required in the pipe to keep the temperature well within the capability of the field coil insulation. An empty pipe will cause overheating and permanent damage to the field coils and, consequently, this type of flowmeter requires an 'empty pipe detection' system to shut down the power to the field coils.

9.9 ELECTRICAL DESIGN

One of the greatest challenges to the practical application of Faraday's Law in EM flowmeters is because of the relatively small flow-induced voltage at the electrodes. For example, a typical value for the induced electromotive force (emf) in an AC flowmeter having a 50 mm ID carrying 50 L/min (producing a velocity of about 4.2 m/s) is only about 2–3 mV. Consequently, the signal voltages are easily distorted by undesirable extraneous voltages – particularly the electrochemical emf.

Dependent on the choice of materials, the metallic electrodes in contact with the flowing liquid create an interfering electrochemical DC voltage. This voltage is dependent on the temperature, the flow rate, the pressure, and the chemical composition of the liquid as well as on the surface condition of the electrodes. Consequently, dependent on the materials used, the electrode potential becomes the CMV (V_{CM}) that needs rejection at the sensor output. Stainless steel electrodes have only a couple of hundred millivolts of V_{CM}, so the common mode can be more easily rejected.

The metallic electrodes in contact with the flowing liquid form a galvanic element that creates an interfering electrochemical DC voltage. This voltage is dependent on the temperature, the flow rate, the pressure, and the chemical composition of the liquid as well as on the surface condition of the electrodes. In practice, the voltage between the liquid and each electrode will be different – giving rise to an unbalanced voltage between the two electrodes. To separate the flow signal from this interfering DC voltage, an AC excitation field is used – allowing the interfering DC voltage to be easily separated from the AC signal voltage by capacitive or transformer coupling. Whilst AC EM flowmeters have been used successfully for many years, the use of an alternating field excitation makes them susceptible to both internal and external sources of errors.

9.9.1 Non-homogeneous Conductivity

Although EM flowmeters are independent of liquid conductivity over a wide range, it is assumed that the conductivity is homogeneous and is thus constant along the cross-section and along the length of the primary head. However, in many sewage and wastewater applications it is often found that, at low flow rates, layers of different densities and conductivities are formed. As a consequence, the eddy current distribution that is created by the time derivative of the induction is completely deformed and therefore interference voltages are produced, which cannot be fully suppressed in the converter.

9.9.2 Fouling of the Electrodes

Fouling of the electrodes by insulating deposits can considerably increase the internal resistance of the signal circuit – changing the capacitive coupling between the field coils and signal circuitry.

9.9.3 Direct Coupling

Because field excitation is derived directly from the mains voltage, it is impossible to separate the signal voltage from external interference voltages. Interference voltages can be transferred by either capacitive or inductive coupling from heavy current carrying cables laid in proximity to the signal cable. Although these interference voltages may be largely suppressed by multiple screening of the signal cable, they might not be completely eliminated.

9.9.4 Axial Currents

Stray currents from other systems are occasionally carried by the pipeline and/or the flowing media that generate voltages at the electrodes, which cannot be distinguished from the signal voltage.

9.9.5 Poor Grounding

Grounding of the primary head as well as the pipeline by grounding rings or properly earthed flanges ensures that the liquid is at zero potential. If the grounding is not symmetrical, earth loop currents give rise to interference voltages – producing zero-point shifts.

9.10 AC FIELD EXCITATION

Early magnetic flowmeters had their field coils connected directly to the AC line voltage operating at 50/60 Hz. Accordingly, the induced signal voltage is easily distinguished from the electrochemical DC voltage.

Unfortunately, the sinusoidal magnetic field also introduces eddy currents (and subsequent interfering voltages) in all the electrical conducting material – including the measuring electrodes. The resultant interference voltages are often superimposed on the signal voltage, making it erroneous. Consequently, AC magnetic flowmeters require periodic zero-point calibration – necessitating the flow to be shut off.

Eddy currents generated in the wall of the measuring tube also generate their own magnetic fields that oppose the signal field of the coils – thus weakening it.

AC magnetic flowmeters are relatively low cost systems having an accuracy of the order of around 2%. They are now rarely found in industry – having been largely replaced by pulsed-DC fields.

9.11 PULSED-DC FIELD

Designed to overcome the problems associated with both AC and DC interference, the pulsed-DC field is periodically switched on and off at specific intervals. The electrochemical DC interference voltage is stored when the magnetic field is switched off and then subtracted from the signal representing the sum of the signal voltage and interference voltage when the magnetic field is switched on.

Figure 9.20 illustrates the voltage at the sensors in which the measured signal voltage V_S is superimposed on the spurious unbalanced CMV V_{CM}. By taking (and storing) samples during the periods A and B, the mean value V_S may be obtained by algebraic subtraction of the two values:

$$V_S = (V_{CM} + V_S) - V_{CM} \tag{9.4}$$

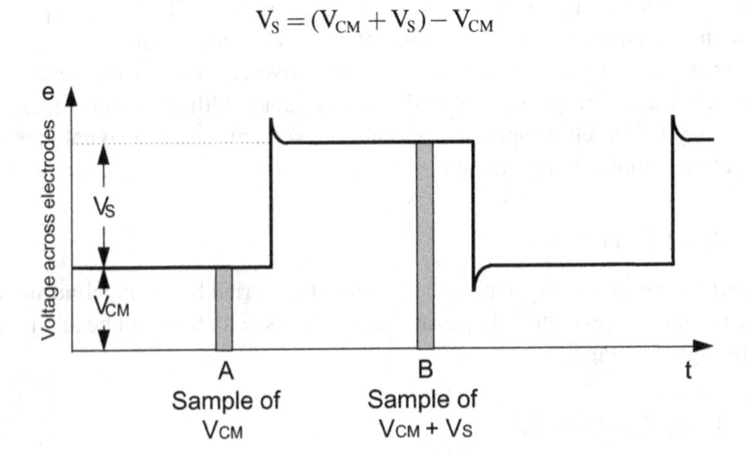

FIGURE 9.20 The voltage at the sensors in which the measured voltage V_m is superimposed on the spurious unbalanced offset voltage V_u.

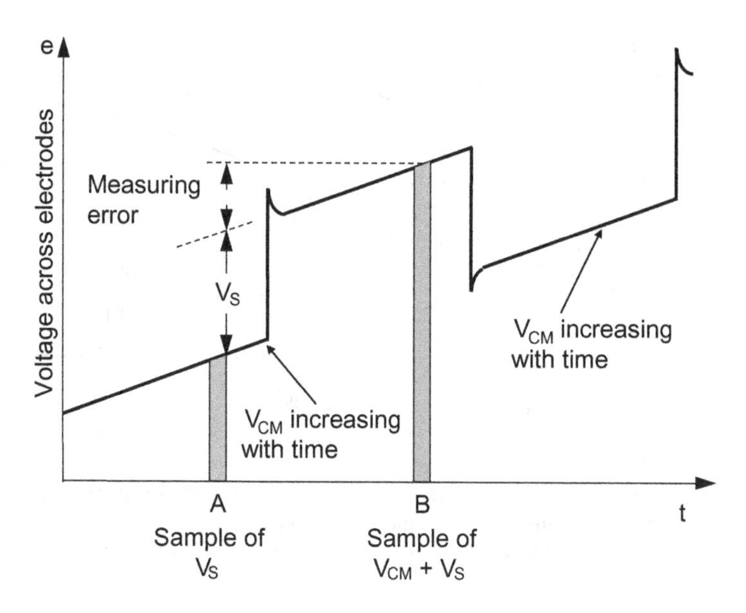

FIGURE 9.21 With the unbalanced offset voltage a steadily increasing ramp, the error is as high as the amount by which the unbalanced voltage has changed during the measuring periods A and B.

This method assumes that the value of the electrochemical interference voltage (V_{CM}) remains constant during this measuring period between samples A and B. However if the interference voltage changes during this period, serious errors are likely to occur. Figure 9.21 shows the unbalanced offset voltage as a steadily increasing ramp. Here, the error is as high as the amount by which the unbalanced voltage has changed during the measuring periods A and B and could result in an induction error of as much as 100%.

One method of overcoming this problem is by a method of linear interpolation as illustrated in Figure 9.22. Prior to the magnetic induction, the unbalanced voltage A is measured. During the magnetic induction phase, the value B (which is the sum of unbalanced voltage and flow signal) is measured and then, after magnetic induction, the changed unbalanced voltage C is measured.

The mean value (A + C)/2 of the balanced voltage prior to and after magnetic induction is electronically produced and subtracted from the sum signal measured during magnetic induction. Therefore, the exact flow signal:

$$V_m = B - \frac{(A + C)}{2} \tag{9.5}$$

is obtained which is free from the unbalanced voltage. This method corrects not only the amplitude of the common mode interference voltage, but also its change with respect to time.

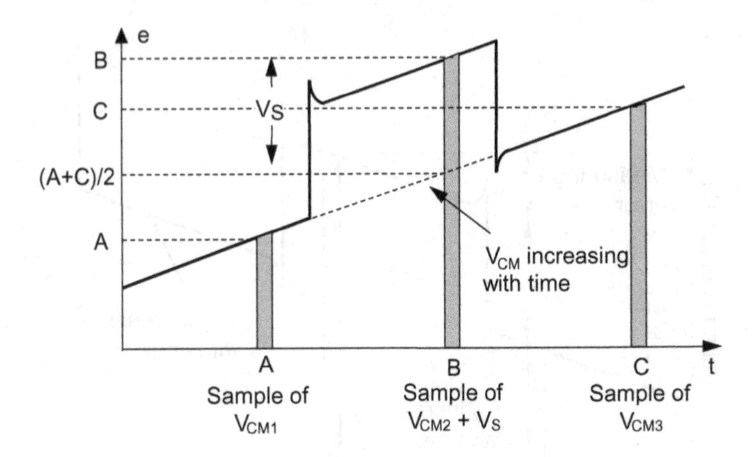

FIGURE 9.22 The unbalanced voltage A is measured prior to the magnetic induction; the value B (the sum of unbalanced voltage and flow signal) is measured during the induction phase, and the changed unbalanced voltage C is measured after magnetic induction.

9.12 BIPOLAR PULSE OPERATION

An alternative method of compensation is shown in Figure 9.23 using an alternating (or bipolar) DC pulse. Under ideal or reference conditions, the values of V_1 and V_2 would be equal and would both have the value V_m, the measured value. Thus:

$$V_1 - V_2 = (V_S) - (-V_S) = 2 \cdot V_S \tag{9.6}$$

If, now, the zero or no-flow signal is off-set by an unbalanced CMV in for example a positive direction (Figure 9.24), then:

$$V_1 = V_{CM} + V_S \tag{9.7}$$

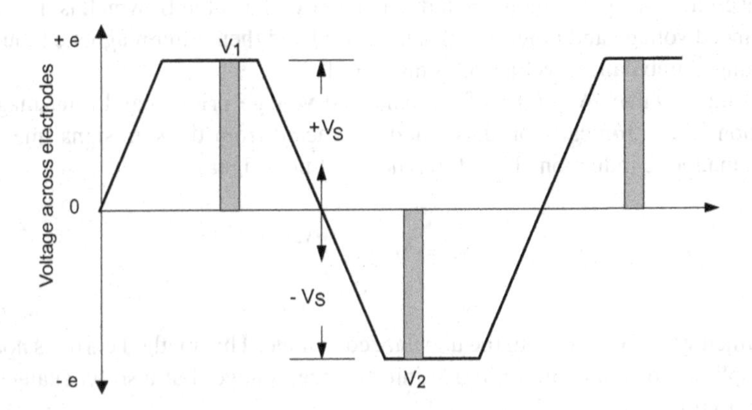

FIGURE 9.23 Bipolar pulse compensation under ideal or reference conditions.

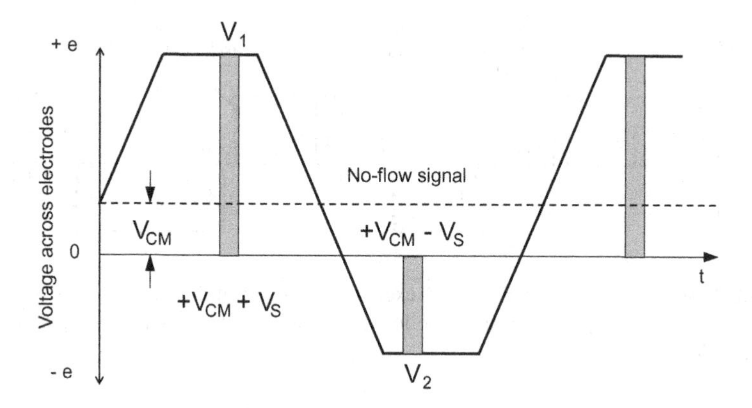

FIGURE 9.24 Bipolar pulse compensation eliminates error due to CMV.

and:

$$V_2 = (V_{CM} - V_S) \tag{9.8}$$

$$V_1 - V_2 = (V_{CM} + V_S) - (V_{CM} - V_S) = 2 \cdot V_S \tag{9.9}$$

Again, linear interpolation methods may be applied as illustrated in Figure 9.25, where five separate samples are taken during each measurement cycle. A zero potential measurement is taken at the commencement of the cycle, a second measurement at the positive peak, a third at zero potential again, a fourth at negative peak, and finally another zero measurement at the completion of the cycle. The result, in this case, will be:

$$2 \cdot V_S = \left(V_1 - \left(\frac{Z_1 - Z_2}{2}\right)\right) - \left(V_2 - \left(\frac{Z_2 + Z_3}{2}\right)\right) \tag{9.10}$$

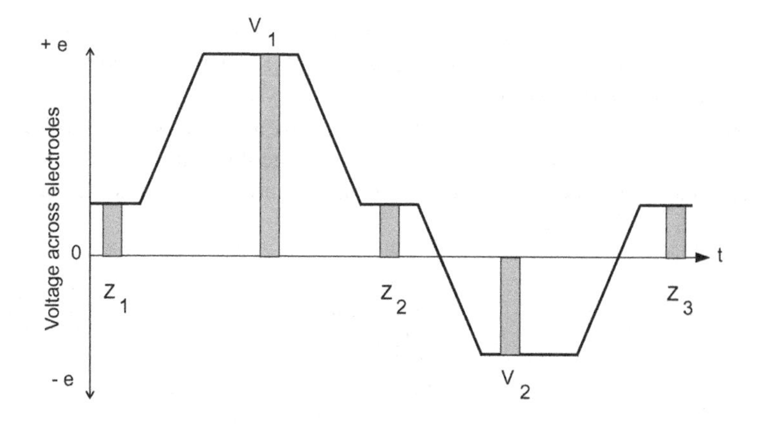

FIGURE 9.25 Bipolar pulse compensation with linear interpolation.

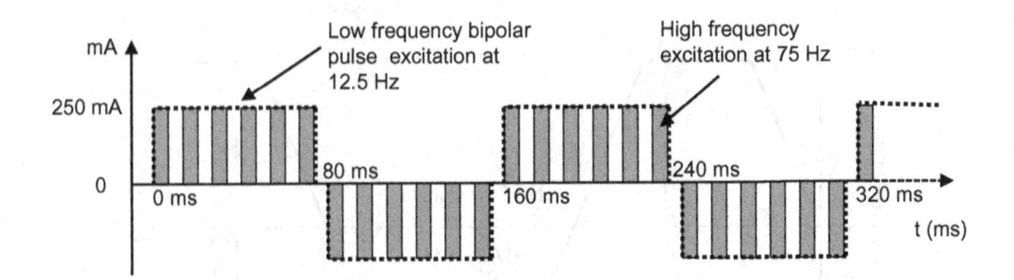

FIGURE 9.26 In this solution, a 250 mA excitation current running at 75 Hz (providing flow data every 13 ms) is superimposed on the low-frequency bipolar signal running at 12.5 Hz. (Courtesy: Yokogawa.)

9.13 DUAL FREQUENCY EXCITATION

Despite the previously mentioned problems of AC excitation in terms of accuracy, zero stability, and high-frequency power consumption, it nonetheless provided a much faster response time than pulsed-DC excitation – thus providing high noise rejection and making it suitable for slurry applications. Conversely, a limitation of low-frequency pulsed-DC design is their relatively low response speed and sensitivity to measurement noise caused by slurry is all low conductivity fluids.

One solution lies with the use of dual frequency excitation in which a high-frequency component is superimposed on the bipolar low-frequency signal (Figure 9.26). In the solution offered by Yokogawa a 250 mA excitation current running at 75 Hz (providing flow data every 13 ms) is superimposed on the low-frequency bipolar signal running at 12.5 Hz. This provides the zero stability of a low-frequency excitation combined with the good noise rejection and high speed response of high frequency excitation.

9.14 INSERTION METERS

Although an ideal candidate for the potable- and wastewater industries the use of EM flowmeters is often precluded because of their high cost in large diameter pipes.

One solution to this problem is through the use of insertion meters (Figure 9.27) in which the EM field is generated by the built-in coil. The voltage (e), proportional to the velocity of the medium flowing through the flux (B), is detected by the built-in sensors separated by distance (d).

Because this type of insertion EM flowmeter only measures the flow velocity in a specific area of the pipe, it is not representative of the flow through the entire cross-sectional area of the pipe. Consequently, accuracies are only of the order of 2–4%.

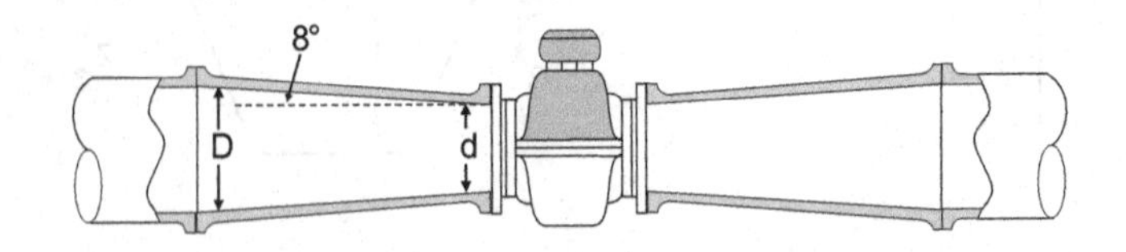

FIGURE 9.27 The voltage (e), proportional to the velocity of the medium flowing through the flux (B), is detected by the built-in sensors separated by distance (d).

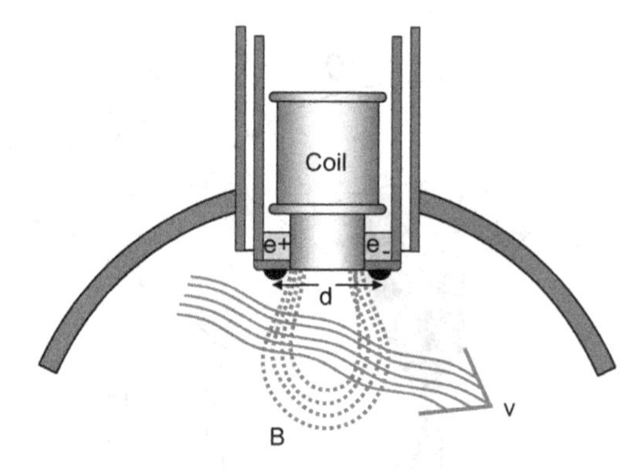

FIGURE 9.28 To avoid air bubbles or solid sedimentation, the sensor must not be installed in the upper or lower parts of the pipe.

In practice, there are quite a few more restrictions to its application. For example, in order to avoid air bubbles or solid sedimentation, very short sensors should not be installed in the upper or lower parts of the pipe (Figure 9.28).

There are also fairly severe restrictions in terms of upstream and downstream discontinuities requiring a minimum of 50D pipe diameters of straight run upstream and 5D downstream.

And finally, it should be remembered that all insertion probe devices are susceptible to vortex shedding that can produce severe probe vibration – resulting in damage and/or measurement instability. This effect is dependent on the insertion length of the probe itself but in any event, it restricts the flow velocity to a maximum of 5 m/s – reducing to as little as 1 m/s for insertion lengths of 1 m.

An improved solution makes use of a full profile insertion probe in which EM coils are installed inside the entire length of the sensor – with electrode pairs installed on the outside of the entire sensor length (Figure 9.29). Consequently, instead of just sampling the flow profile over a small portion, this solution averages the profile across the full diameter of the pipe to provide an accuracy of 0.5%.

9.15 2-WIRE OPERATION

A major feature of early AC field-excited magnetic flowmeters was their high power consumption – ranging from 200 VA to even higher for large pipe diameters. And whilst systems based on pulsed-DC fields require relatively low power to the field coils – typically between 14 and 20 VA – this was still too high to cater for loop powered devices where the requirement is for less than 0.1 VA.

Referring to the Faraday equation below:

$$e = B \cdot d \cdot v \tag{9.11}$$

where

e = induced voltage (V)
B = magnetic flux density (Wb/m^2)
d = distance between sensing electrodes (m)
v = velocity of flow medium (m/s)

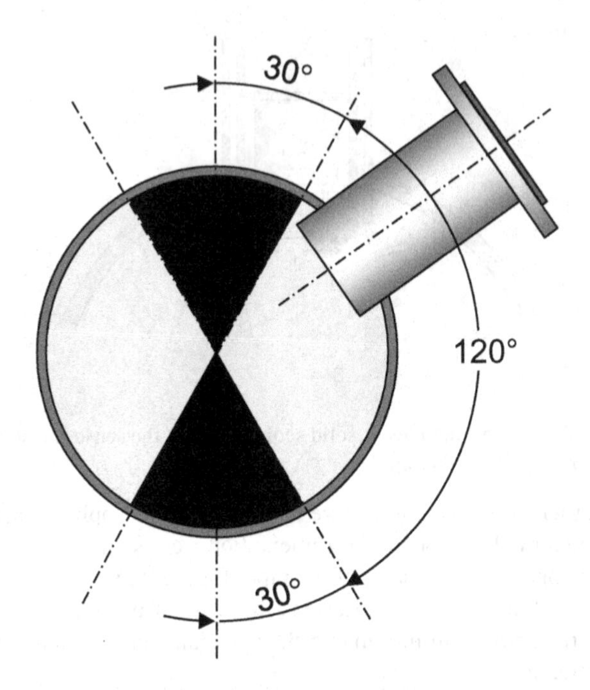

FIGURE 9.29 An improved accuracy solution makes use of a full profile insertion probe in which EM coils are installed inside the entire length of the sensor – with electrode pairs installed on the outside of the entire sensor length. (Courtesy: McCrometer.)

This shows that the induced voltage (e) can only be maximised by increasing the magnetic flux density (B) – determined by the number of windings, the coil length, and the magnetising current.

At first glance this might appear to be a relatively easy problem to solve. Nonetheless, a number of problems needed to be overcome.

Firstly, although increasing the distance between the sensing electrodes would, on the face of it, help in maximising the voltage, the problem arises that as the pipe diameter increases maximising the flux density over an increased area becomes increasingly difficult. Consequently, loop powered 2two-wire operation is generally constrained to pipe diameters of 200 mm (8 in) and less.

A further problem lies in the fact that the resultant lower field strength produces much lower signal-to-noise ratios – making 2-wire systems more susceptible to background noise produced by electrically active fluids such as slurries and paper and pickling liquors.

To conserve power between readings 2-wire systems tend to slow down and their response speed. Consequently their use in batching operations, that require faster response, produces totalisation errors.

2-wire systems also make use of power conversion circuitry to store the coil operating current. Unfortunately, this generally precludes their use in Class I Division 1 (Zones 0 and 1) applications and even in some Class I Division 2 (Zone 2) hazardous areas.

Finally, not only does the performance of 2-wire systems fall short of conventional 4-wire systems in terms of accuracy, but they also tend to be priced substantially higher (as much as 25%) over their traditional 4-wire counterparts.

9.16 METER SIZING

Generally the size of the primary head is matched to the nominal diameter of the pipeline. However, it is also necessary to ensure that the flow rate of the medium lies between the minimum and maximum full scale ranges of the specific meter. Typical values of the minimum and maximum full scale ranges are 0.3 and 12 m/s, respectively.

Experience has also shown that the optimum flow velocity of the medium through an EM flowmeter is generally 2–3 m/s – dependent on the medium. For example, for liquids having solid content, the flow velocity should be between 3 and 5 m/s to prevent deposits and to minimise abrasion.

Knowing the volumetric flowrate of the medium in for example cubic metres per hour, and knowing the pipe diameter, it is easy to calculate and thus check to see if the flow velocity falls within the recommended range. Most manufacturers supply nonograms or tables that allow users to ascertain this data at a glance.

Occasionally, in such cases where the calculated meter size needs to be smaller than that of the media pipe size, a transition using conical sections can be installed. The cone angle should be 8° or less and the pressure drop resulting from this reduction can, again, be determined from manufacturers' tables (Figure 9.30).

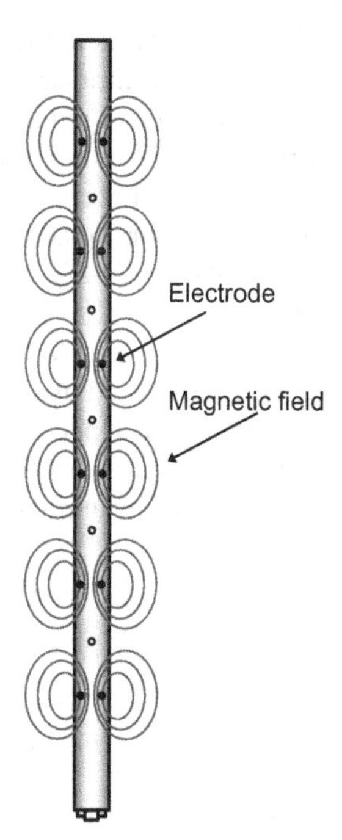

FIGURE 9.30 Conical section used to cater for reduced meter size.

9.17 CONCLUSION

The EM flowmeter is regarded by many users (outside the oil and gas industries) as the universal answer to more than 90% of all flow-metering applications. Some of the many benefits offered by the EM flowmeter include:

- No pressure drop
- Short inlet/outlet sections (5D/2D)
- Relationship is linear (not square root)
- Insensitive to flow profile changes (laminar to turbulent) including many non-Newtonian liquids
- Rangeability of 40:1 or better
- Inaccuracy of better than ±0.1% of actual flow over full range
- No recalibration requirements
- Bi-directional measurement
- No taps or cavities
- No obstruction to flow
- Not limited to clean fluids
- High-temperature capabilities
- High-pressure capabilities
- Volumetric flow
- Can be installed between flanges
- Can be made from corrosion resistance materials at low cost

But, there is one major drawback – EM flowmeters require a conductive fluid. This precludes their use to measure gases, steam, ultrapure water, and all hydrocarbons.

Nonetheless, huge strides have been made in reducing the limit for conductivity. For most modern DC field driven instruments, the minimum conductivity is about 1 μS/cm. However many instruments employing capacitively coupled sensors can be used in liquids with conductivity levels down to 0.05 μS/cm. Although this is very close to the upper hydrocarbon limit of 0.0017 μS/cm for crude oil, it falls far short of even jet fuel (150–300 pS/cm).

As discussed previously, another disappointment is in the flow measurement of pure and ultrapure water which, with a conductivity extending down to 0.1 μS/cm, would therefore appear to be covered by an extended range down to 0.05 μS/cm.

Unfortunately water, being a bipolar vibratory molecule, produces relatively large amplitudes of electrical noise that tends to swamp the amplifiers used to gain this sensitivity.

10 Ultrasonic Flowmeters

10.1 INTRODUCTION

First introduced in 1963 by Tokyo Keiki, ultrasonic flowmeters are suitable for both liquids and gases and are increasingly being used in custody transfer applications.

Unfortunately, although originally hailed as a general panacea for the flow measurement industry, lack of knowledge and poor understanding of the limitations of early instruments (especially the Doppler method) often lead to its use in unsuitable applications. Nonetheless, the ultrasonic meter is probably the only meter capable of being used on large diameter pipes (above 3 m bore) at a reasonable cost and performance.

10.2 ULTRASONIC TRANSDUCERS

The term 'ultrasonic' (or often simply US), applies to frequencies from as low as 5 kHz through to 10 MHz or even higher – but is generally applied to frequencies above the normal hearing range of human beings (20 kHz).

Ultrasonic technology is centred on transducers employing piezoelectric ceramic crystals that transmit and receive the acoustic signal. Pioneered by the brothers Pierre Curie and Jacques Curie, the phenomenon is known as the piezoelectric effect (piezo from the Greek *piezein* to 'squeeze'). When a mechanical stretching or compressing force is applied to an asymmetrical crystalline material, such as quartz (SiO_2), equal and opposite electrical charges appear across it (Figure 10.1).

Equally, when a voltage is applied across the material, the polarised molecules will align themselves with the resultant electric field. This alignment of molecules causes the material to change dimensions (Figure 10.2). This phenomenon is known as electrostriction.

In practice, a high-frequency electrical signal is applied to the electrostrictive material – with the small mechanical movement converted, through a membrane, into an acoustic signal. Conversely, when a received acoustic signal is transmitted through a membrane, it converts this small mechanical movement into an electrical signal.

Earlier piezoelectric transducers were fabricated from ceramic materials based on barium titanate ($BaTiO_3$). However, these devices had a limited temperature range due to their low Curie temperature (around 120°C) – the temperature at which the material loses its piezoelectric characteristics. Consequently, most modern transducers now make use of lead zirconate titanate ($PbZrTiO_3$) (PZT) due to its higher Curie temperature of up to 250°C.

Higher temperature applications up to 300°C or 400°C generally make use of some form of an isolating mechanism that couples the acoustic signal through for example a coupling rod (Figure 10.3). Even higher temperature applications are possible to 600°C making use of an acoustic waveguide.

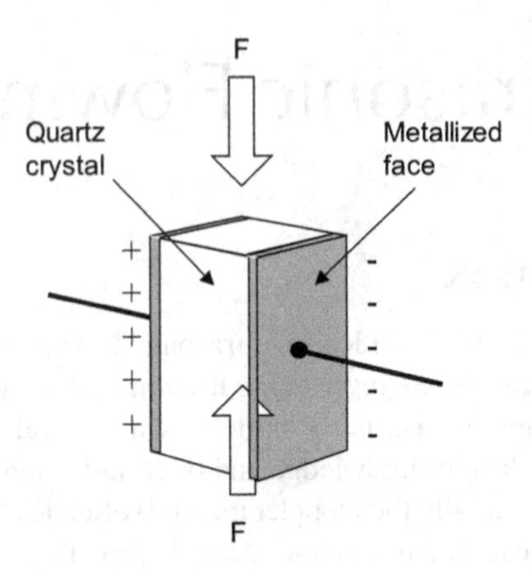

FIGURE 10.1 When a mechanical stretching or compressing force is applied to an asymmetrical crystalline material, equal and opposite electrical charges appear across it.

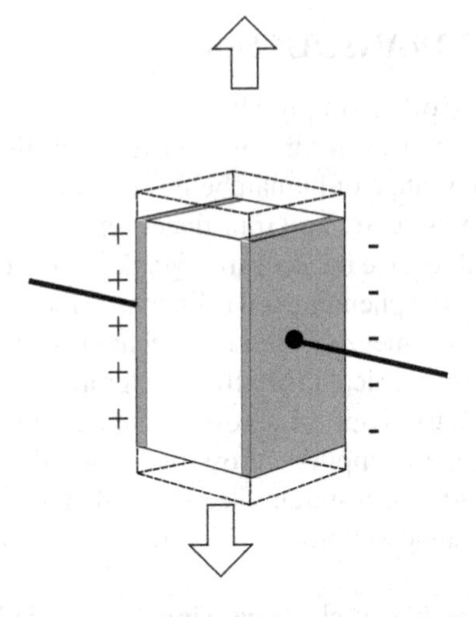

FIGURE 10.2 When a voltage is applied across an electrostrictive material, the polarised molecules align themselves with the resultant electric field to produce a change in dimensions.

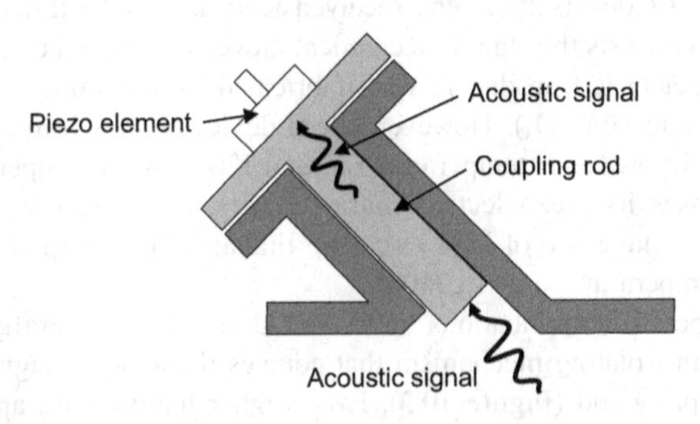

FIGURE 10.3 Higher temperature applications make use of an isolating mechanism such as a coupling rod.

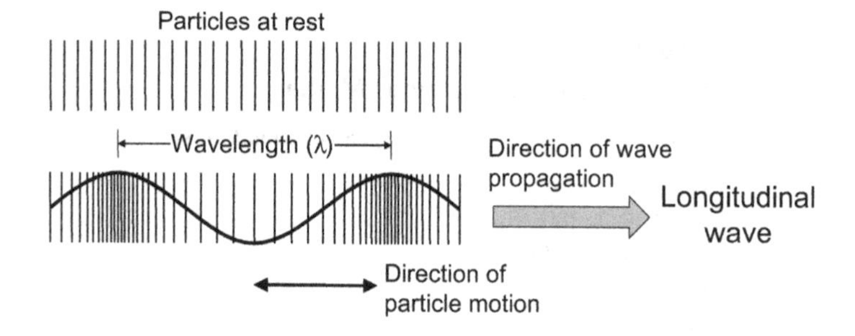

FIGURE 10.4 In air, sound travels by the compression and rarefaction of air molecules in the longitudinal direction – the direction of travel.

10.3 ACOUSTIC PROPAGATION

In air, sound travels by the compression and rarefaction of air molecules in the longitudinal direction – the direction of travel (Figure 10.4). These are called longitudinal waves.

However, in solids, molecules can support vibrations in other directions and thus a number of different types of acoustic waves are possible. One such mode of propagation is called a transverse or shear wave (Figure 10.5).

The speed of sound varies considerably according to the medium (Table 10.1).

In gases for example the speed of sound is largely determined by:

$$c = \sqrt{\frac{\gamma \cdot R \cdot T}{M}} \tag{10.1}$$

where

c = velocity sound (m/s)
γ = adiabatic index (1.4)*

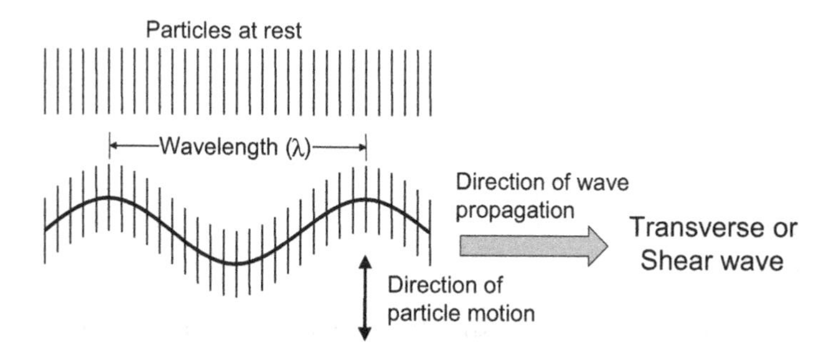

FIGURE 10.5 In solids, molecules can support vibrations in other directions and one such mode of propagation is called a transverse or shear wave.

* This is a typical value but can be as high as 1.6667 for monatomic noble gases and 1.3 for triatomic molecule gases.

TABLE 10.1
Speed of Sound in Various Media

Medium	Speed of Sound (m/s)
Gases (20°C)	
Air (21% O_2, 78% N_2)	343
Carbon dioxide (CO_2)	268
Ethylene (C_2H_4)	336
Methane (CH_4)	445
Nitrogen (N_2)	349
Liquids (25°C)	
Glycerol	1,904
Kerosene	1,324
Methyl alcohol	1,103
Water	1,493
Solids	
Aluminium	5,100
Iron	5,960
Stainless steel	5,800

R = universal molar gas constant (8,314.5 J/kmol)
T = temperature (K)
M = molecular weight (kg/kmol)

Thus, in dry air at 20°C (68°F), the speed of sound is 343 m/s (1,125 ft/s) – equivalent to 1,236 km/h (768 mph) or about 1 km in 3 s. This figure is heavily dependent on air temperature, but is nearly independent of the pressure or density of air.

Since fluids do not sustain shear forces, the speed of sound in a fluid is given by:

$$c_{fluid} = \sqrt{\frac{K}{\rho}} \tag{10.2}$$

where
K = bulk modulus
ρ = density

In solids, it is possible to generate sound waves with different velocities depending on the deformation mode.

In essence, there are three basic principles used in ultrasonic metering: the Doppler method, the time-of-flight method, and the frequency difference method. However all are centred on the use of ultrasonic transducers based on the piezoelectric effect.

10.4 DOPPLER-BASED METERS

Doppler flowmeters are based on the Doppler effect – the change in frequency that occurs when a sound source and receiver move either towards or away from each other. The classic example is that of an express train passing through a station. To an observer, standing on the platform, the sound of the train appears to be higher as the train approaches and then falls as the train passes through the station and moves away. This change in frequency is called the Doppler shift.

In the Doppler-ultrasonic flowmeter, an ultrasonic beam (usually of the order of 1–5 MHz) is transmitted, at an angle, into the liquid (Figure 10.6). Assuming the presence of reflective particles (dirt, gas bubbles, or even strong eddies) in the flowstream, some of the transmitted energy will be reflected back to the receiver. Because the reflective particles are moving towards the sensor, the frequency of the received energy will differ from that of the transmitted frequency (the Doppler effect).

This frequency difference, the Doppler shift, is directly proportional to the velocity of the particles.

► Assuming that the media velocity (v) is considerably less than the velocity of sound in the media (c), the Doppler frequency shift (Δf) is given by:

$$\Delta f = \frac{2 \cdot f_t \cdot v \cdot \cos\theta}{c} \tag{10.3}$$

where f_t is the transmitted frequency.

From this it can be seen that the Doppler frequency, Δf, is directly proportional to the flow rate.

The velocity of sound in water is about 1,500 m/s. If the transmitted frequency is 1 MHz, with transducers at 60°, then for a media velocity of 1 m/s, the Doppler shift is around 670 Hz. ◄

Since this technique requires the presence of reflecting particles in the medium, its use in ultra-clean applications or, indeed with any uncontaminated media, is generally, precluded. Although some manufacturers claim to be able to measure 'non-aerated' liquids, in reality such meters rely on the presence of bubbles due to micro-cavitation originating at valves, elbows, or other discontinuities.

In order for a particle to be 'seen', it needs to be approximately 1/10 larger than the wavelength of the acoustic frequency in the liquid. Again, using water as an example,

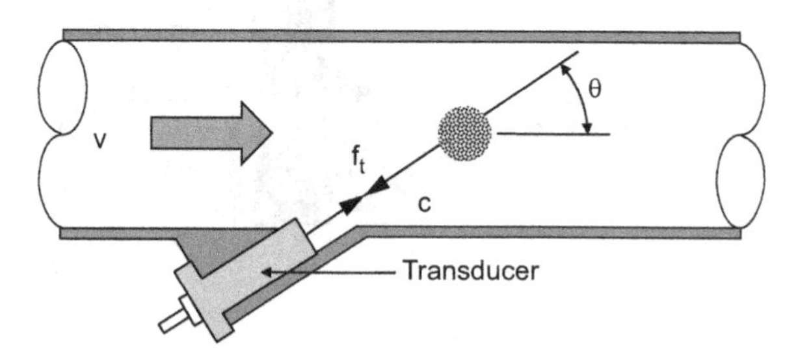

FIGURE 10.6 In the Doppler-ultrasonic flowmeter, an ultrasonic beam is transmitted, at an angle, into the liquid.

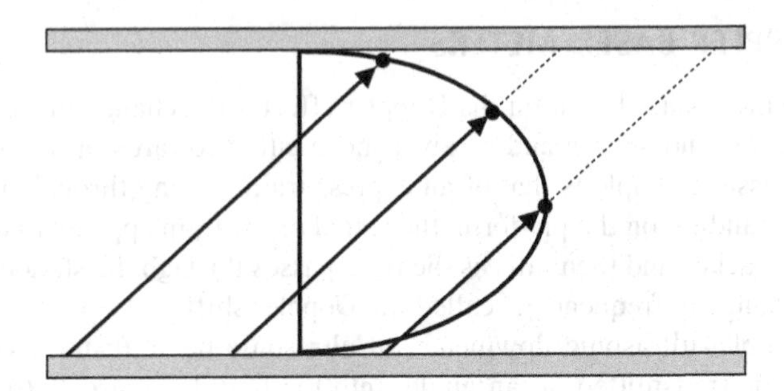

FIGURE 10.7 Because there can be no guarantee as to the particles' position (and therefore velocity) there may be several different frequency shifts – each originating at different positions in the pipe.

a 1 MHz ultrasonic beam would have a wavelength of about 1.5 mm and so particles would need to be larger than 150 μm in order to reflect adequately.

Whilst air, oil particle, and sand are excellent sonic reflectors, the presence of too many particles can attenuate the signal so that very little of the signal reaches the receiving transducer.

Probably the single biggest drawback of this technology is that in multiphase flows, the particle velocity may bear little relationship to the media velocity. Even in single-phase flows, because the velocity of the particles is determined by their location within the pipe, and there can be no guarantee as to the particles' position (and therefore velocity), there may be several different frequency shifts – each originating at different positions in the pipe (Figure 10.7).

As a result, the Doppler method often involves a measurement error of 10% or even more.

In the insertion type probe shown in Figure 10.8, the reflective area is, to a large extent, localised and the potential source of errors is thereby reduced.

Generally, Doppler meters should not be considered as high-performance devices and are cost effective when used as a flow monitor. They work well on dirty fluids and typical applications include sewage, dirty water, and sludge. Doppler meters are sensitive to velocity profile effects and are temperature sensitive.

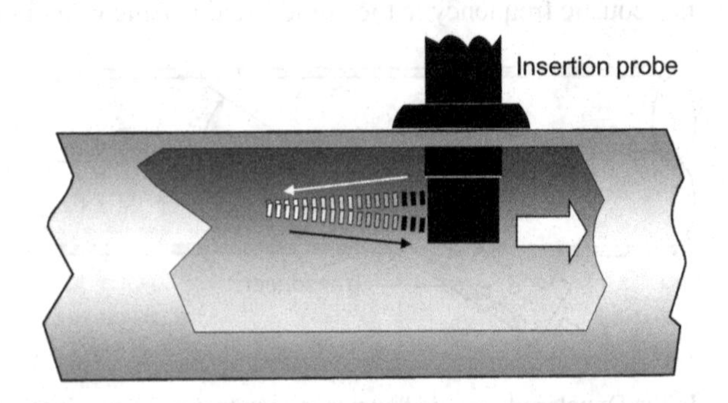

FIGURE 10.8 Insertion type probe Doppler probe. (Courtesy: Dynasonics.)

10.5 TRANSIT TIME METERS

The ultrasonic transit time measuring method is based on the fact that, relative to the pipe and the transducers, the propagation speed of an ultrasonic pulse travelling against the media flow will be reduced by a component of the flow velocity. Similarly, the speed of propagation of the pulse travelling downstream is increased by the fluid velocity. The difference between these two transit times can be directly related to the flow velocity.

In practice, the meter comprises two transducers (A and B) mounted at an angle to the flow and having a path length L (Figure 10.9) – with each acting alternately as the receiver and transmitter. The transit time of an ultrasonic pulse, from the upstream to the downstream transducer, is first measured and then compared with the transit time in the reverse direction.

▶ Mathematically:

$$T_{AB} = \frac{L}{c - v \cdot \cos\theta} \tag{10.4}$$

and:

$$T_{BA} = \frac{L}{c + v \cdot \cos\theta} \tag{10.5}$$

where

T_{AB} = upstream travel time
T_{BA} = downstream travel time
L = path length through the fluid
c = velocity of sound in medium
v = velocity of medium

The difference in transit time ΔT is:

$$\Delta T = T_{AB} - T_{BA} \tag{10.6}$$

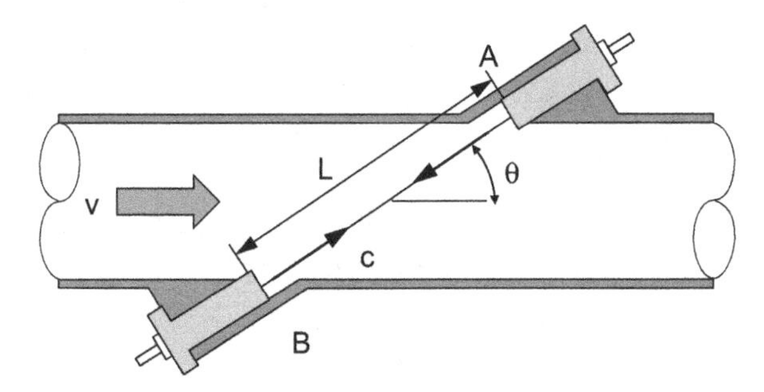

FIGURE 10.9 In the transit time meter, two transducers (A and B) each act alternately as the receiver and transmitter.

$$\Delta T = \frac{L}{c - v \cdot \cos\theta} - \frac{L}{c + v \cdot \cos\theta} \quad (10.7)$$

$$\Delta T = \frac{2 \cdot L \cdot v \cdot \cos\theta}{c^2 - v^2 \cdot \cos^2\theta} \quad (10.8)$$

Since the velocity of the medium is likely to be much less than the velocity of sound in the medium itself (e.g. 15 m/s compared with 1,500 m/s), the term $v^2 \cos^2 \theta$ will be very small compared with c^2 and may thus be ignored for all practical flow velocities.

Thus:

$$\Delta T = \frac{2 \cdot L \cdot v \cdot \cos\theta}{c^2} \quad (10.9)$$

$$v = \frac{\Delta T \cdot c^2}{2 \cdot L \cdot \cos\theta} \quad (10.10)$$

This shows that the flow velocity v is directly proportional to the transit time difference ΔT. This also illustrates that v is directly proportional to c^2 (the square of the speed of sound) which will vary with temperature, viscosity, and material composition.

Fortunately, it is possible to eliminate the variable c^2 from the equation:

$$c = \frac{L}{T_M} \quad (10.11)$$

where T_M is the mean transit time given by:

$$T_M = \frac{(T_{AB} + T_{BA})}{2} \quad (10.12)$$

therefore:

$$c = \frac{2L}{(T_{AB} + T_{BA})} \quad (10.13)$$

and:

$$c^2 = \frac{4L^2}{(T_{AB} + T_{BA})^2} \quad (10.14)$$

now:

$$v = \frac{\Delta T \cdot 4 \cdot L^2}{2 \cdot L \cdot \cos\theta (T_{AB} + T_{BA})^2} \quad (10.15)$$

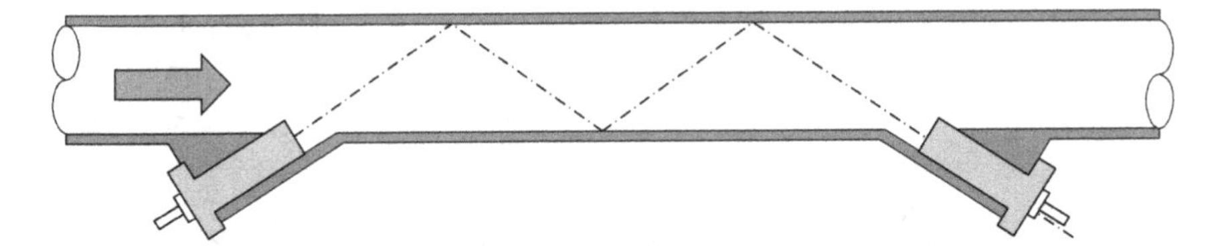

FIGURE 10.10 Increasing the path length using a double traverse, single 'V' path on the centre line.

or:

$$v \propto \frac{k \cdot \Delta T}{(T_{AB} + T_{BA})^2} \tag{10.16}$$

◀

Since both length L and angle θ are likely to remain constant, it is only necessary to calculate the sum and difference of the transit times in order to derive the flow rate independent of the velocity of sound in the medium.

As distinct from Doppler meters, transit time meters work better on clean fluids and typical applications include water, clean process liquids, liquefied gases, and natural gas pipes.

The accuracy of measurement is determined by the ability of the instrument to measure accurately the transit time. In a 300 mm diameter pipe for example with the transducers set at 45° and the media flowing at 1 m/s, the transit time is about 284 μs and the time difference ΔT is less than 200 ns. This means that in order to measure the velocity with a full scale accuracy of 1%, the time resolution must be, at the very least, down to 2 ns. With smaller diameter pipes, the measurement accuracy would need thus to be in the picosecond range.

Obviously, with longer path lengths, this stringent time measurement requirement becomes easier to meet. The performance thus tends to be better with large bore pipes. Alternatively, multiple traverses, as illustrated in Figure 10.10, can increase the path length.

These arrangements are frequently used for gas measurement in lines and gas flow measurement. The double traverse, single path flow meter is frequently used for low-cost liquid measurement and accurate real-time measurement of hazardous and non-hazardous gas flows in lines from 100 to 900 mm DN.

The 'U'-form meter as shown in Figure 10.11 can be used for very low flows.

10.6 FLOW PROFILE

▶ The average velocity along an ultrasonic path (Figure 10.12) is given by:

$$V_{average} = \int_0^D V \cdot dx \tag{10.17}$$

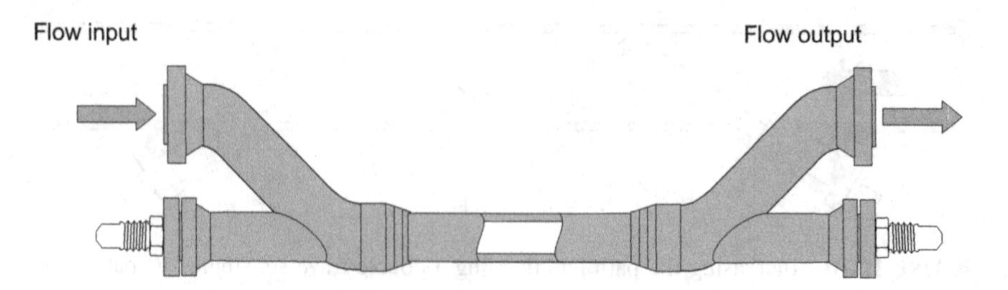

FIGURE 10.11 The 'U'-form meter can be used for very low flows.

where

D = pipe internal diameter

x = distance across the pipe

Thus, with a single path across the flow, the average flow is made up of the sum of the instantaneous velocities at each point across the diameter of the pipe. ◀

The transit time meter thus provides a picture of the total flow profile along the path of the beam. However, the validity of the measurement can only be relied on if the flow profile is not subjected to an asymmetric velocity profile or symmetric swirl. In addition it is important to know the flow profile. If for example the flow profile is not fully developed, then, as shown in Figure 10.13, the laminar-to-turbulent error can be up to 33%.

Using a dual path as shown in Figure 10.14, the laminar-to-turbulent error can be reduced to 0.5%.

10.6.1 Problems with Swirl

One of the effects of swirl is illustrated in Figure 10.15. In this example the clockwise swirl aids the instantaneous acoustic signal transmission from transducer A to transducer B and thus makes it go too fast. Conversely, the same swirl opposes the instantaneous signal transmission from transducer C to D and causes it to go too slow. As a result the transit time difference becomes too large and the meter over reads.

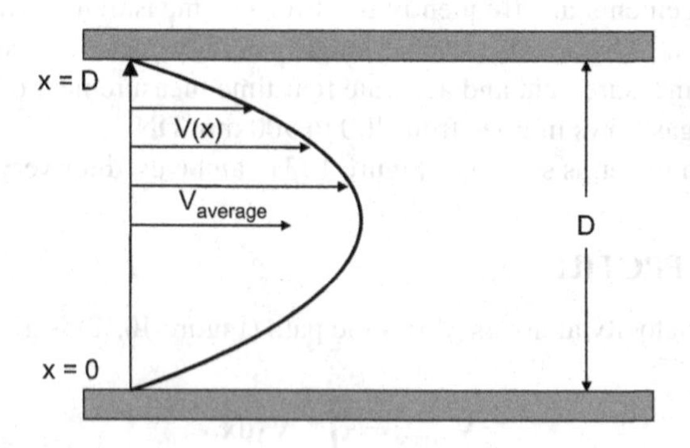

FIGURE 10.12 Average velocity along an ultrasonic path.

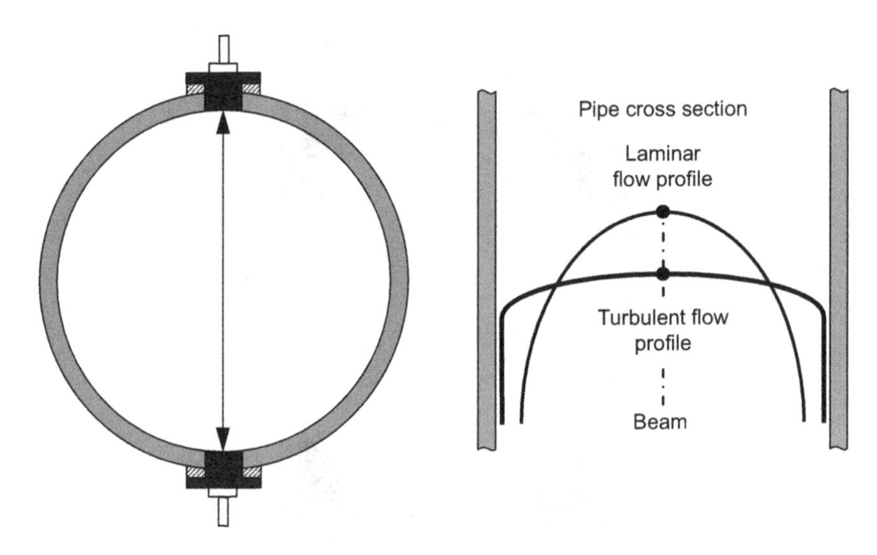

FIGURE 10.13 A single path produces a laminar-to-turbulent error up to 33%. (Courtesy: Krohne.)

10.6.2 5-Path Solution

In the Krohne multi-channel custody transfer ultrasonic flowmeter, 10 sensors form five measurement paths located in the cross-section of the flow tube (Figure 10.16).

This approach provides a wealth of information on the flow profile under laminar and turbulent flow conditions and provides highly accurate flow even in the presence of non-symmetric flow profiles and swirl – thus providing a measurement that is

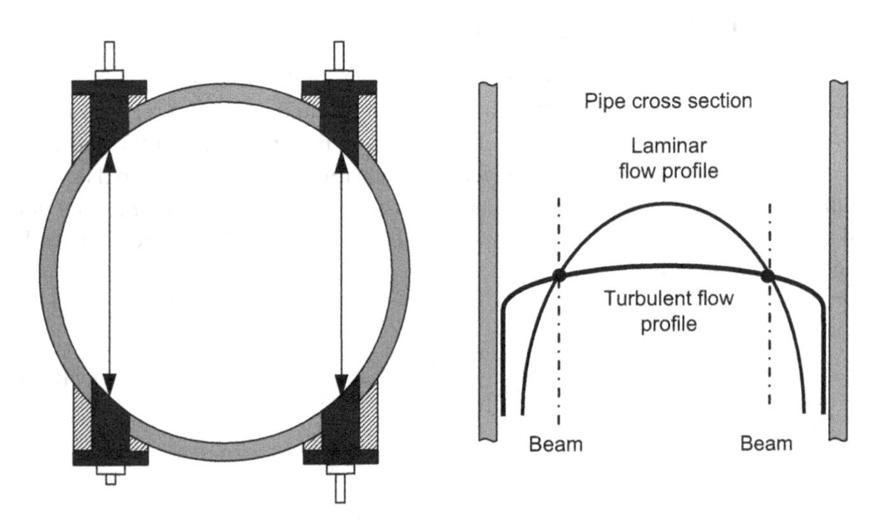

FIGURE 10.14 A dual path reduces the laminar-to-turbulent error to 0.5%. (Courtesy: Krohne.)

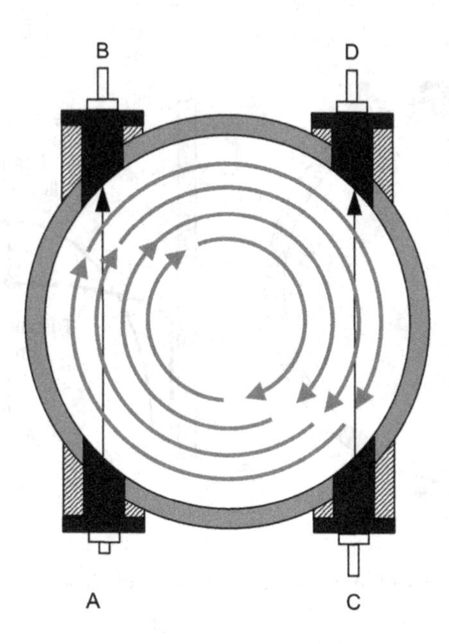

FIGURE 10.15 In this example, the clockwise swirl aids the instantaneous acoustic signal transmission from transducer A to transducer B and thus makes it go too fast. Conversely, the same swirl opposes the instantaneous signal transmission from transducer C to D and causes it to go too slow.

essentially independent of the flow profile – with accuracies to 0.15% and repeatability down to 0.02%.

10.7 FREQUENCY DIFFERENCE

The frequency difference or 'sing-around' flowmeter makes use of two independent measuring paths – with each having a transmitter (A_1 and A_2) and a receiver (B_1 or B_2) (Figure 10.17). Each measuring path operates on the principle that the arrival of a transmitted pulse at a receiver triggers the transmission of a further pulse. As a result, a pair of transmission frequencies is set up – one for the upstream direction and another for the downstream direction. The frequency difference is directly proportional to the flow velocity.

▶ Thus:

$$F_1 = \frac{(c - v \cdot \cos\theta)}{L} \tag{10.18}$$

and:

$$F_2 = \frac{(c + v \cdot \cos\theta)}{L} \tag{10.19}$$

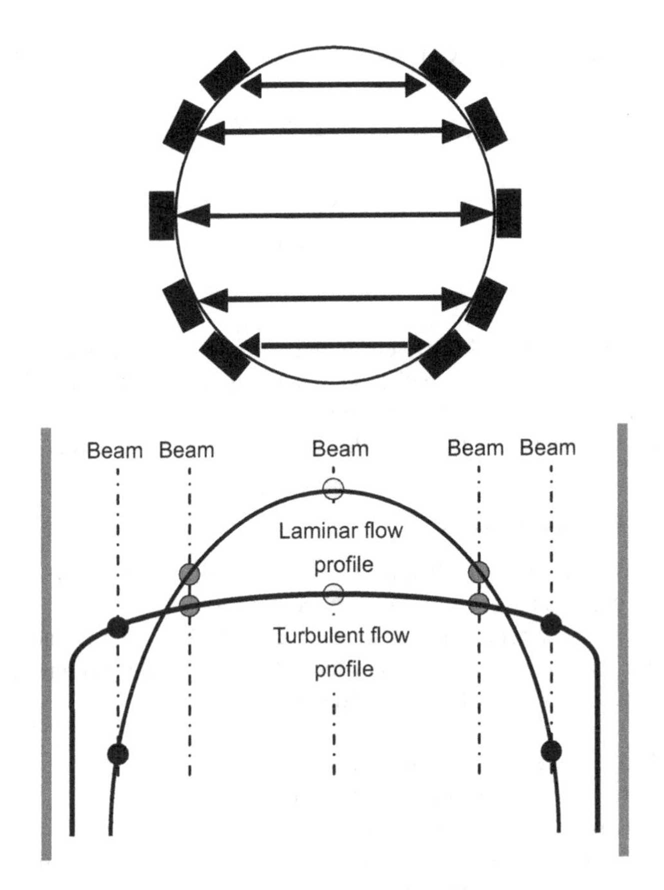

FIGURE 10.16 Five measurement paths provide a measurement that is essentially independent of the flow profile. (Courtesy: Krohne.)

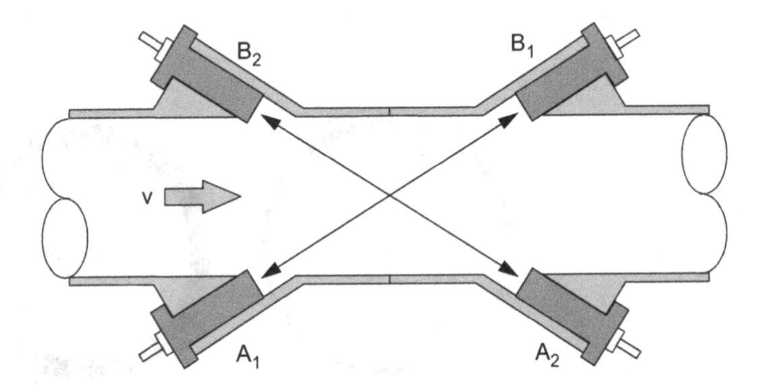

FIGURE 10.17 'Sing-around' flowmeter makes use of two independent measuring paths each having a transmitter (A_1 and A_2) and a receiver (B_1 or B_2).

The frequency difference ΔF is given by:

$$\Delta F = F_1 - F_2 = \frac{2 \cdot v \cdot \cos\theta}{L} \quad (10.20)$$

$$v = \frac{\Delta F \cdot L}{2 \cdot \cos\theta} \quad (10.21)$$

◀

The main advantage of this system is that because the frequency difference is directly proportional to flow, no maths function is required. Further, the measurement is independent of the velocity of sound in the medium.

Despite these advantages, however, this technology is rarely used.

10.8 GAS APPLICATIONS

The technology used for gas applications is essentially the same as that used for liquid except for the use of a lower frequency – typically 100 kHz.

However, gas measurement poses some unique problems that are not found with liquids. One of the major concerns associated with gas of flow measurement is fouling and condensate that may be classified into several categories:

- Evenly distributed coating on the inside of the pipe
- Condensate flow in the bottom of the pipe
- Intermittent sticking to the pipe wall
- Dirt build-up on the transducers (especially on those facing upstream)
- Liquid build-up in the transducer pockets

10.8.1 EVENLY DISTRIBUTED COATING

One of the immediate effects of fouling is that it reduces the internal diameter of the meter (Figure 10.18) and, therefore, the calculated volumetric flow rate is no longer valid.

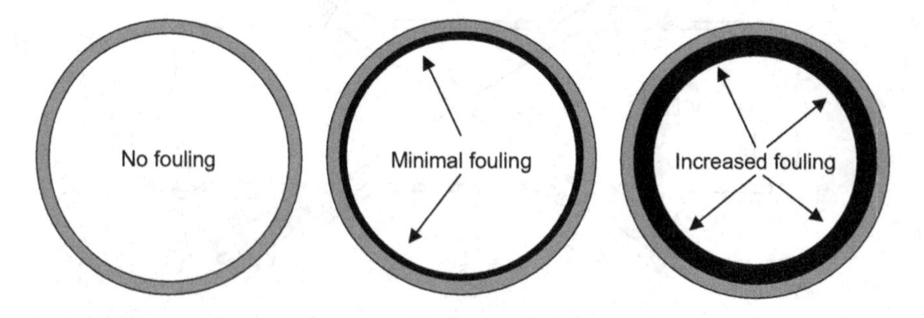

FIGURE 10.18 Fouling reduces the internal diameter of the meter and therefore the calculated volumetric flow rate is no longer valid.

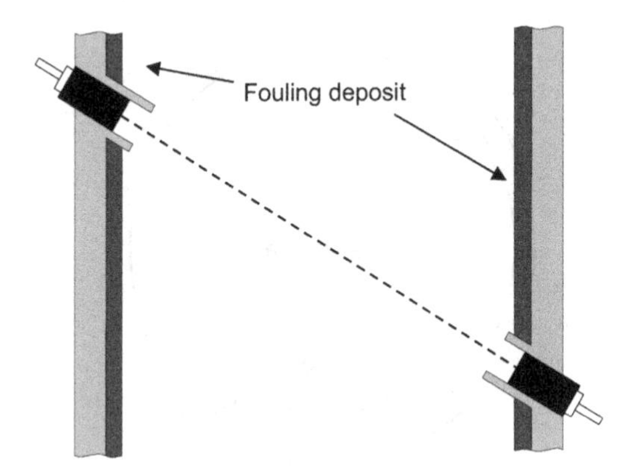

FIGURE 10.19 When fouling is present, the use of direct-path metering can lead to serious measurement errors since the cross-sectional area is reduced.

When fouling is present (Figure 10.19) the use of direct-path metering can lead to serious measurement errors since, although the measured distance remains correct due to the self-cleaning action of the ultrasonic probes, the cross-sectional area is reduced.

This may be overcome by using a single-traverse reflective system (Figure 10.20) in which any reduction in the cross-sectional area is immediately detected and compensated for.

10.8.2 Intermittent Fouling

Intermittent fouling increases the wall roughness and provides intermittent attenuation and absorbance of the acoustic signal.

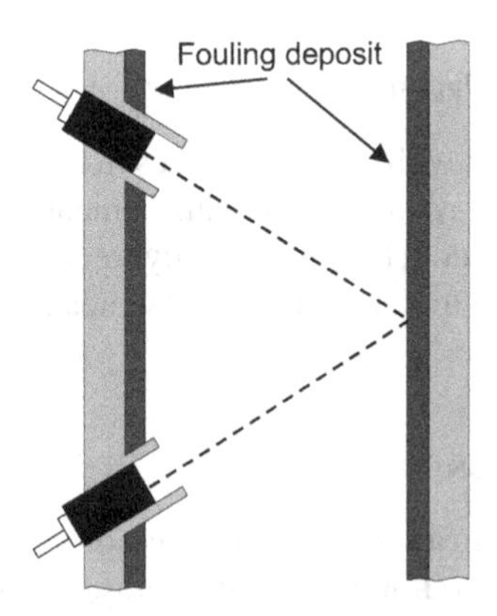

FIGURE 10.20 By using a single-traverse reflective system, any reduction in the cross-sectional area is immediately detected and compensated for.

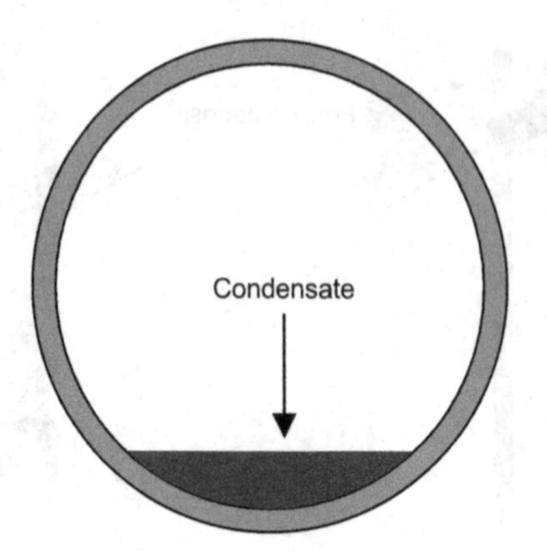

FIGURE 10.21 Condensates forming on the bottom of the pipe reduces the cross-sectional area.

10.8.3 Condensate Flow

In a similar manner, condensates forming on the bottom of the pipe will also reduce the cross-sectional area (Figure 10.21).

10.8.4 Dirt Build-up on Transducers

Dirt build-up on the transducers may produce attenuation of the acoustic signal due to absorbance. However, as noted previously, this problem is largely offset by the self-cleaning action of the ultrasonic probes.

10.8.5 Liquid in Transducer Pockets

Where possible this should be avoided owing to the possibility of increased cross-talk.

10.8.6 Non-Uniform Profiles

Another problem associated with a gas flow measurement lies with the fact that gas velocity profiles are not always uniform – with elements of swirl and asymmetrical flow profile within the meter. This, accordingly, requires the use of an increased number of chords in order to compute the bulk mean velocity – the average gas velocity across the meter area.

10.8.7 Gas Composition

Since sound cannot travel in a vacuum, transmission between the sensors requires that the gas has a defined minimum density. Subsequently, details including gas type and the minimum/maximum pressure and temperature must be clarified in advance.

10.8.8 Vortex Formation

The edges of any components built into the pipeline, such as a valves or dampers, can generate vortices – producing frequencies far above the 100 kHz frequency range generally required for gas measurement. Consequently, straight unimpeded inlet piping is essential – with no internals likely to cause gas flow separation.

10.8.9 Speed of Sound

Earlier, we had seen that the speed of sound (c) may be determined by computing the reciprocal of the mean transit time and multiplying it by the path length (Equation 10.13). Obviously, if the transit time measurement is incorrect so too will be the speed of sound calculation.

It was also seen that there was a necessity to determine the gas composition – typically carried out using an in-line gas chromatograph (GC). A GC will also be capable of providing the compositional data necessary to additionally calculate the speed of sound.

It is thus possible to compare the computed values of the speed of sound from the meter and from the independently computed value from the GC. This provides an important diagnostic feature.

10.9 MULTIPATH REFLECTIVE SYSTEMS

To overcome the problems of fouling, condensate, and non-uniform profiles, several companies, including Cameron, Emerson Daniel, Honeywell-Ester, and Krohne have introduced multipath reflective systems having three or more paths.

The technology is typified by Krohne's Altosonic V12 meter in which five transducer pairs provide 10 acoustic paths in horizontal planes perpendicular to the pipe axis (Figure 10.22). The data from these five horizontal planes is used to calculate the bulk gas velocity.

A further two chords, located vertically, are used to detect the presence, and thickness, of any condensate layer in the bottom of the pipe.

10.10 CLAMP-ON SYSTEMS

Clamp-on ultrasonic flowmeters employ external transducers that are attached to the walls of the pipe to provide portable non-intrusive flow measurement systems that can be installed within a few minutes to virtually any pipe (Figure 10.23). Pipe materials include metal, plastic, ceramic, asbestos cement, and internally and externally coated pipes.

Clamp-on transducers are also often used in permanent installations that cannot justify a permanent in-line meter but nonetheless require periodic metering.

A problem that may be encountered in small diameter pipes is that the path length may be insufficient to provide good resolution. This problem may be overcome by using a single traverse (Figure 10.24) to increase the path length and thus increase the resolution.

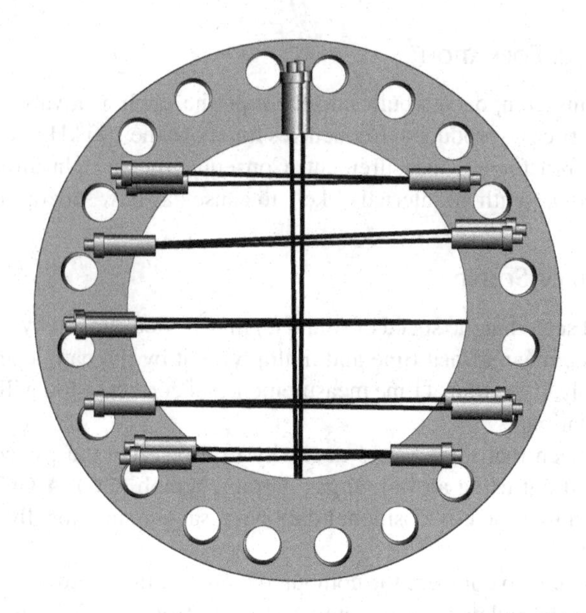

FIGURE 10.22 Diagrammatic representation of the arrangement of acoustic transducers and chords. (Courtesy: Krohne.)

One of the major problems associated with clamp-on transducers is refraction. Refraction occurs at virtually all the interfaces such as transducer-to-pipework, pipework-to-fluid, fluid-to-pipework, and pipework-to-transducer. In addition, there may be further refraction due to external coating (e.g. paint) and internal deposits on the inside pipe surface. Coatings may also affect the transmitted signal strength.

Consequently, because the ultrasonic signal is 'bent', the precise exit point at which the downstream sensor needs to be located is often quite difficult to determine. Furthermore, any change in the characteristics of the liquid that affects the speed of

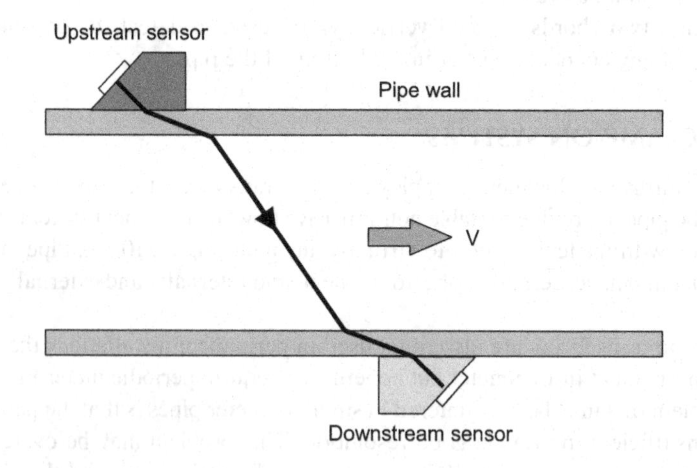

FIGURE 10.23 Clamp-on transducers must take into account the thickness and material of construction of the pipe wall.

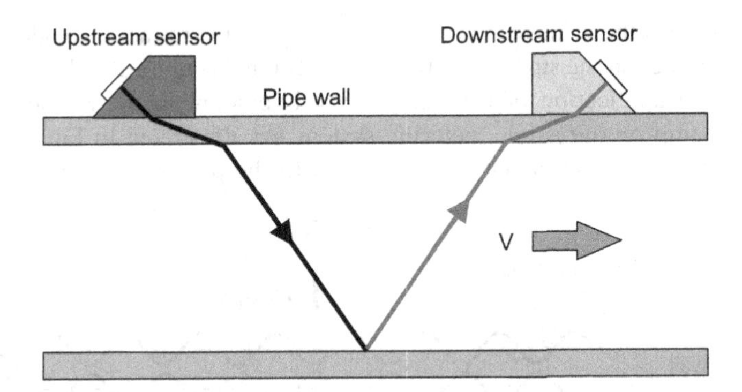

FIGURE 10.24 A single traverse may be used to increase the path length and thus increase the resolution.

sound will have a direct effect on the refraction angle. With sufficient change in the refraction angle, the signal from the upstream transducer will not be received by the downstream sensor.

Despite these obstacles, many modern clamp-on ultrasonic meters, incorporating microprocessor technology that allows the transducer mounting positions and calibration factors to be calculated for each application, provide measuring accuracies of 1–3% – depending on the application.

In these single-path conventional designs use is made of shear wave injection to produce a single-ultrasonic beam. And, as we've seen, changes in the refraction angle may prevent the signal from the transmitting transducer reaching the downstream sensor.

An alternative scheme lies with the use of axial injection using Lamb waves. Often referred to as 'plate waves', Lamb waves can be generated in a plate – producing both symmetric and asymmetric displacements within the layer. The symmetric mode (Figure 10.25) is also called longitudinal because the wave is 'stretching and compressing' the plate in the wave motion direction.

In the anti-symmetric mode (often called the 'flexural mode') a large portion of the motion moves in a normal direction to the plate – with little motion occurring in the direction parallel to the plate but with the plate body bending as the two surfaces move in the same direction (Figure 10.26). This is the strongest and most easily excited mode.

Using the flexural mode the pipe wall may be incorporated into the signal transmission system in which the pipe itself becomes the launching point of the acoustic signal and allows a much wider signal beam to be transmitted from one

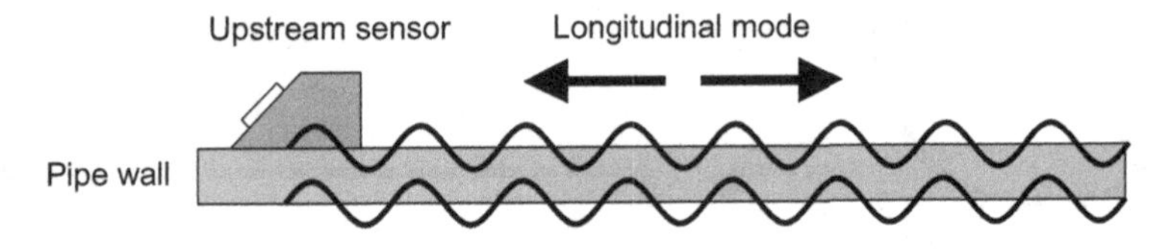

FIGURE 10.25 In the symmetric mode, also called longitudinal, the wave is 'stretching and compressing' the plate in the wave motion direction.

transducer to the other. The result is that any change in the refraction angle will have a negligible effect on the strength of the received signal (Figure 10.27).

A practical application of this technology is evident in the Siemens Sitrans FUH1010 clamp-on ultrasonic metering system. As illustrated in Figure 10.28, a dual-path single-traverse layout is employed with the path separated by 90°. This

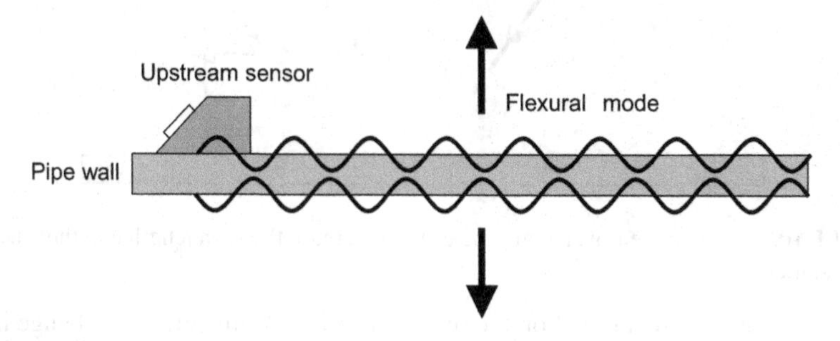

FIGURE 10.26 In the flexural mode the plate body bends as the two surfaces move in the same direction.

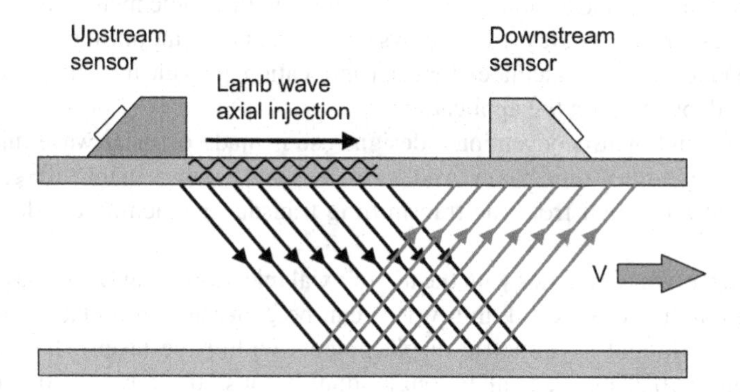

FIGURE 10.27 Using the flexural mode the pipe becomes the launching point of the acoustic signal and allows a much wider signal beam to be transmitted. Consequently any change in the refraction angle will have a negligible effect on the strength of the received signal.

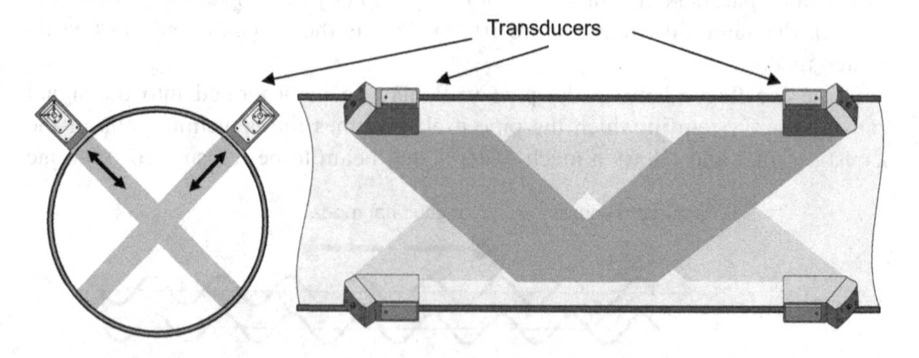

FIGURE 10.28 A dual-path single-traverse layout divides the pipe into four equal sections – with the sonic beam crossing in the centre of the pipe at right angles. This arrangement allows much higher accuracies to be achieved – down to ±0.15%. (Courtesy: Siemens.)

divides the pipe into four equal sections – with the sonic beam crossing in the centre of the pipe at right angles. This arrangement allows much higher accuracies to be achieved – down to ±0.15%.

10.11 SUMMARY OF ULTRASONIC METERING

10.11.1 Advantages

- Suitable for high accuracy custody transfer applications
- Suitable for large diameter pipes
- No obstructions, no pressure loss
- No moving parts, long-operating life
- Not affected by corrosion, erosion, or viscosity
- Fast response
- Multi-beam systems can be used to eliminate the effects of profile disturbances
- Not affected by fluid properties
- Most ultrasonic flowmeters are bi-directional

10.11.2 Disadvantages

- In single-beam meters, the accuracy is dependent on the flow profile
- Fluid must be acoustically transparent
- Expensive
- Pipeline must be full

11 Coriolis Mass Flowmeters

11.1 INTRODUCTION

Most chemical reactions are based largely on their mass relationship. Consequently, by measuring the mass flow of the product it is possible to control the process more accurately. Furthermore, the components can be recorded and accounted for in terms of mass.

Mass flow is a primary unit of flow measurement and is unaffected by viscosity, density, conductivity, pressure, and temperature. As a result it is inherently more accurate and meaningful for measuring material transfer.

Traditionally, mass flow has been measured inferentially. Electromagnetic, orifice plate, turbine, ultrasonic, Venturi, vortex shedding, etc., all measure the flow of the medium in terms of its velocity through the pipe (e.g. metres per second). However, because the dimensions of the pipe are fixed, we can also determine the volumetric flow rate (e.g. litres per second). Furthermore, by measuring the density and multiplying it by the volumetric flow rate, we can even infer the mass flow rate. However, such indirect methods commonly result in serious errors in measuring the mass flow.

Possibly the most significant advancement in flow measurement over the past few years has been the introduction of the Coriolis mass flowmeter. This technology not only allows mass flow to be measured directly but is also readily able to cope with the extremely high densities of for example dough, molasses, asphalt, liquid sulphur, etc., found in many industries.

11.2 THE CORIOLIS EFFECT

Probably one of the earliest recorded studies of, what is now termed the Coriolis effect, was by the Italian scientist Giovanni Riccioli in 1651. If, in the astronomical model of the solar system the Earth rotated around the Sun (as heliocentrics believed), a cannon ball is fired towards the north pole it should deflect to the east. Since, at that time, this was not an observable effect, it was used as an argument for geocentrism, which placed the Earth at the centre.

Although the Coriolis acceleration equation was derived by Euler in 1749 it was not until Gaspard-Gustav de Coriolis published a paper in 1835, in connection with the theory of water wheels, that the terms *Coriolis Effect* and *Coriolis Force* were actually coined.

In recent years, the major impact of the Coriolis effect has been in the field of meteorology – describing how it affects global wind patterns – with cyclonic rotation having an anti-clockwise direction in the northern hemisphere and clockwise direction in the southern hemisphere.

But the practical application of the Coriolis effect in the field of engineering has been in mass flow measurement.

11.3 THE CORIOLIS FORCE

Consider two children, Anne and Belinda, playing on a children's roundabout (Figure 11.1) which is rotating at a constant angular velocity (ω). Anne is situated mid-way between the axis and the outer edge of the platform while Belinda is sat at the outer edge itself. If Anne now throws a ball directly to Belinda, Belinda will fail to receive the ball!

The fact is that whilst Anne and Belinda both have the same angular velocity (ω) their tangential velocities are different (Figure 11.2). Anne's tangential velocity (V_A) is only half of Belinda's tangential velocity (V_B). In fact, the peripheral speeds of each are directly proportional to the radius, that is:

$$v = r \cdot \omega \tag{11.1}$$

where

v = peripheral velocity
r = radius
ω = angular speed

Consequently, to move the ball from Anne to Belinda, its tangential velocity needs to be accelerated from V_A to V_B. This acceleration is a result of what is termed the Coriolis force, named after the French scientist who first described it, and is directly

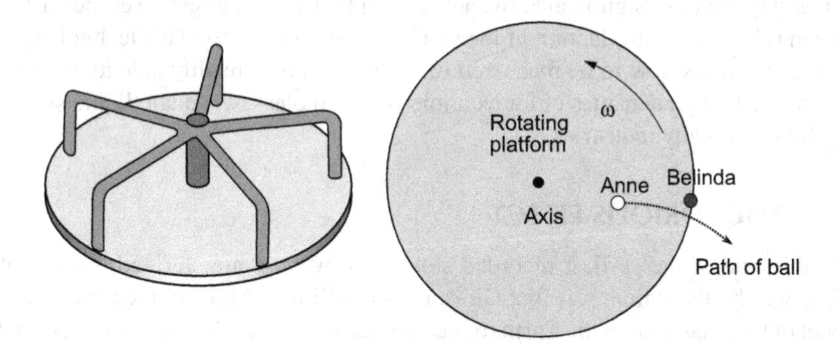

FIGURE 11.1 Anne and Belinda are playing on a children's roundabout. If Anne throws a ball directly to Belinda, Belinda will fail to receive the ball due to the Coriolis effect.

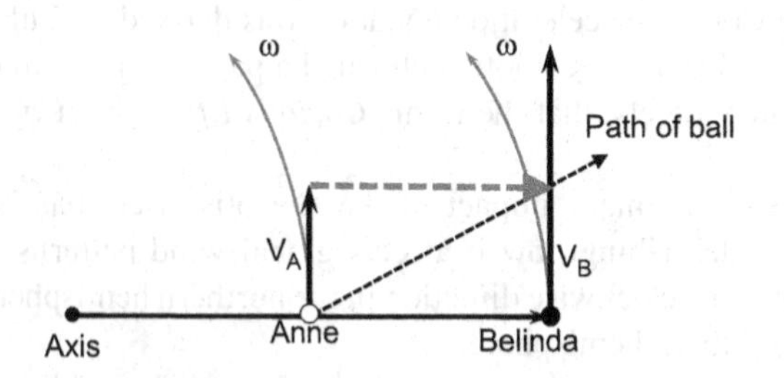

FIGURE 11.2 Belinda, at the edge of the platform will have a peripheral speed of twice that of Anne and thus the ball's peripheral speed needs to be accelerated from V_A to V_B.

proportional to the product of the mass in motion, its speed, and the angular velocity of rotation:

$$F_{cor} = 2 \cdot m \cdot \omega \cdot v \quad (11.2)$$

where

F_{cor} = Coriolis force
v = tangential velocity
ω = angular speed
m = the mass of the ball

Looking at this from another point, if we could measure the Coriolis force (F_{cor}), knowing the tangential velocity (v) and the angular velocity (ω) we could determine the mass (m) of the ball.

How does this relate to mass measurement of fluids? Consider a liquid-filled pipe sealed at both ends, rotating about an axis at an angular velocity (ω).

The tangential velocity (v) of any individual particle of the fluid is simply the angular velocity (ω) times the distance (r) from the centre of rotation (Figure 11.3). Thus, at distance r_1, the tangential velocity of a particle would be $r_1 \cdot \omega$, whilst at double the distance r_2, the tangential velocity would also double to $r_2 \cdot \omega$.

If now, the liquid flows in a direction away from the axis (Figure 11.4), then as each mass particle moves for example from r_1 to r_2 it will be accelerated by an

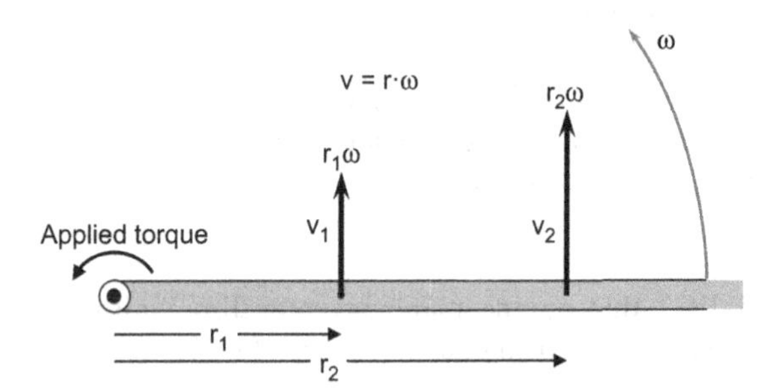

FIGURE 11.3 At distance r_1, the tangential velocity of a particle would be $r_1 \cdot \omega$ whilst at double the distance r_2, the tangential velocity would also double to $r_2 \cdot \omega$.

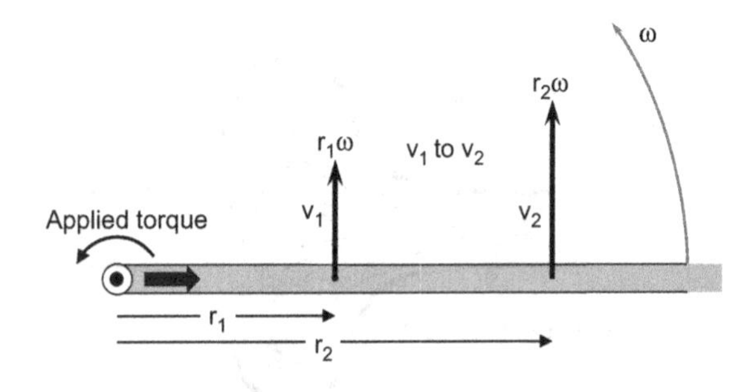

FIGURE 11.4 As the liquid flows away from the axis, each mass particle will be accelerated by an amount equivalent to its movement along the axis from a low to a higher orbital velocity.

amount equivalent to its movement along the axis from a low to a higher tangential velocity. This increase in velocity is in opposition to the mass inertial resistance and is felt as a force opposing the pipe's direction of rotation – that is it will try to slow down the rotation of the pipe. Conversely, if we reverse the flow direction, particles in the liquid flow moving towards the axis are forced to slow down from a high velocity to a lower velocity and the resultant Coriolis force will try to speed up the rotation of the pipe.

Thus, if we drive the pipe at a constant torque, the Coriolis force will produce either a braking torque or an accelerating torque (dependent on the flow direction) that is directly proportional to the mass flow rate. In other words, the torque required to rotate the pipe will increase in direct proportion to the actual mass flow of the liquid.

11.4 INITIAL IMPLEMENTATION

The possibility of applying the Coriolis effect to measure the mass flow rate was recognised many years ago and initial implementations were typically based on rotating systems typified by the radial-vane type meter (Figure 11.5).

Here, the flowing fluid enters the centre of a centrifugal pump-type impeller driven at a constant angular velocity (ω). As the fluid's tangential velocity ($r \cdot \omega$) increases, it reacts on the impeller with a force directly proportional to the mass flow rate. This Coriolis force is measured by the torque-measuring tube.

Obvious problems include:

- Maintaining the motor speed to provide a constant angular velocity (ω)
- Slip rings for power and measurement of torque
- High unrecoverable pressure loss
- Wear and tear

It was not until 1977 that engineer and inventor Jim Smith, founder of Micro Motion, patented the first practical system (Figure 11.6). Rather than using rotational movement this system was based on a vibrating tube system to provide oscillatory movement.

The basic principle is illustrated in Figure 11.7 in which a tubular pipe, carrying the liquid, is formed in a loop and vibrated around the *z*-axis. The straight parts of the

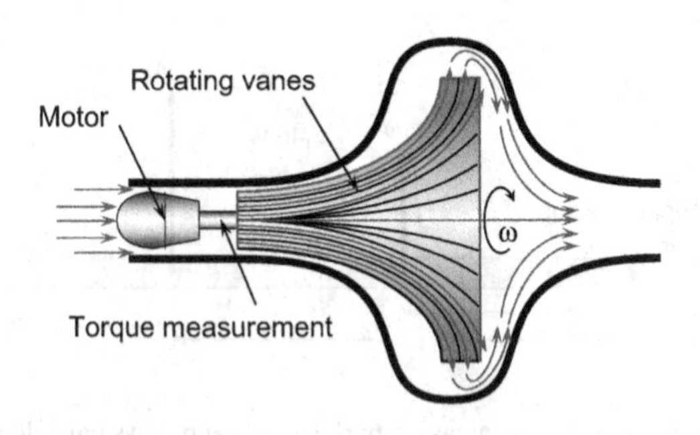

FIGURE 11.5 Radial-vane type meter.

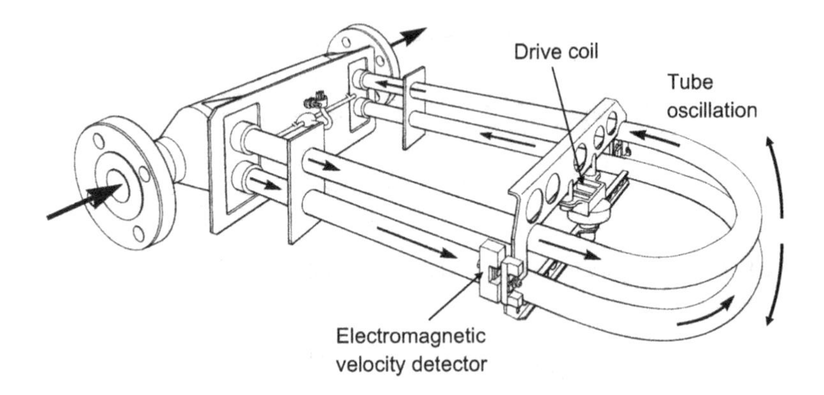

FIGURE 11.6 First practical system from Micro Motion. (Courtesy: Emerson.)

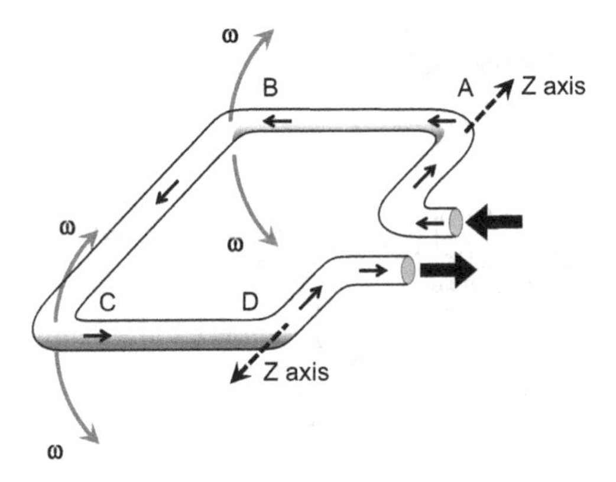

FIGURE 11.7 Pipe, formed in a loop, is vibrated around the *z*-axis so that the straight parts of the pipe, A-B and C-D, oscillate on the arcs of a circle.

pipe, A-B and C-D, oscillate on the arcs of a circle and without any flow will remain parallel to each other throughout each cycle.

When a liquid flows through the tube in the direction shown, then the fluid particles in section A-B will move from a point having a low tangential velocity at A to a point having a high tangential velocity at B. This means that each mass particle must be accelerated in opposition to the mass inertial resistance. This opposes the pipe's direction of rotation and produces a Coriolis force in the opposite direction.

Conversely, in section C-D, the particles move in the opposite direction – from a point having a high tangential velocity at C to a point having a low tangential velocity at D.

The resultant effect of these Coriolis forces is to delay the oscillation in section A-B and accelerate it in section C-D. As a result section A-B tends to lag behind the undisturbed motion whilst section C-D leads this position. Consequently, the complete loop is twisted by an amount that is directly and linearly proportional to the mass flow rate of the fluid – with the twisting moment lent to the pipe arrangement being measured by sensors.

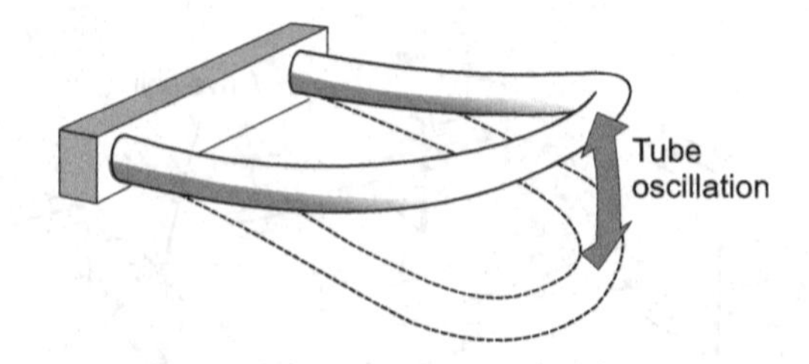

FIGURE 11.8 Oscillatory motion applied to a single tube.

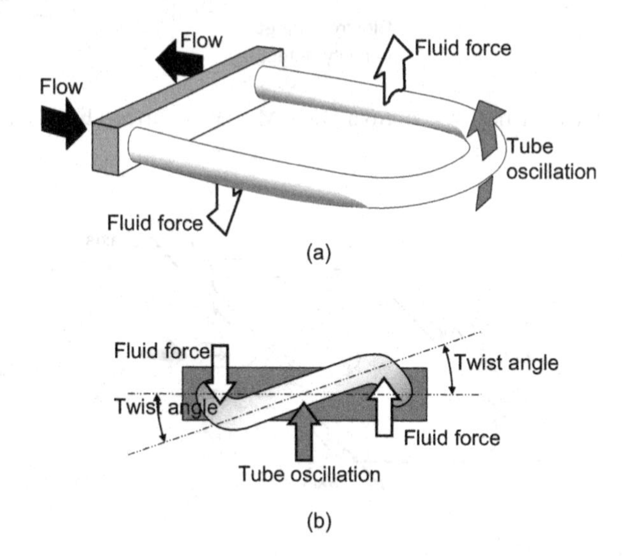

FIGURE 11.9 (a) Forces acting on the tube in which there is fluid flow. (b) Complete loop is twisted by an amount that is directly and linearly proportional to the mass flow rate of the fluid.

Figure 11.8 shows the oscillatory motion applied to a single tube. Figure 11.9a shows the forces acting on the tube in which there is fluid flow. As a result, the complete loop is twisted by an amount that is directly and linearly proportional to the mass flow rate of the fluid (Figure 11.9b) – with the flexure of the pipe arrangement being measured by sensors.

To reduce the risk of stress fractures the oscillation amplitude is limited between 0.1 and 1 mm which, in an optimally designed system, is about 20% of the maximum permitted value.

The distortion caused by the Coriolis forces is about 100 times smaller (a magnitude of about 10 μm). And in order to provide a measurement resolution to meet an accuracy of for example ±0.1% magnitudes of the order of a few nanometres would need to be resolved. The reality is we don't actually measure the amplitude but the phase difference (i.e. time) with flexure of the pipe arrangement measured by velocity sensors (Figure 11.10).

With no fluid flow (and thus no flexure) the sinusoidal outputs of the two velocity sensors would be in-phase – with no time difference (Figure 11.11a).

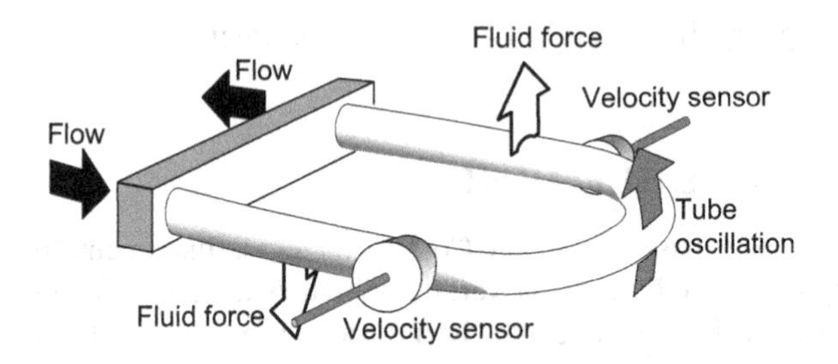

FIGURE 11.10 Flexure of the pipe arrangement measured by velocity sensors.

However, with fluid flow, the flexure of the pipe produces a phase difference directly proportional to the mass flow (Figure 11.11b).

Measurement resolution of ±0.1% amounts to only a few nanoseconds. And in order to achieve an accuracy of ±0.1%, where the resolution needs to be typically

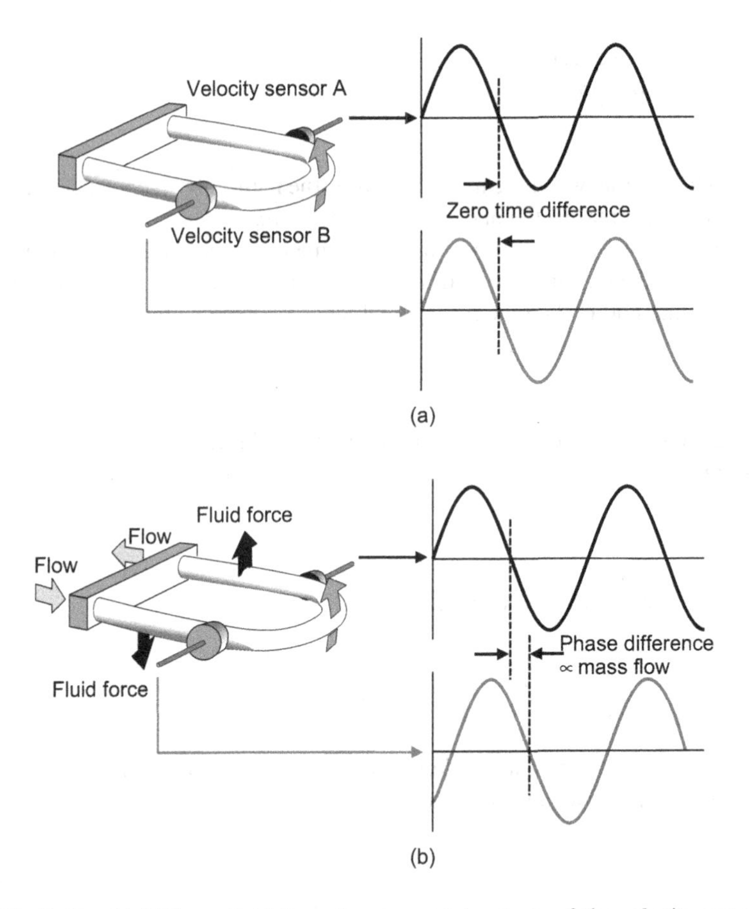

FIGURE 11.11 (a) With no fluid flow, the sinusoidal outputs of the velocity sensors are in-phase with no time difference. (b) With fluid flow, the flexure of the pipe produces a phase difference directly proportional to the mass flow.

5–10 times higher, time shift difference measurements are required down to picosecond levels.

11.5 DENSITY MEASUREMENT

The measurement of mass flow by the Coriolis meter is, fundamentally, independent of the density of the medium. However, the vibratory action of the oscillating tubes employed in the Coriolis meter can be harnessed to provide an independent measure of the medium density.

Hooke's Law spring equation states that a mass suspended on a spring will oscillate at a resonant frequency:

$$f = \frac{1}{2\pi} \cdot \sqrt{\frac{k}{m}} \tag{11.3}$$

where

f = resonant frequency
k = spring constant
m = mass

This indicates that when the mass increases, the natural frequency decreases, and when the mass decreases, the natural frequency increases.

► How does Hooke's Law relate to density measurement in a vibrating tube?

The mass of the system (m) equals the mass of the tube (m_{tube}), which is fixed, plus the mass of the fluid in the tube (m_{fluid}) – which is variable with the process:

$$m = m_{tube} + m_{fluid} \tag{11.4}$$

In turn, the fluid mass (m_{fluid}) is determined by the volume of the tube (V_{tube}), which is fixed, multiplied by the density of the fluid (ρ_{fluid}), which is variable:

$$m_{fluid} = V_{tube} \cdot \rho_{fluid} \tag{11.5}$$

Rearranging Equations 11.3, 11.4 and 11.5:

$$\rho_{fluid} = \frac{k}{f^2 \cdot V_{tube} \cdot 4\pi^2} - \frac{m_{tube}}{V_{tube}} \tag{11.6}$$

With all other values constant, the density of the fluid (ρ_{fluid}) is inversely proportional to the square of the resonant frequency:

$$\rho_{fluid} \propto \frac{1}{f^2} \tag{11.7}$$

◄

Consequently, by measuring this frequency, we calculate the fluid's density.

Note: It is important to recognise that the density measurement is not based on the Coriolis effect but on the effect of the vibrating tube.

Therefore, in addition to providing a direct indication of mass flow the oscillating pipe system also, independently, provides a direct indication of the density by tracking the resonant oscillation frequency.

Obviously the value of k in Equation 11.6 hides a lot of dependencies – one of which is temperature. Consequently, the temperature must be measured as an independent quantity and used as a compensating variable. The temperature is also available as a measured output.

With knowledge of both the mass flow (Q_m) and the density (ρ) of the fluid, it is now possible to also calculate the volumetric flow rate (Q) since:

$$Q_m = Q \cdot \rho \tag{11.8}$$

therefore:

$$Q = \frac{Q_m}{\rho} \tag{11.9}$$

From the foregoing we can see the importance of vibrating the tubes at their resonant frequency. Furthermore, excitation at the resonant frequency requires less drive energy and ensures that excitation is in the primary resonant mode at all times.

This is accomplished through a simple feedback system from the pick-up coils (Figure 11.12).

11.6 MOVING FORWARD

At the beginning of the 1980s, there was only one Coriolis meter manufacturer. But by 1990, there were more than 13 different manufacturers with a plethora of new designs – each one seeking to overcome the shortcomings of previous versions.

So what exactly were the major perceived shortcomings of this technology?

Not necessarily in order of importance:

- The bent tube design
- Maximum pipe diameter

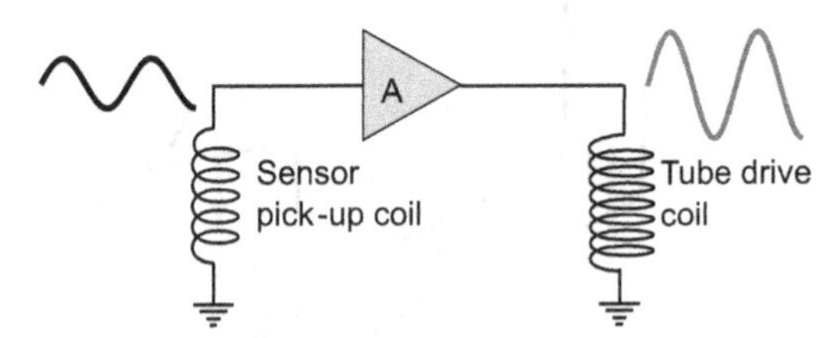

FIGURE 11.12 Excitation at the resonant frequency is accomplished through a simple feedback system from the pick-up coils.

- Entrained gas
- Liquefied natural gas (LNG) measurement
- Latency

11.7 TUBE CONFIGURATIONS

Maximising sensitivity, whilst simultaneously minimising the effects of extraneous noise and vibration, led to a variety of different bent-tube designs. But the bent-tube arrangement, itself, produced problems.

In any arrangement requiring the tube to be bent, the outside wall is stretched and becomes thinner whilst the inner wall becomes thicker. And, when the flowmeter requires two such convoluted tubes, it becomes difficult to balance them both dimensionally and dynamically. Furthermore, if the fluid is abrasive, this already weakened part of the flowmeter is likely to be most severely stressed. Abrasive materials can also cause erosion that will change the stiffness of the resonant elements and so cause measurement errors.

And for many liquids, the resultant pressure drop, due to the bends, could result in flashing or even cavitation damage. Furthermore, some of the bent-tube configurations did not cater for self-draining – an important consideration in many industries including food and beverage, pharmaceuticals, and chemical.

A typical dual bent-tube design, as shown in Figure 11.13, featured a high total cross-sectional area combined with the flexibility of two pipes. On the negative side, the flow divider introduced a high pressure drop. In addition the flow may not be equally divided, and the dual tube arrangement does not allow clean-in-place (CIP) to be implemented.

And whilst the continuous loop configuration (Figure 11.14) caters for CIP, a larger cross-sectional area is required to reduce the pressure loss. This leads to increased rigidity – making it less sensitive at low rates.

Other designs proliferated (Figure 11.15).

However, the next major milestone occurred with the first straight-tube design introduced in 1986 by Endress+Hauser (Figure 11.16).

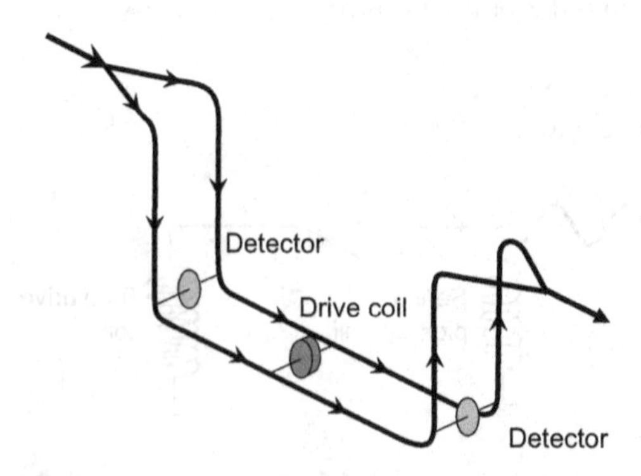

FIGURE 11.13 Typical dual bent-tube design.

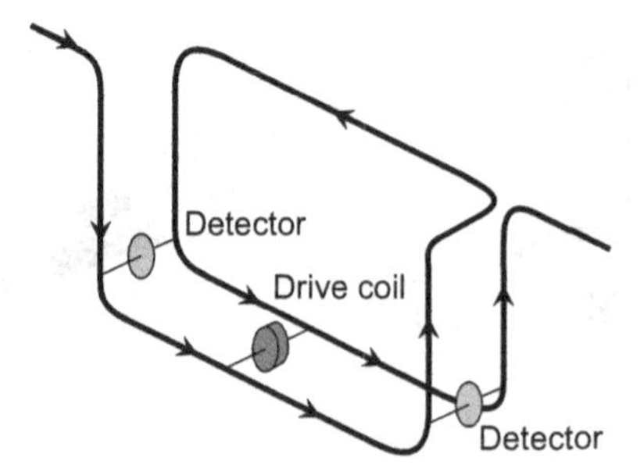

FIGURE 11.14 Continuous loop configuration.

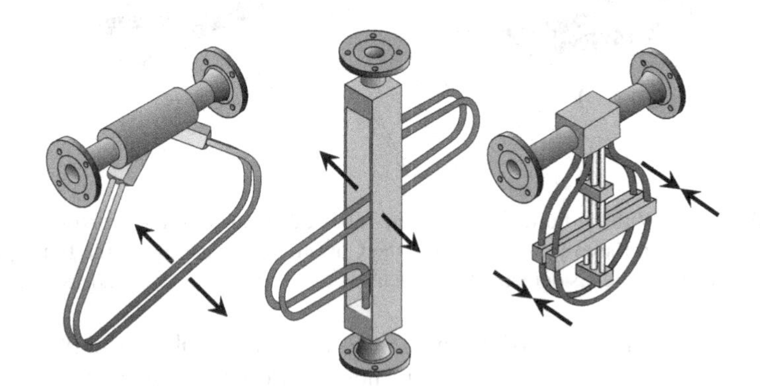

FIGURE 11.15 Variety of dual bent-tube designs.

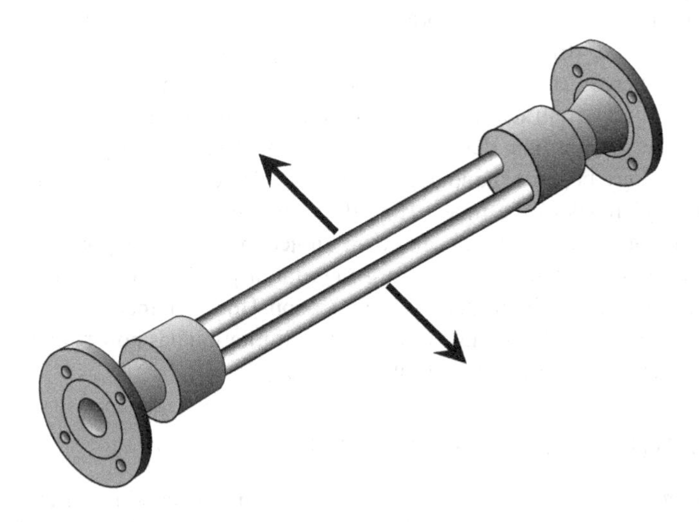

FIGURE 11.16 Straight-tube design employing dual measuring tubes.

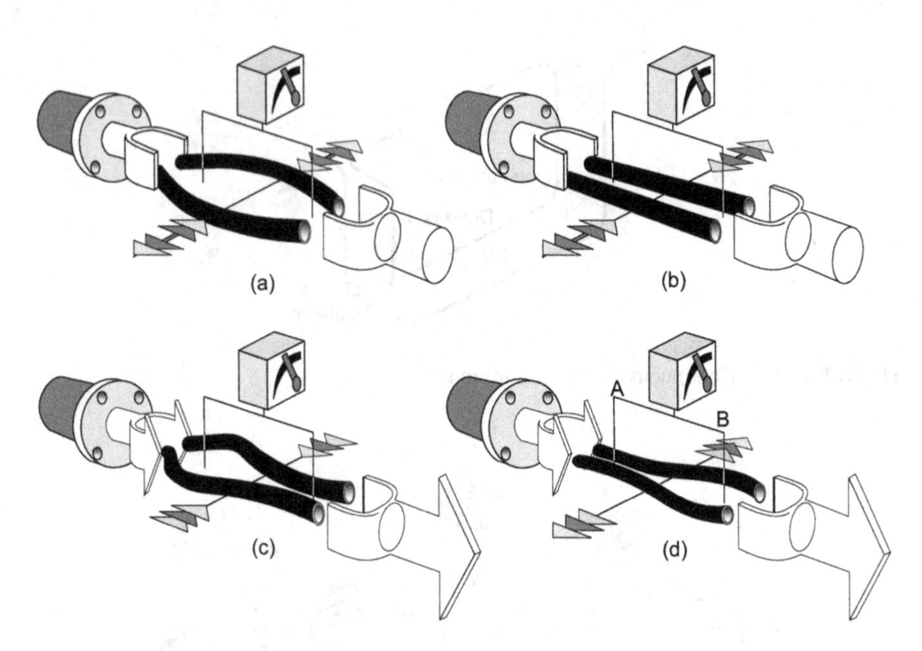

FIGURE 11.17 With no flow, flexure of the tubes takes place in the vibrational plane (a and b). However, in the event of fluid flow the Coriolis forces acting on the tubes produce a distorted flexure which is detected by the sensors (c and d). (Courtesy: Endress+Hauser.)

With no flow, the flexure of the tubes takes place in the vibrational plane (Figure 11.17a and b). However, in the event of fluid flow, the Coriolis forces acting on the tubes produce a distorted flexure which is detected by the sensors (Figure 11.17c and d).

But whilst this design overcame the problems associated with having a bent tube (weakening of the tubes at the bends, erosion, and flashing) it still employed dual measuring tubes with the need to split the flow. This limitation was surmounted when, in 1994, Krohne introduced the world's first industrial single straight tube meter.

However, these straight tube designs also brought in their wake several other problems. Whereas the flexibility of the bent tube arrangement easily allowed for expansion and contraction due to temperature and/or pressure variations, the rigidity of the straight tube design is less forgiving. Consequently, the use of strain gauge technology became mandatory in order to detect the slight variations in the dimensional changes of the tubes. How these measurements were incorporated to provide accurate compensation became the subject of complex algorithms.

Early designs were constrained to a maximum pipeline diameter of typically 25–50 mm which severely limited their application. However, incrementally this limit has been extended until, only late last year, this limitation has now been extended to 400 mm pipeline diameter straight tube systems.

11.8 MEMS-BASED TECHNOLOGIES

At the other end of the scale there has been a growing demand in for micro-dosing in the pharmaceutical and silicon fabrication industries as well as in the laboratory

FIGURE 11.18 MEMS-based integrated Coriolis system. (Courtesy: Integrated Sensing Systems.)

sample analysing and processing. Whilst most Coriolis meters find their application in measuring flow rates greater than 1 kg/h, these industries require dosing and a measuring rates in the order of 1 g/h and lower.

These challenges have been met through the application of Micro Electro-Mechanical System (MEMS)-based technologies in which the entire mechanical Coriolis fluidic-sensing system is integrated into a single silicon microchip (Figure 11.18).

Whilst some of these designs have focussed on flow measurement at very low flow rates (down to 10 mg/h or better) other designs have been mainly developed to measure gas density and concentration.

At first glance, the use of silicon to fabricate the actual vibrating-sensing tubes might seem at odds with its inherent rigidity, as compared with metals that can bend. Nevertheless, whilst a metal tube will plastically deform over time when exposed to fatigue, when a silicon structure is bent it either goes back to its original position or it breaks. Consequently, because it never deforms, it is virtually free of fatigue, hysteresis, and drift.

Another benefit of silicon is the resonant frequency of the vibrating tube. Conventional metal tube Coriolis mass flowmeters resonate at 100–1,500 Hz leaving them susceptible to the spectrum of common external mechanical vibration and shock frequencies that are under 2,000 Hz.

The micro-tube resonant frequency of silicon, on the other hand, is high (typically of the order of 20 kHz) making it virtually impervious to any form of extraneous vibration.

A third advantage of silicon over steel is that its density is 3.4 times lower. Thus, whilst conventional Coriolis flow sensors manufactured from steel are typically not sensitive enough to accurately measure gas density at low pressures, a significantly higher sensitivity can result from the use of silicon.

11.9 ENTRAINED GAS

Two-phase flow is notoriously problematic for most flow technologies.

Gas bubbles can form for a variety of reasons:

- Degassing
- Leaks upstream of or in a negative pressure area
- Excessive cavitation
- Supply container levels falling below minimums
- Empty-full-empty applications
- Tank agitation
- Top filling tanks with long drops
- Process-control status transitions, such as system start-up, shut down, or cleaning

Conversely, there are processes in which the entrainment of air in the liquid is critical to achieve the right quality.

Examples include the production of:

- Yoghurt
- Mayonnaise
- Ice cream

Assume a 5% gas by volume entrained in a liquid stream. Velocity based meters (e.g. turbine, ultrasonic, etc.) calculate the volume flow (Q) by multiplying velocity (v) by the cross-sectional area (A). This means the desired liquid measurement would be overstated by the amount of the 'gas void fraction' resulting in an error of about 5%.

Unfortunately, a two-phase flow is often not even recognised as a problem. And even when it is identified as the culprit, it often requires 'workarounds'.

One solution involves the use of venting devices upstream in the process. However, this does not capture 100% of the air and also exposes the process to the atmosphere – possibly compromising the product quality.

In pharmaceutical, food and beverage, and other sanitary applications, requiring closed-loops, the use of venting devices is not an option at all.

Another workaround involves removing air from the process liquid. However, because air can enter from a wide variety of points eliminating so many sources is costly.

Flow regimes in the horizontal flow are dependent on what is termed the superficial liquid* and gas velocities and are also affected by density, viscosity, and surface tension (Figure 11.19).

Multiphase flow regimes have no sharp boundaries – often changing smoothly from one regime to another (Figure 11.20).

* Superficial velocity is the velocity that the (liquid or gas) phase would have, in case it was the only phase flowing through the pipe.

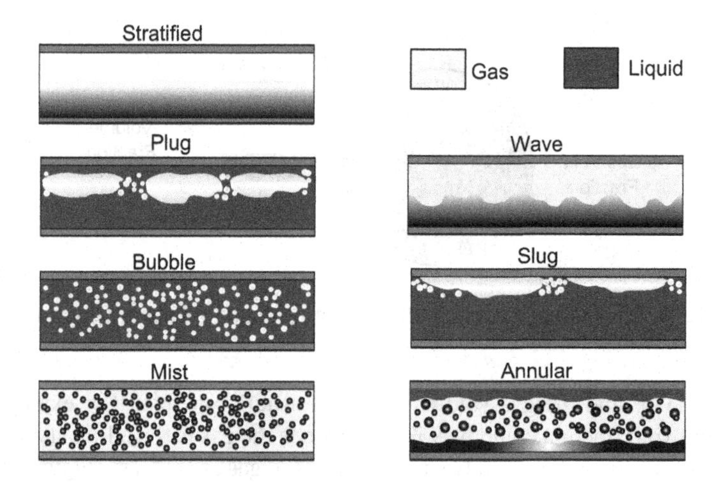

FIGURE 11.19 Flow regimes in horizontal flow are dependent on liquid and gas velocities, and also by density, viscosity, and surface tension.

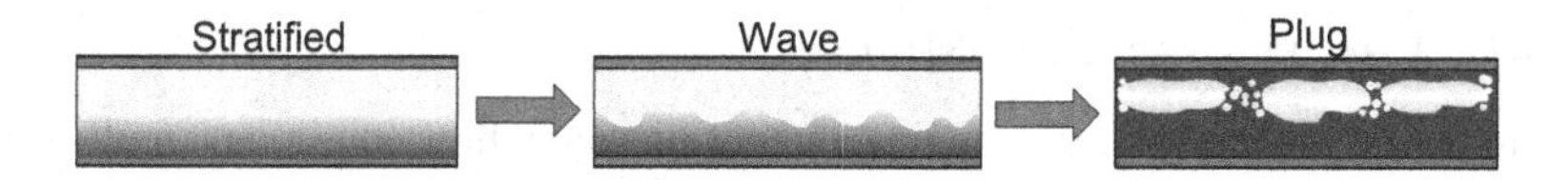

FIGURE 11.20 Multiphase flow regimes often change smoothly from one regime to another.

GAS VOLUME FRACTION VS. GAS VOID FRACTION

With both sharing terms sharing the same initials, it's important to distinguish between the Gas Volume Fraction (GVF) and the Gas Void Fraction.

The GVF is the ratio of the gas volumetric flow rate to the total volumetric flow rate of the total stream volume of oil, water, and gas. The gas void fraction is based on local areas – the actual cross-sectional areas occupied by the gas and liquid, respectively (Figure 11.21). The gas void fraction is usually described as the percentage of the total volume occupied by the gas under the conditions being considered. Consequently, it is dimensional and the volumes of each fluid phase need to be calculated in the same volume units for example gallons, SCF, BBL, etc.

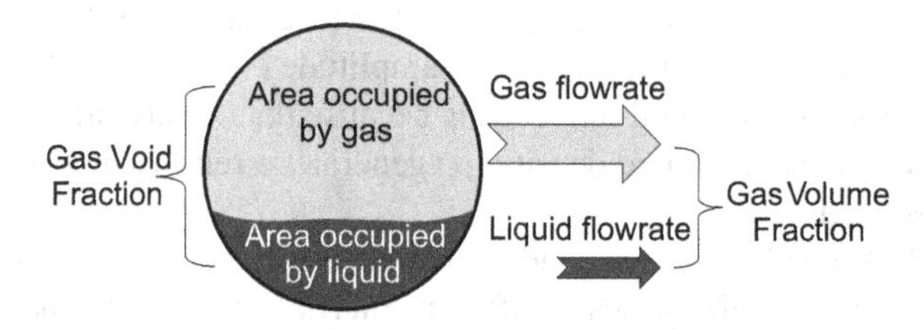

FIGURE 11.21 GVF is based on flow rates whilst the gas void fraction is based on local areas.

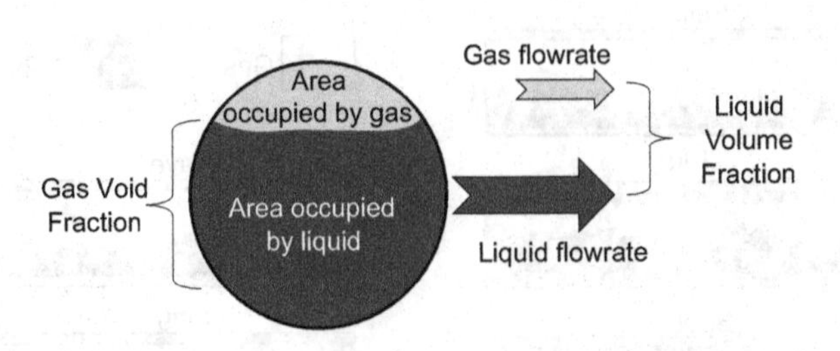

FIGURE 11.22 LVF is defined as 'the ratio of the liquid volumetric flow rate to the total volumetric flow rate'.

The GVF and the gas void fraction are usually unequal. For example, a 70% gas void fraction could be 95% GVF since the gas is travelling at a higher velocity. Where only relatively small amounts of gas are present (Figure 11.22) the term liquid volume fraction (LVF) is sometimes used and is defined as: 'the ratio of the liquid volumetric flow rate to the total volumetric flow rate'.

11.9.1 Challenges of Entrained Gas

Due to the large variety of flow regimes under unsteady flow conditions, multiphase flow measurements can be a challenging for traditional Coriolis mass flowmeters.

At first glance the measurement of liquid mass flow would not appear to be a problem. Remember, Coriolis meters measure the mass flow – the combined mass of the liquid plus the gas:

$$\dot{m}_{mixture} = \dot{m}_{gas} + \dot{m}_{liquid} \tag{11.10}$$

Since the mass of the gas is negligible the measurement is virtually identical to the liquid mass flow.

11.9.2 Decoupling

However, as the GVF increases errors arise due to what is termed 'decoupling' which occurs when gas bubbles move relative to the liquid during vibration of the flow tube.

A gas bubble in the measuring tube does not strictly follow the oscillation of the surrounding liquid with the same amplitude if the liquid cannot 'hold' the bubble. As illustrated in Figure 11.23 the oscillation amplitude of the gas bubble (u_g) will not be the same as that of the liquid (u_l). This is because the density difference between the gas density (ρ_g) and the liquid density (ρ_l) generates a relative motion between the bubble and the liquid.

This induced secondary flow around the bubble is usually in the opposite direction to the tube vibration. This causes a different inertial effect from the one that the flowmeter uses to sense mass flow rate since the oscillation amplitude of the bubble u_g is greater but in-phase with that of the measuring tube or the liquid u_l. Consequently,

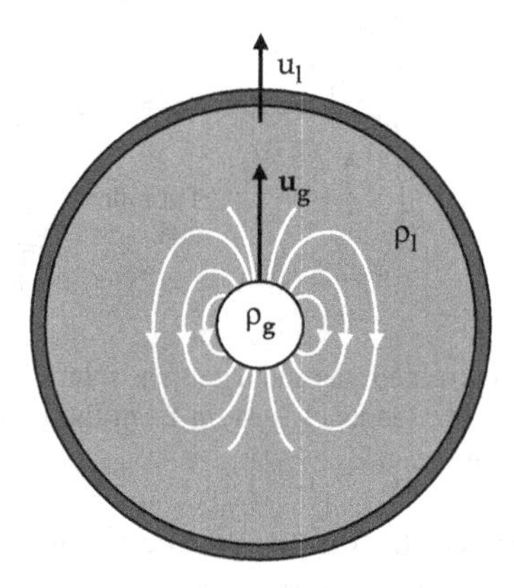

FIGURE 11.23 Density difference between the gas density ρ_g and the liquid density ρ_l generates a relative motion between the bubble and the liquid.

a part of the inertia of the liquid that should be felt by the tube wall for a particular flow is lost – thus leading to an underestimation of the real density and mass flow rate of the liquid phase.

We have already seen that density measurement is performed by tracking the resonant frequency in which lower density fluids have higher resonant frequencies and vice versa. Water thus has a lower resonant frequency than pure gas (Figure 11.24). When the gas is entrained in the liquid it produces a decoupling between the measurement tube and the flow. This absorbs some of the available excitation energy manifesting itself as a reduced signal amplitude.

A second effect of entrained gas is that the fluid density fluctuates swiftly. Consequently, the resonance frequency changes rapidly over time making it increasingly difficult to track.

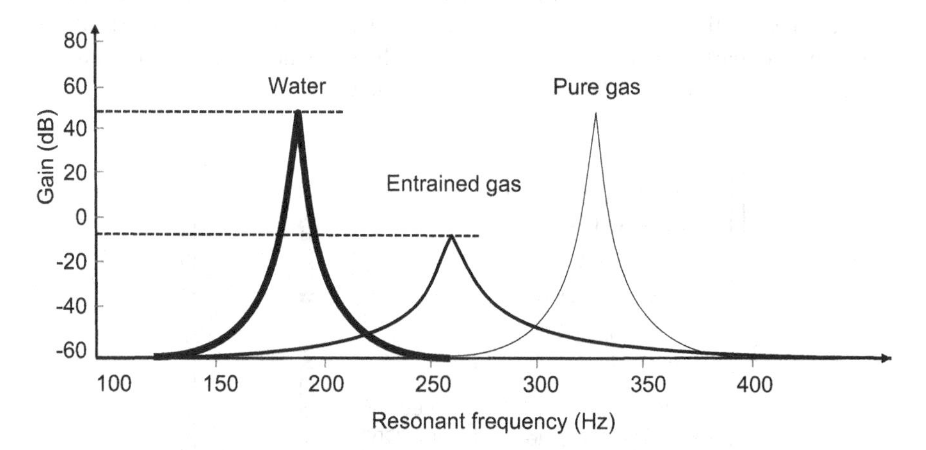

FIGURE 11.24 Water has a lower resonant frequency than pure gas. When the gas is entrained in the liquid some of the available excitation energy is absorbed producing a reduced signal amplitude. (Courtesy: Krohne.)

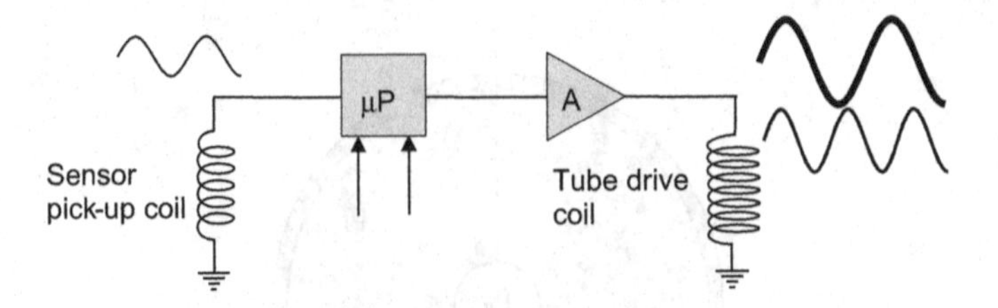

FIGURE 11.25 Typically entrained gas management solutions are centred on the use of a synthesised drive in which the feedback system from the pickup coils is modified by a microprocessor driven algorithm.

And even further problem is that the mass flow error margins change with different gas flow fractions, different flow velocities, and pressure variations.

11.9.3 Entrained Gas Management Systems

A variety of entrained gas management solutions are now available. Typically such systems are centred on the use of a synthesised drive (Figure 11.25) in which the feedback system from the pickup coils is a modified by a variety of microprocessor driven sophisticated algorithms. This has allowed several manufacturers to handle GVFs ranging from 0% to 100%.

11.10 LNG MEASUREMENT

As illustrated in Figure 11.26 it is important to draw distinction between natural gas liquids (NGLs) and Liquefied Natural Gas (LNG).

NGLs comprise those hydrocarbon components in a produced gas stream that can be extracted and sold in their respective markets. NGLs are the lighter condensable hydrocarbon fractions – the result of fractionation that separates the raw mixture into its component parts. Although methane is the lightest it cannot be condensed under normal processes, because the boiling point is so low.

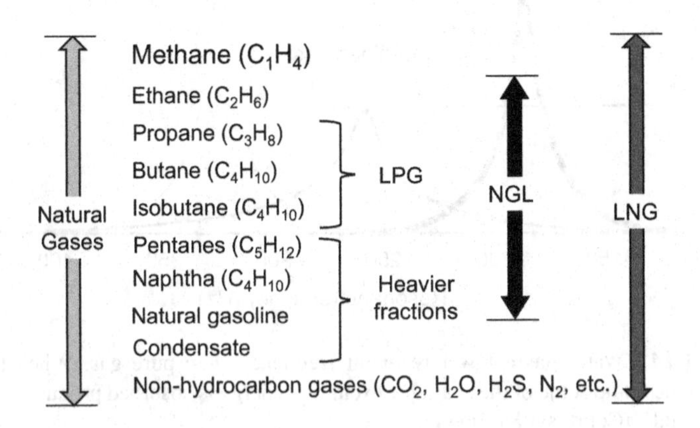

FIGURE 11.26 Distinction between NGLs and LNG.

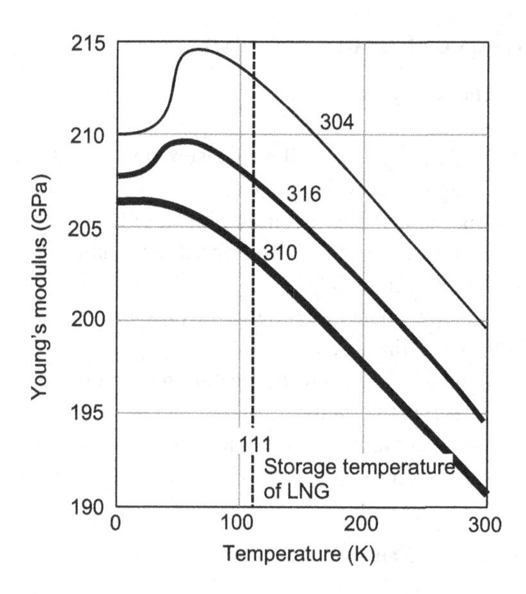

FIGURE 11.27 Young's modulus of the measuring tube (and therefore the natural resonant frequency) changes with temperature.

If the effluent gas from an NGL plant is totally liquefied it is called LNG which is a sale or pipeline quality gas in a liquid form.

LNG is a clear, colourless, non-toxic, non-corrosive liquid formed when natural gas (including methane) is cooled to around –153.1°C at close to atmospheric pressure.

This shrinks the volume of the gas 600 times, making it easier to store and transport.

Because of the cost of keeping LNG cold enough to remain in its liquid state it is normally kept at a temperature only slightly above its boiling point for example –162°C. As a consequence even a small pressure drop can cause flashing to occur.

Furthermore, moving mechanical parts and wetted seals can be adversely affected by the cold and cease to function or fail. Consequently, flow measurement is constrained to instruments capable of handling non-conductive cryogenic liquids.

One of the issues with regards to Coriolis-based equipment relates to the Young's modulus of the measuring tube since this (and therefore the natural resonant frequency) changes with the temperature (Figure 11.27).

11.11 SUMMARY OF CORIOLIS MASS MEASUREMENT

Coriolis meters may well supplant the electromagnetic flowmeter as the answer to the majority of flowmetering applications. For critical control, the mass flow rate is the preferred method of measurement and, because of their accuracy, Coriolis meters are becoming common for applications requiring very tight control. Apart from custody transfer applications, they are used for chemical processes and expensive fluid handling.

11.11.1 Advantages of Coriolis Metering

Some of the many benefits include:

- Direct, in-line, and accurate mass flow measurement of both liquids and gases
- Turndown ratios up to 500:1
- Accuracies as high as ±0.05% accuracy for liquid and ±0.35% for gases
- Mass flow measurement ranges cover from less than 5 g/m to more than 4,600,000 kg/h (350 tonnes/h)
- Measurement is independent of temperature, pressure, viscosity, conductivity, and density of the medium
- Direct, in-line, and accurate density measurement of both liquids and gases down to as little as 0.0005 g/cm^3
- Mass flow, density, and temperature can be accessed from one sensor
- Can be used for almost any application irrespective of the density of the process

11.11.2 Drawbacks of Coriolis Meters

On the downside, despite tremendous strides in the technology, some of the drawbacks include:

- Expensiveness
- Many models are affected by vibration
- Current technology limits the upper pipeline diameter to 400 mm
- Secondary containment can be an area of concern

11.11.3 Application Consideration

In the petrochemical industries the Coriolis meter has earned an unjust reputation for producing flashing and cavitation.

Properly sized, a straight-tube Coriolis meter produces virtually no unrecoverable pressure drop – and therefore no flashing or cavitation results. However, in many early applications, with pipe diameters limited to 100 DN, many hundreds of instruments were undersized – resulting in their misapplication and their subsequent high unrecoverable pressure drop.

12 Thermal Mass Flowmeters

12.1 INTRODUCTION

Thermal mass flow measurement, which dates back to the 1930s, is a quasi-direct method, suited, above all, for measuring the gas flow. Thermal mass flowmeters infer their measurement from the thermal properties of the flowing medium (such as specific heat and thermal conductivity) and hence are capable of providing measurements which are proportional to the mass of the medium.

In the ranges normally encountered in the process industry, the specific heat c_p of the gas is essentially independent of pressure and temperature and is proportional to density and therefore to mass.

The two most commonly used methods of measuring the flow using thermal techniques are either to measure the rate of heat loss from a heated body in the flow stream or to measure the rise in temperature of the flowing medium when it is heated.

12.2 HEAT LOSS OR 'HOT WIRE' METHOD

In its simplest form a hot body (a heated wire, thermistor, or resistance temperature detector [RTD]) is placed in the main stream of the flow (Figure 12.1). According to the first law of thermodynamics, heat may be converted into work and vice versa. Thus, the electrical power (I^2R) supplied to the sensor is equal to the heat convected away from it.

Since it is the molecules (and hence mass) of the flowing gas that interact with the heated boundary layer surrounding the velocity sensor and convect away the heat, the electrical power supplied to the sensor is a direct measure of the mass flow rate.

► The rate of heat loss of a small wire is given by:

$$P = h \cdot A \cdot (T_w - T_f) \tag{12.1}$$

where

P = heat loss in watts
h = heat transfer coefficient
A = surface area of the wire
T_w = wire temperature
T_f = fluid temperature

The heat transfer coefficiency depends on the wire geometry, specific heat, thermal conductivity, and density of the fluid as well as the fluid velocity in the following way:

$$H = C_1 + C_2 \cdot \sqrt{\rho \cdot v} \tag{12.2}$$

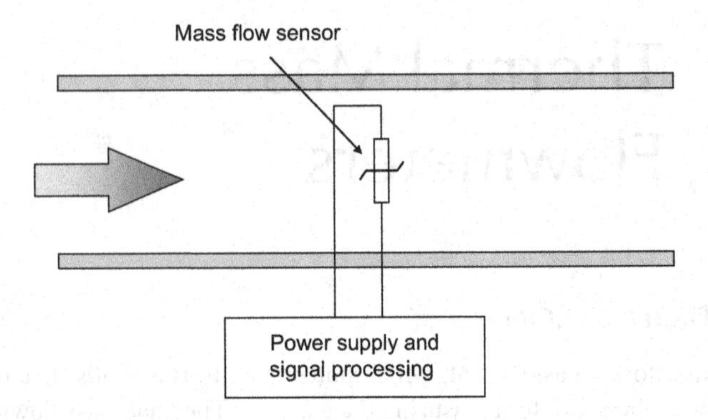

FIGURE 12.1 Basic schematic diagram of the 'hot wire' method.

where C_1 and C_2 are constants that depend on the wire geometry and gas properties.

The term $\sqrt{\rho \cdot v}$ indicates that the output of the hot wire flowmeter is related to the product of density and velocity, which can be shown to be proportional to the mass flow rate. ◀

In practice, this device can be used only if the medium temperature is constant, since the measured electrical resistance of the hot wire cannot determine whether the change in resistance is the result of a change in flow speed or of a change in medium temperature. To solve this problem, the temperature of the medium must be used as a reference value and a second temperature sensor must be immersed in the flow to monitor the medium temperature and correct for temperature changes (Figure 12.2).

The mass measuring RTD has a much lower resistance than the temperature RTD and is self-heated by the electronics. In a constant temperature system, the instrument measures I^2R and maintains the temperature difference between the two sensors at a constant level.

Complete hot wire mass flowmeters (Figure 12.3) are available for pipes up to 200 mm diameter (size DN 200). Above this size, insertion probes, which incorporate a complete system at the end of a rod, are used.

The main limitation of this method is that by its very 'point' measurement it is affected by the flow profile within the pipe as well as by the media viscosity and

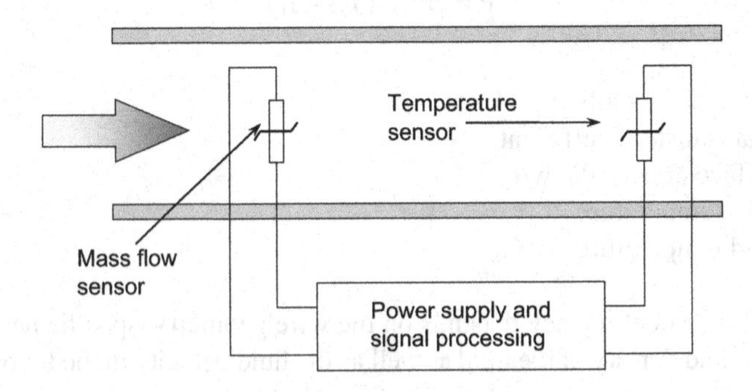

FIGURE 12.2 Second 'temperature sensor' monitors the gas temperature and automatically correct temperature changes.

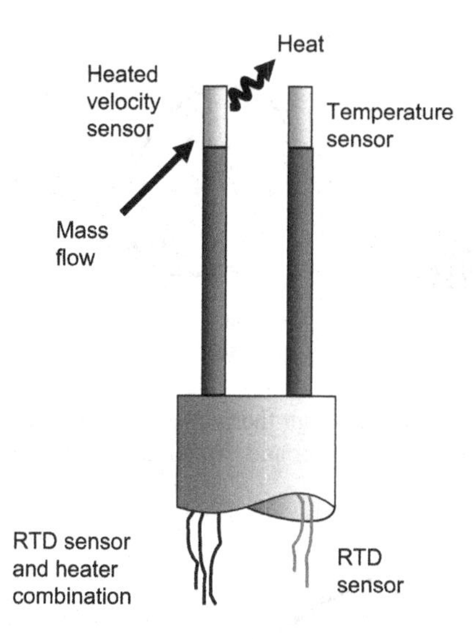

FIGURE 12.3 Typical in-line hot wire mass flowmeter. (Courtesy: Sierra Instruments Ltd.)

pressure. Further, since the measurement is determined by the thermal characteristics of the media, the system must be calibrated for each particular gas – with each mass flow/temperature sensor pair individually calibrated over its entire flow range.

The measured value, itself, is primarily nonlinear and thus requires relatively complex conversion. On the positive side, however, this inherent nonlinearity is responsible for the instrument's wide rangeability (1,000:1) and low speed sensitivity (60 mm/s).

Such instruments also have a fast response to velocity changes (typically 2 s) and provide a high level signal, ranging from 0.5 to 8 W over the range of 0–60 m/s.

One of the limitations of many conventional hot wire systems is that they soon reach their performance limits when higher mass flow speeds need to be detected. The thermal current into the medium depends on the flow speed and thus a constant heat input would mean that when the flow speed is low there would be a build-up of heat and the corresponding temperature increases. And at high flow speeds the temperature difference would be around zero. To overcome this problem, the heat input may be adapted to the flow speed. This is achieved in the sensor shown in Figure 12.4 which consists of a high thermal-conductive ceramic substrate upon which are deposited a thick film heating resistor (Rh) and two temperature-dependent thick film resistors (T_1 and T_2) (Figure 12.5).

As the process medium flows along the front of the ceramic substrate, the thermal current produced by the heating resistor forms a temperature gradient as illustrated in Figure 12.5. The temperature difference between the two resistors is then used to regulate the current controlling the heating resistor.

12.3 TEMPERATURE RISE METHOD

In this method, the gas flows through a thin tube in which the entire gas stream is heated by a constantly powered source – with the change in temperature being measured by

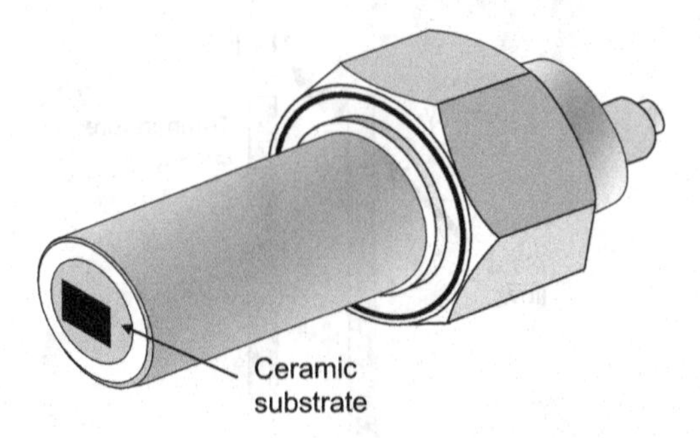

FIGURE 12.4 Sensor consists of a high thermal-conductive ceramic substrate upon which are deposited a thick film heating resistor and two temperature-dependent thick film resistors. (Courtesy: Weber Sensors Group.)

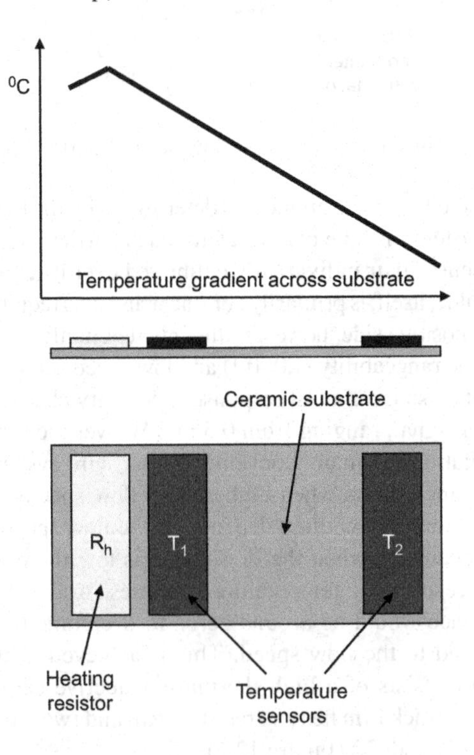

FIGURE 12.5 As the process medium flows along the front of the ceramic substrate the thermal current produced by the heating resistor forms a temperature gradient.

RTDs located upstream and downstream of the heating element (Figure 12.6). Because of the heat requirements this method is used for very low gas flows.

► Here, the mass flow rate q_m is:

$$q_m = \frac{k \cdot q_Q}{c_p \cdot \Delta T} \tag{12.3}$$

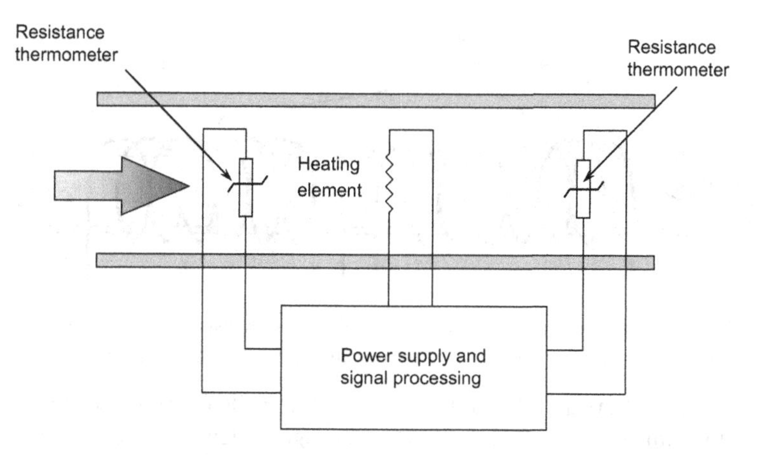

FIGURE 12.6 Basic schematic diagram of the 'temperature rise' method.

where

k = constant,
q_Q = the heat input (W),
c_p = specific heat capacity of the gas (J/kg K)
ΔT = temperature difference (°C) ◄

The main disadvantages of this method are that it is only suitable for low gas flows, the sensors are subject to erosion and corrosion and the multiple tapping points increase the chances of leakage.

12.4 EXTERNAL TEMPERATURE RISE METHOD

An alternative arrangement places the heating element and temperature sensors external to the pipe. In the arrangement shown in Figures 12.7 and 12.8, the heating elements and temperature sensors are combined so that the RTD coils are used to

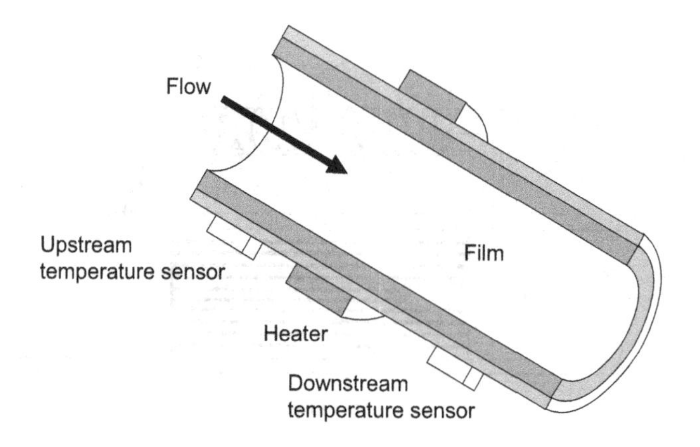

FIGURE 12.7 Thermal flowmeter with external elements and heater.

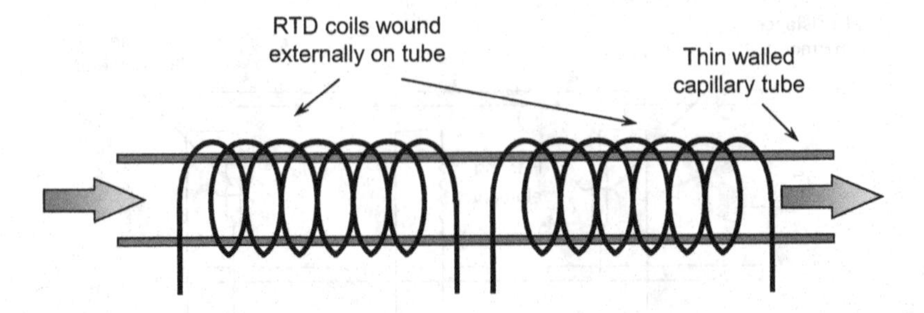

FIGURE 12.8 In the capillary tube meter the RTD coils are used to direct a constant amount of heat through the thin walls of the sensor tube into the gas. (Courtesy: Sierra Instruments.)

direct a constant amount of heat through the thin walls of the sensor tube into the gas. At the same time, the RTD coils sense changes in temperature through changes in their resistance.

The main advantage of this method is that it provides non-contact, non-intrusive sensing with no obstruction to flow.

12.5 CAPILLARY TUBE METER

In a typical capillary tube thermal mass flowmeter, the medium divides into two paths, one (m_2) through the bypass and the other (m_1) through the sensor tube (Figure 12.9).

As the name implies, the role of the bypass is to bypass a defined portion of the flow so that a constant ratio of bypass flow to sensor flow (m_2/m_1) is maintained. This condition will only apply if the flow in the bypass is laminar so that the pressure drop across the bypass is linearly proportional to the bypass flow. An orifice bypass for example has non-laminar flow so that the ratio of the total flow to sensor flow is non-linear.

One solution lies in the use of multiple discs or sintered filter elements. Another solution is the bypass element used by *Sierra* (Figure 12.10) which comprises a single-machined element having small rectangular passages with a high length-to-width ratio. This element provides pure laminar flow and is easily removed and cleaned.

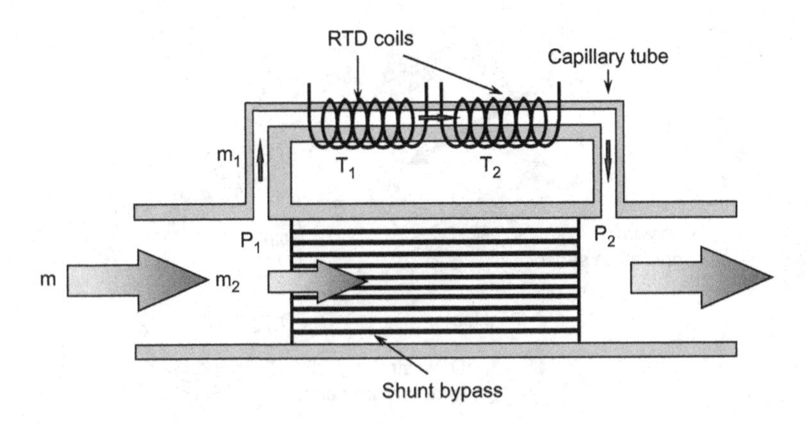

FIGURE 12.9 Typical capillary tube thermal mass flowmeter. (Courtesy: Sierra Instruments.)

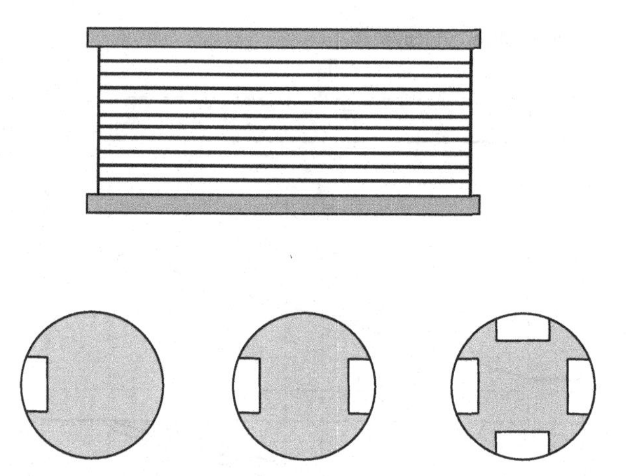

FIGURE 12.10 Single-machined elements having small rectangular passages with a high length-to-width ratio provide pure laminar flow and are easily removed and cleaned. (Courtesy: Sierra Instruments.)

With a linear pressure drop ($P_1 - P_2$) maintained across the sensor tube, a small fraction of the mass flow passes through the sensor tube. The sensor tube has a relatively small diameter and a large length-to-diameter ratio in the range of 50:1 to 100:1 – both features being characteristic of capillary tubes.

These dimensions reduce the Reynolds number to a level less than 2,000 to produce a pure laminar flow in which the pressure drop ($P_1 - P_2$) is linearly proportional to the sensor's mass flow rate (m_1).

In operation, the long length-to-diameter ratio of the tube ensures that the entire cross-section of the stream is heated by the coils – with the mass flow carrying heat from the upstream coil to the downstream coil. This means the first law of thermodynamics can be applied in its simplest form.

This method is largely independent of the flow profile and the medium viscosity and pressure. It means that the flow calibration for any gas can be obtained by multiplying the flow calibration for a convenient reference gas by a constant K-factor. K-factors are now available for over 300 gases, giving capillary tube meters almost universal applicability.

Although the output is not intrinsically linear with mass flow, it is nearly linear over the normal operating range. Accurate linearity is achieved with multiple-breakpoint linearisation (e.g. at 25%, 50%, 75%, and 100% of full scale).

In addition to its applicability to very low gas flows, the capillary tube method can also be used for larger flows by changing the bypass to affect a higher or lower value of the bypass ratio (m_2/m_1).

12.6 LIQUID MASS FLOW

Although the main application of the thermal mass flowmeter lies with gases, the same technology can also be applied to the measurement of very low liquid flows for example down to 30 g/h. A typical meter is shown in Figure 12.11.

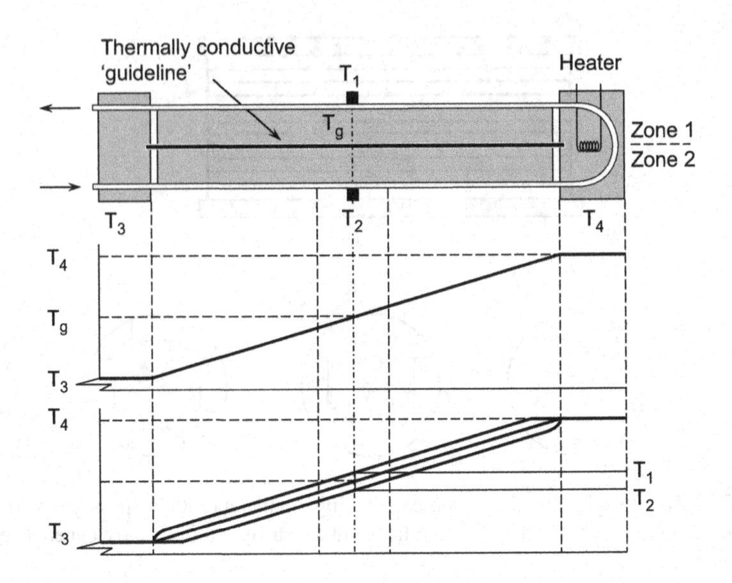

FIGURE 12.11 Typical liquid thermal mass flowmeter. (Courtesy: Brookes–Rosemount.)

Here, the inlet and outlet of the sensor tube are maintained at a constant temperature by a heat sink – with the mid-point of the sensor tube heated to a controlled level for example 20°C above the temperature of the inlet–outlet heat sink. These two locations, together with the flow tube, are mechanically connected by a thermally conductive path.

In this manner, the flowing fluid is slightly heated and cooled along the sensor zones, 1 and 2, respectively, to create an energy flow perpendicular to the flow tube. Two RTDs (T_1 and T_2), located at the mid-point of the sensor tube, determine the temperature difference. This temperature difference is directly proportional to the energy flow and is, therefore, directly proportional to the mass flow times the specific heat of the fluid.

13 Open Channel Flow Measurement

13.1 INTRODUCTION

In many applications, liquid media is distributed in open channels. Open channels are found extensively in water irrigation schemes, sewage processing and effluent control, water treatment, and mining beneficiation.

The most commonly used method of measuring flow in an open channel is through the use of a hydraulic structure (known as a *primary measuring device*) that changes the level of the liquid. By selecting the shape and dimensions of the primary device (a form of restriction) the rate of flow through or over the restriction will be related to the liquid level in a known manner. In this manner, a secondary measuring element may be used to measure the upstream depth and infer the flow rate in the open channel.

In order that the flow rate can be expressed as a function of the head over the restriction, all such structures are designed so that the liquid level on the upstream side is raised to make the discharge independent of the downstream level. The two primary devices in general use are the weir and the flume.

13.2 THE WEIR

A weir (Figure 13.1) is essentially a dam mounted at right angles to the direction of flow, over which the liquid flows.

The dam usually comprises a notched metal plate – with the three most commonly used being the rectangular weir, the triangular (or V-notch) weir, and the trapezoidal (or Cipolletti) weir – each having an associated equation for determining the flow rate over the weir that is based on the depth of the upstream pool. The crest of the weir, the edge, or surface over which the liquid passes, is usually bevelled – with a sharp upstream corner.

For the associated equation to hold true and accurate flow measurement determined, the stream of water leaving the crest (the nappe) should have sufficient fall (Figure 13.2). This is called free or critical flow, with air flowing freely beneath the nappe so that it is aerated. If the level of the downstream water rises to a point where the nappe is not ventilated, the discharge rate may be inaccurate and dependable measurements cannot be expected.

13.2.1 Rectangular Weir

The rectangular weir was probably the earliest type in use and, due to its simplicity and ease of construction, is still the most popular type.

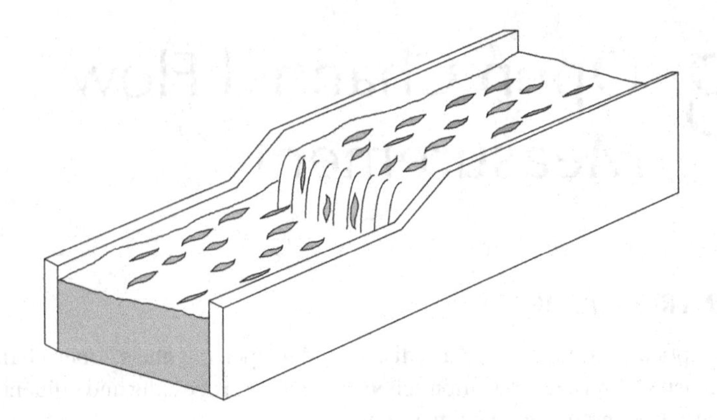

FIGURE 13.1 Basic weir – a dam mounted at right angles to the direction of flow.

In its simplest form (Figure 13.3a), the weir extends across the entire width of the channel with no lateral contraction.

► The discharge equation (head vs. flow rate), without end contractions, is:

$$Q = k \cdot L \cdot h^{1.5} \tag{13.1}$$

where

Q = flow rate
k = constant
L = length of crest
h = the head ◄

Generally, this means that for a 1% change in flow, there will be a 0.7% change in the level.

A problem with rectangular weirs without contraction is that the air supply can become restricted and the nappe clings to the crest. In such cases, a contracted rectangular weir (Figure 13.3b) is used where end contractions reduce the width

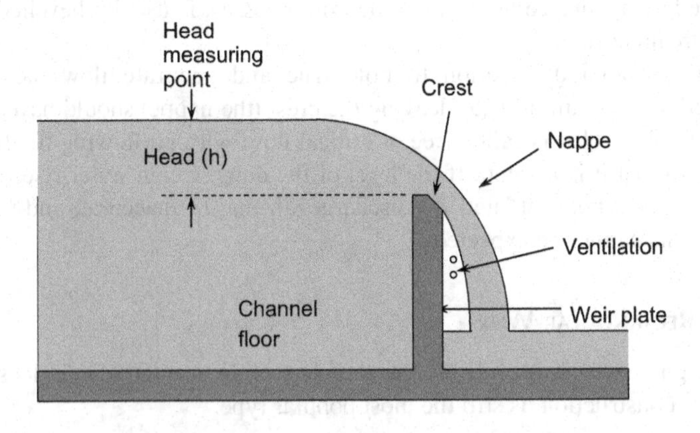

FIGURE 13.2 For accurate flow measurement, the nappe should have sufficient fall.

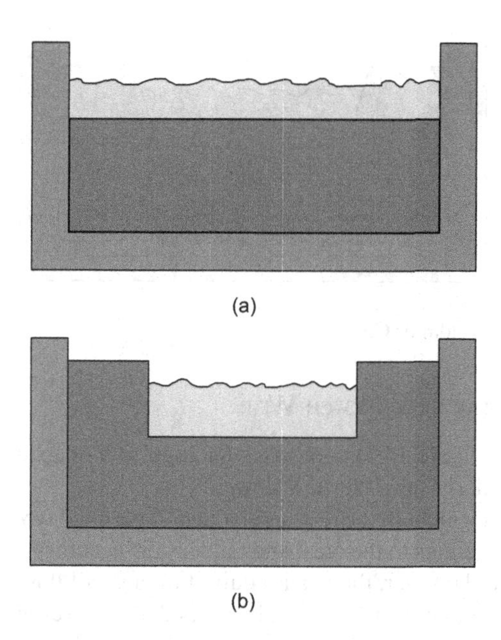

FIGURE 13.3 Rectangular weir (a) with no contraction and (b) with lateral contraction.

and accelerate the channel flow as it passes over the weir and provides the needed ventilation.

► In this case the discharge equation of such a restriction, with end contractions, becomes:

$$Q = k \cdot (L - 0.2 \cdot h) \cdot h^{1.5} \tag{13.2}$$

where
Q = flow rate
k = constant
L = length of crest
h = the head ◄

The rectangular weir can normally handle flow rates in the range of 1:20 from about 0–15 L/s up to 10,000 L/s or more (3 m crest length).

13.2.2 Trapezoidal (Cipolletti) Weir

In the trapezoidal type of weir (Figure 13.4) the sides are inclined to produce a trapezoidal opening. When the sides slope one horizontal to four vertical the weir is known as a Cipolletti weir and its discharge equation (head vs. flow rate) is similar to that of a rectangular weir with no end contractions:

$$Q = k \cdot L \cdot h^{1.5} \tag{13.3}$$

The trapezoidal type of weir has the same flow range as a rectangular weir.

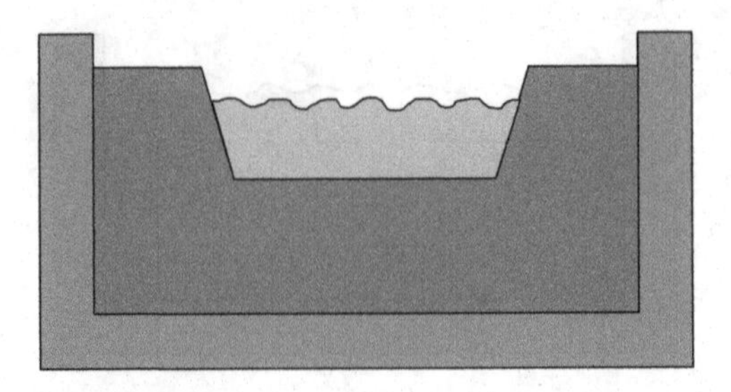

FIGURE 13.4 Trapezoidal or Cipolletti weir.

13.2.3 Triangular or V-Notch Weir

The V-notch weir (Figure 13.5) comprises an angular V-shaped notch – usually of 90° – and is particularly suited for low flows.

A major problem with both the rectangular and trapezoidal type weirs is that at low flow rates the nappe clings to the crest and reduces the accuracy of the measurement. In the V-notch weir, however, the head required for a small flow is greater than that required for other types of weirs and freely clears the crest – even at small flow rates.

► The discharge equation of the V-notch weir is given by:

$$Q = k \cdot h^{2.5} \tag{13.4}$$

where

Q = flow rate
k = constant
h = the head

This equates to a 0.4% change in height for a 1% change in flow. ◄

V-notch weirs are suitable for flow rates between 2 and 100 L/s and, for good edge conditions, the flow range is 1:100. Higher flow rates can be obtained by placing a number of triangular weirs in parallel.

There is a high unrecoverable pressure loss with weirs, which may not be a problem in most applications. However with the operation of a weir, it is required that the flow

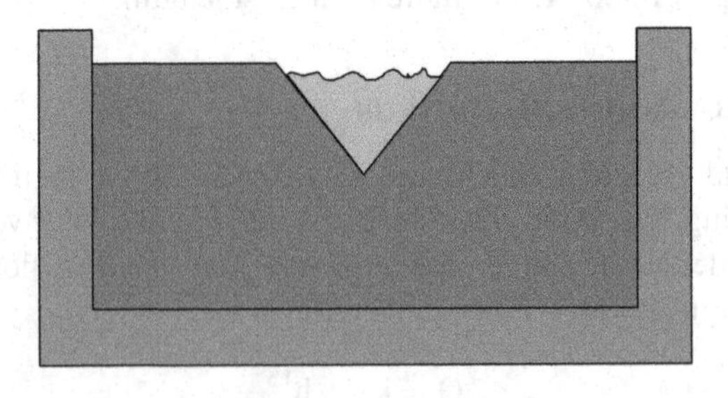

FIGURE 13.5 Triangular or V-notch weir.

clears the weir on departure. If the liquid is not free flowing and there is back pressure obstructing the free flow, then the level over the weir is affected and hence the level and flow measurement.

13.2.4 Advantages of Weirs

- Simple operation
- Good rangeability (for detecting high and low flows)

13.2.5 Disadvantages of Weirs

- Damning action can cause changes in the inflow region
- Damning action can produce silt build-up
- Accuracy of about 2%

13.3 THE FLUME

The second class of primary devices in general use is the flume (Figure 13.6). The main disadvantage of flow metering with weirs is that the water must be dammed, which may cause changes in the inflow region. Further, weirs suffer from the effects of silt build-up on the upside stream. In contrast, a flume measures the flow in an open channel in which a specially shaped flow section restricts the channel area and/or changes the channel slope to produce an increased velocity and a change in the level of the liquid flowing through it.

Major benefits offered by the flume include a higher flow rate measurement than for a comparably sized weir, a much smaller head loss than a weir, and better suitability for flows containing sediment or solids because the high flow velocity through the flume tends to make it self-cleaning.

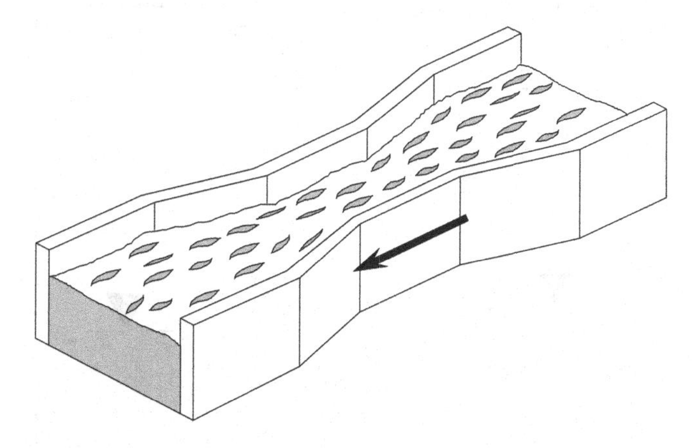

FIGURE 13.6 Basic flume in which a specially shaped flow section produces an increased velocity and a change in the liquid level.

The major disadvantage is that a flume installation is typically more expensive than a weir.

13.3.1 Flume Flow Consideration

An important consideration in flumes is the state of the flow. When the flow velocity is low and is due mainly to gravity, it is called tranquil or sub-critical. Under these conditions, it is necessary to measure the head in both the approach section and in the throat in order to determine the discharge rate.

As the flow velocity increases and the inertial forces are equal to or greater than the gravitational force, the flow is termed critical or supercritical. For both critical and supercritical states of flow, a definitive head/discharge relationship can be established and measurement can be based on a single-head reading.

13.3.2 Venturi Flume Meter

The most common flume is the Venturi flume (Figure 13.7) whose interior contour is similar to that of a Venturi flow tube with the top removed, normally consisting of a converging section, a throat section, and a diverging section.

The rectangular Venturi flume, with constrictions at the side, is the most commonly used since it is easy to construct. In addition, the throat cross-section can also be trapezoidal or U-shaped. Trapezoidal flumes are more difficult to design and construct, but provide a wide flow range with low pressure loss. A U-shaped

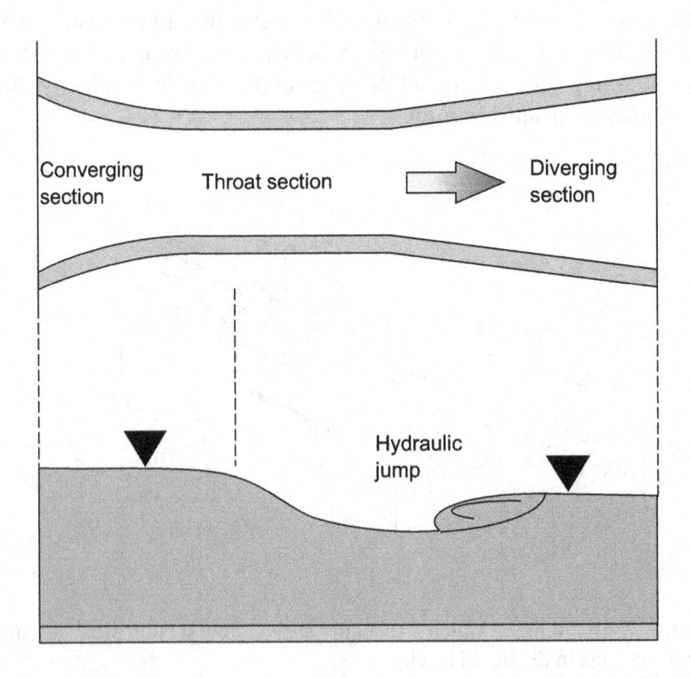

FIGURE 13.7 Rectangular Venturi flume with constrictions at the side.

section is used where the upstream approach section is also U-shaped and gives higher sensitivity – especially at low (tranquil) flows.

► Although the theory of operation of flumes is more complicated than that of weirs, it can be shown that the volume flow rate through a rectangular Venturi flume is given by:

$$q = k \cdot h^{1.5} \tag{13.5}$$

where

q = the volume flow
k = constant determined by the proportions of the flume
h = the upstream fluid depth ◄

13.3.3 Parshall Venturi Flume

The Parshall Venturi flume (Figure 13.8) differs from the conventional flat-bottomed Venturi flume in that it incorporates a contoured or stepped floor that ensures the transition from sub-critical to supercritical flow. This allows it to function over a wide operating range whilst requiring only a single-head measurement. The Parshall Venturi flume also has better self-cleaning properties and relatively low head loss.

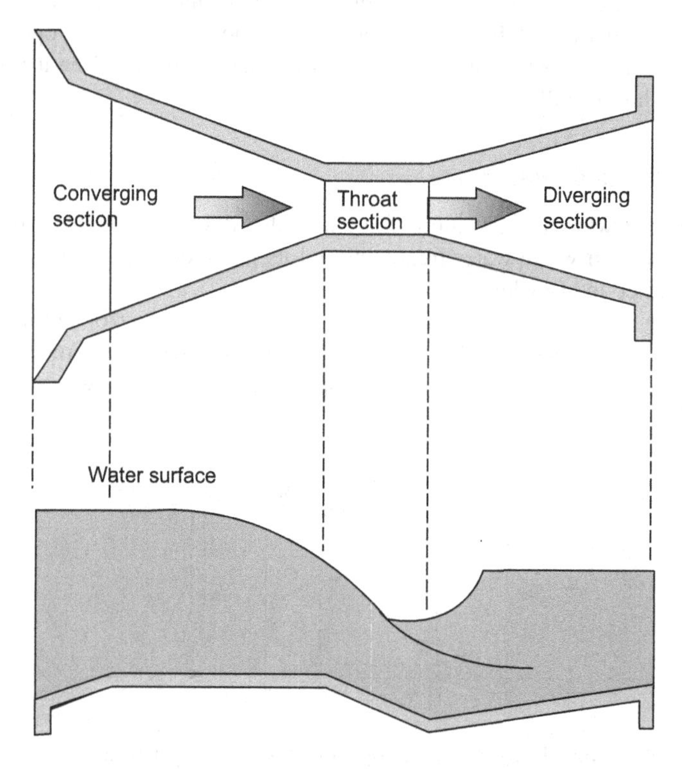

FIGURE 13.8 Parshall Venturi flume incorporates a contoured or stepped floor.

Parshall Venturi flumes are manufactured in a variety of fixed sizes and are usually made of glass fibre-reinforced polyester. The user only needs to install it in the existing channel.

► Because of its slightly changed shape, the discharge equation of the Parshall Venturi flume changes slightly to:

$$q = k \cdot h^{n} \tag{13.6}$$

where

q = flow rate
h = the head
k and n = constants determined by the proportions of the flume

Generally, the exponent n varies between 1.522 and 1.607, determined mainly by the throat width. ◄

13.3.3.1 Application Limits of Venturi Flume Meter

Providing excellent self-cleaning properties, the Venturi flume has replaced the weir in most applications, and the Parshall flume is, at present, possibly the most accurate open channel flow-measuring system with flow ranges from 0.15 to 4,000 L/s.

The advantages include reliable and repeatable measurements, no erosion, insensitive to dirt and debris, low head pressure loss, and simple operation and maintenance. However it is more expensive than the rectangular Venturi flume and more difficult to install.

13.3.4 Palmer Bowlus

The Palmer Bowlus flume (Figure 13.9) was also developed in the United States in 1936 for use in waste-water treatment and its name derives from the inventors, Messrs. Palmer and Bowlus.

As shown it comprises a U-sectioned channel having a trapezoidal throat section and a raised invert. Its main advantage is its ability to match up to circular pipes and it can be fitted inside the existing pipes in special applications; flow ranges from 0.3 to 3,500 L/s.

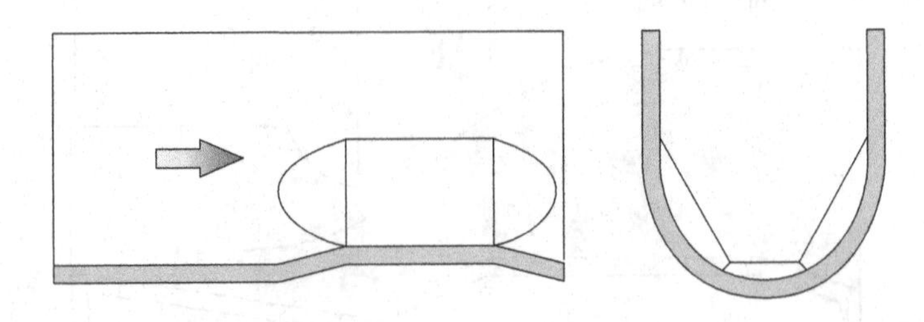

FIGURE 13.9 Palmer Bowlus flume comprises a U-sectioned channel having a trapezoidal throat section and a raised invert. (Courtesy: Neuplast.)

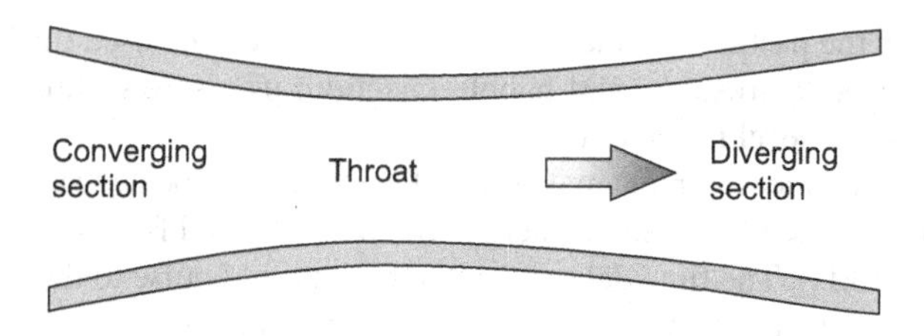

FIGURE 13.10 Khafagi flume does not have a parallel throat section.

13.3.5 Khafagi Flume

Similar to the Venturi flume the Khafagi flume (Figure 13.10) does not have a parallel throat section. Instead, the throat section is that point at which the inlet section meets the curve of the divergent discharge section. The floor is horizontal throughout its length. The flow range is from 0.25 to 1,500 L/s.

13.4 LEVEL MEASUREMENT

Whilst a weir or a flume restricts the flow and generates a liquid level which is related to the flow rate, a secondary device is required to measure this level.

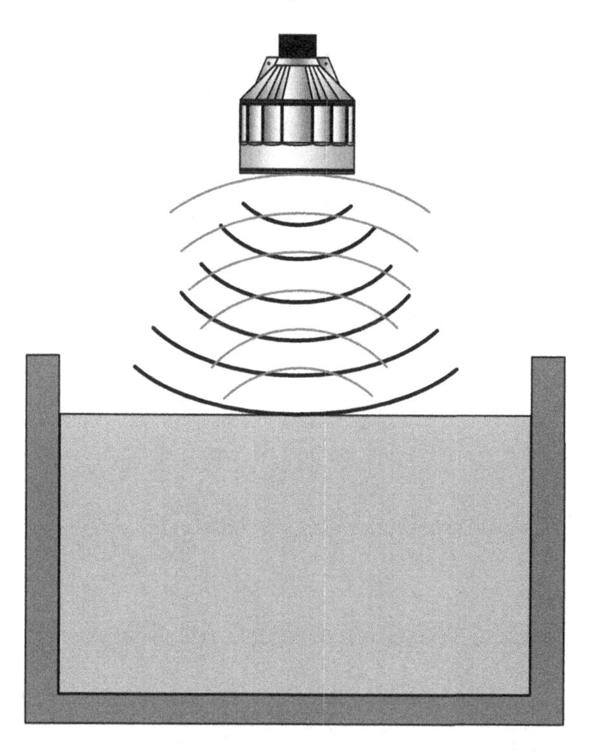

FIGURE 13.11 Ultrasonic level measurement uses a transducer mounted above the channel, which transmits a burst of ultrasonic energy that is reflected from the liquid surface. (Courtesy: Milltronics, Siemens.)

Although in the past several measuring methods have been used (including floats, capacitive probes, hydrostatic, and bubble injection) use is now almost universally made of ultrasonic level measurement.

Ultrasonic level measurement makes use of a transducer, located above the channel that transmits a burst of ultrasonic energy which is reflected from the surface of the water (Figure 13.11). The time delay from the transmitted pulse to the received echo is converted into distance and hence determines the liquid level.

Ultrasonic sensors have no contact with the liquid, are easy to install, require minimal maintenance, and are not affected by grease, suspended solids, silt, and corrosive chemicals in the flow stream.

Modern ultrasonic systems are also capable of providing very high level measuring accuracies (down to ±0.25%).

Open channel flow measurement does not end with the measurement of level, since it still remains to convert the measured liquid level into a corresponding flow rate.

In modern level measuring instruments, linearisation is carried out in software in which a wide range of different compensating curves are stored in the instrument's memory. During commissioning of the system, users may then access the correct curve – dependent on the type and dimensions of the weir or flume.

14 Multiphase Flow Metering

14.1 INTRODUCTION

Multiphase flow metering (MPFM), along with water-cut metering, is a technology that is very specific to the oil and gas industry.

However, the term 'multiphase' is somewhat misleading because it covers both multi-component and multiphase. Thus dirty gas, air-in-water, cavitational, and steam flows may all be referred to as multiphase flows within this broad use of the term.

During its life the output components of an oil well change dramatically. This is illustrated in Figure 14.1 which shows how the delivery of oil and gas decreases over the life of the well with a commensurate increase in the delivery of water.

In its early life a typical well might deliver 90% oil and less than 10% water. Ideally the gas cap forces the produced fluids out under pressure for some period of time.

In midlife, as the gas cap pressure falls, pumps may be required to bring the fluids to the surface, which might now comprise about 50% oil and 50% water.

In later life, with almost no gas, require artificial lift in order to raise fluids to the surface and obtain maximum oil recovery which may now possibly comprise less than 20% oil and 80% water. Artificial lift systems include rod pumping using a pumping jack (nodding donkey) as a prime mover, gas lifting (gas injection), hydraulic pumping (water injection), and centrifugal pumping. A combination of gas and/or water injection and pumps might be required to bring the fluids to the surface.

14.2 VERTICAL THREE-PHASE FLOW

Consider first the flow from an oil well – making the assumption that this is vertical* (Figure 14.2).

The crude oil will reach the well-head, having flowed up a pipe of about 100 mm bore for distances of several kilometres or about 30,000 pipe diameters. The flow should, therefore, be fully developed.

Initially, the flow will be single-phase – essentially oil only. As oil is removed from the well, the pressure in the reservoir will decrease, and the gas fraction in the well flow line will increase and appear as gas bubbles. This is known as bubbly flow.

With further ageing, the gas bubbles will become larger and the water content, present as droplets, is likely to increase. Yet further ageing will result in slugs of gas that travel up to the centre of the pipe leaving a slower layer of moving liquid on the

* Although not strictly true, it will provide some basic concepts.

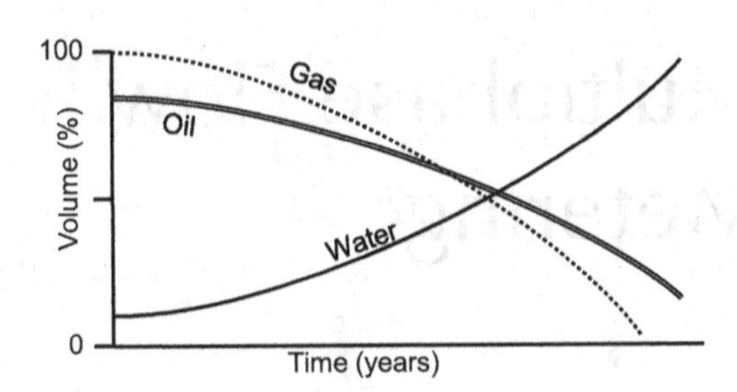

FIGURE 14.1 Decrease in the delivery of oil and gas over the life of the well with a commensurate increase in the delivery of water.

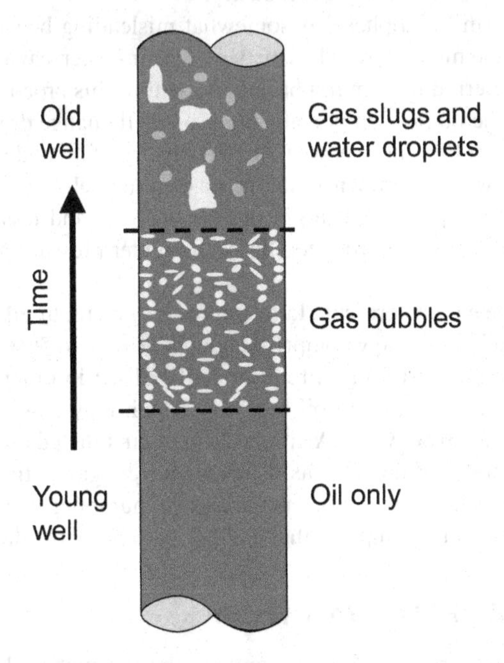

FIGURE 14.2 Initially, the flow will be single-phase oil only. Later, as the gas fraction increases, gas bubbles will appear. Yet further ageing will result in slugs of gas combined with water droplets.

wall. These slugs tend to overtake each other forming larger slugs, typically up to 20 m long.

14.3 MULTIPHASE FLOW

A multiphase mixture of three different components (oil, water, and gas) is a complex phenomenon producing a variety of flow regimes whose distribution, in both space and time, differ from each other. Several different types of flow regimes are illustrated in Figure 14.3 and can be broadly grouped into dispersed flow (bubble flow), separated (stratified and annular), and intermittent (piston and slug flow). Because they are

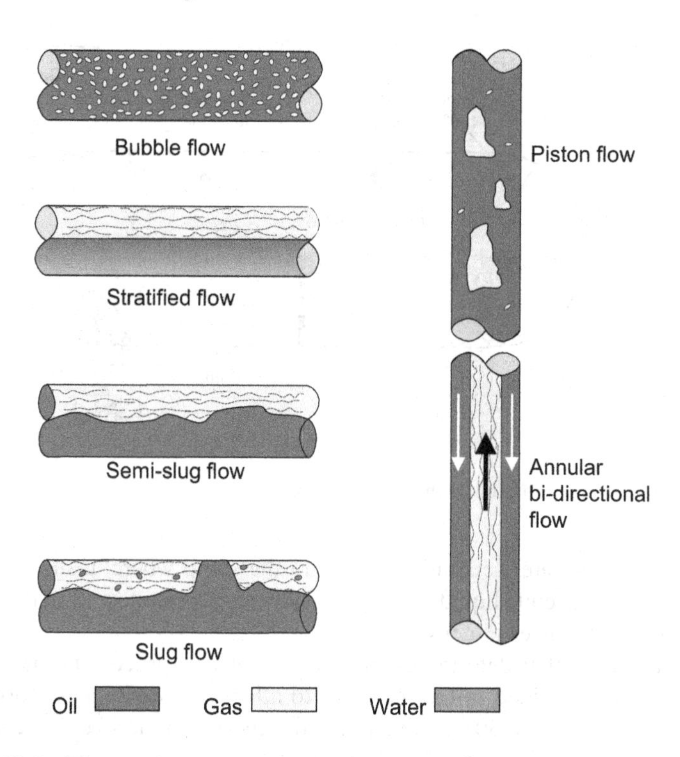

FIGURE 14.3 Flow regimes can be broadly grouped into dispersed flow (bubble flow), separated (stratified and annular), and intermittent (piston and slug flow).

dynamic, these flow regimes fall completely outside the control of an engineer or operator and are difficult to predict or model.

14.4 SEPARATION

At the well-head surface, these three constituents (oil, gas, and water) are separated into their respective components using a series of two, three, or more separators. A typical separator (as illustrated in Figure 14.4) involves three principles to realise physical separation of the oil, gas, and water, such as momentum, gravity settling, and coalescing.

In the representative system shown, the flow through the inlet nozzle impinges on a diverter baffle that produces an abrupt change of direction. Since the momentum of the heavier oil/water phase particles is greater than that of the gas, they cannot turn as rapidly and separation occurs. Enhanced separation of the entrained droplets occurs in the gravity section where the gas moves at a relatively low velocity – with little turbulence. If the gravitational force acting on the droplet is greater than the drag force of the gas flowing around the droplet, liquid droplets will settle out of a gas phase.

In the coalescing section, a mist extractor (normally comprising a series of vanes or a knitted wire mesh pad) removes the very small droplets of liquid from the gas by impingement on a surface where they coalesce.

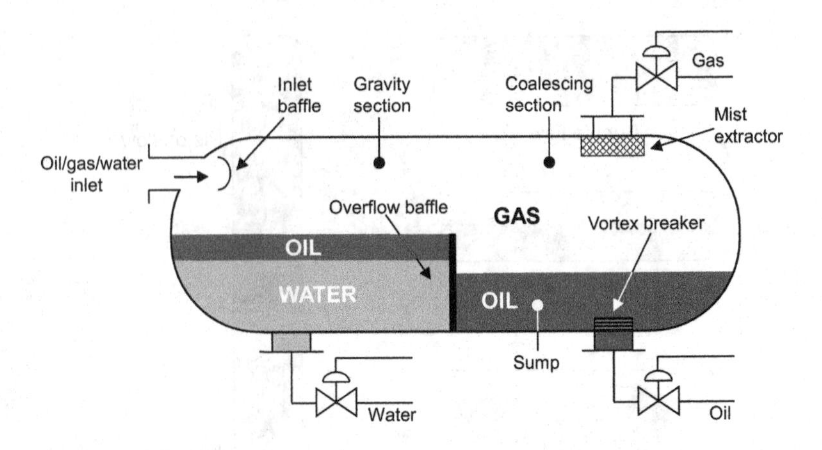

FIGURE 14.4 Typical horizontal three-phase separator.

And finally, oil–water separation occurs in the sump where, using an overflow baffle, the two immiscible liquid phases (oil and water) separate within the vessel by virtue of their differences in density.

A critical factor that determines the efficacy of this process is that sufficient retention time be provided in the separator to allow for gravity separation to take place. This can vary from 30 s up to 3 min – depending on the size of the separator, the inflow, the gas/oil ratio, and the oil/water ratio.

The oil output from a first stage separator should typically contain 1–3% water with (possibly) up to 0.5% gas. The water output should typically contain less than 300 ppm of oil. To achieve this balance, tight control of the three component outflows must be applied which must be balanced against the inflow. And this, quite simply, is where the problem lies – accuracy and reliability, simultaneous measurement of the input multiphase flow.

14.5 SEPARATION-TYPE FLOW METERING

The technology required to measure any one of these regimes is already complex. A further complicating consideration is that the multiphase mixture pressure may vary from almost 0 to 2,000 bar and the temperature can vary from −40°C to 200°C.

For a single technology to cover them all, borders on the insurmountable. On this basis, the traditional method is to split the multiphase steam into discrete phases, using a test separator (Figure 14.5), with the measurement taking place on a single-phase (or in the case of oil–water mixtures, two phase) stream. The compact upstream separation device provides a relatively liquid-free gas stream and a liquid stream – with each metered separately.

The metering methods most commonly used are:

Gas: Orifice plate and turbine
Liquid: Orifice plate, turbine, and Coriolis

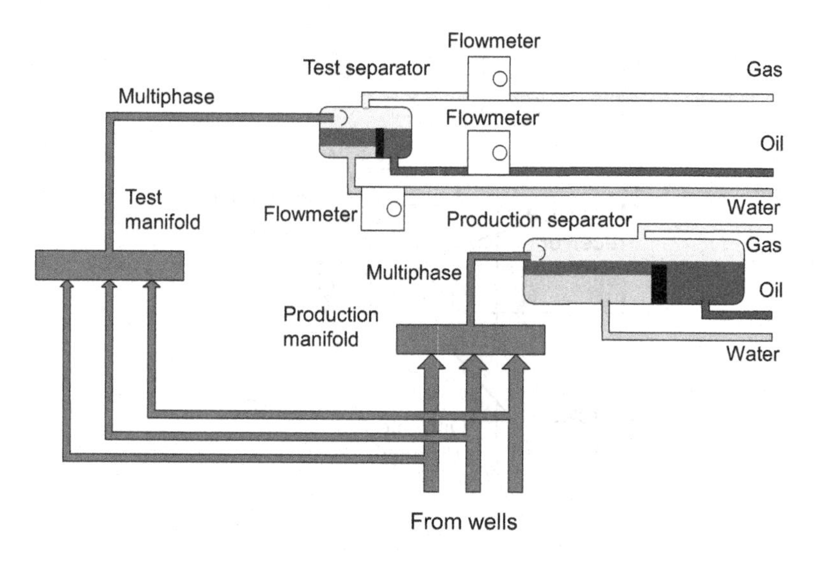

FIGURE 14.5 Traditional method is to split the multiphase steam into discrete phases using a test separator.

A major problem with this method lies in separating the production stream into its component parts. Indeed, rarely can complete phase separation occur and entrapment of phases within each other is common.

14.6 IN-LINE MULTIPHASE FLOW METERING (MPFM)

Generally, full multiphase stream-measuring systems make use of two or more sensors that combine the data to yield individual phase flow rates. A major difficulty in exploring the technologies used in MPFM is that manufacturers are inclined towards a high degree of secrecy in order to preserve their competitive edge in what could prove to be a highly rewarding and lucrative market. Estimates indicate that installed costs of a single MPFM can range from US$100,000 to US$500,000 – depending on the size and application.

One such system, PhaseWatcher, from Schlumberger, comprises a Venturi tube combined with a nuclear dual-energy fraction meter (Figure 14.6). The Venturi tube is fitted with differential pressure, static pressure, and temperature-measuring transducers for temperature-corrected flow measurement. The nuclear-measuring section comprises a barium source and a dual-energy fraction detector to measure two different photon energy levels – high energy for measuring the density and low energy to calculate the water–oil ratio.

Other technologies include a combination of a microwave sensor and positive displacement flowmeter (Agar), capacitance and inductance sensors (FlowSys, Roxar), and Venturi tube and nuclear densitometers (Schlumberger, Haimo, Roxar).

Extensively tested by a large number of oil companies, conventional wisdom indicates that, despite claims by the vendors, no one solution appears to be totally satisfactory. The author's spot opinion poll, conducted amongst more than 20

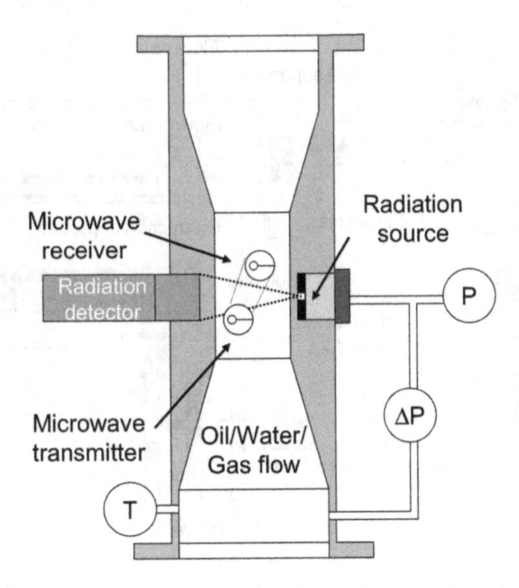

FIGURE 14.6 PhaseWatcher comprises a Venturi tube combined with a nuclear dual-energy fraction meter. (Courtesy: Schlumberger.)

current, and former facility (topside offshore platforms) engineers and operational staff, indicates a deep cynicism regarding the claims by many vendors and that often calibration is referenced back to both the empirical experience, and sometimes even priori feelings and instincts of the operational staff.

14.7 WATER-CUT METERING

The water-cut is the ratio of the water which is produced in a well compared with the volume of the total liquids produced. As the field fills with water, a mix of oil and water flows out of the well. The percentage of water at these wells is called the water-cut.

Water-cut meters measure the water content (cut) of an oil/water mixture, expressed in % by volume, and are typically used in the oil and gas industry to measure the water-cut of oil flowing from a well, produced oil from a separator, crude oil transfer in pipelines, and in loading tankers. As with MPFM, several technologies are available including oscillatory, capacitive, microwave dielectric, and infrared (IR).

One such system, from Phase Dynamics, makes use of 'oscillator load pull'* in which a radio-frequency (RF) oscillator (150–500 MHz) sets up a standing wave within a resonant chamber through which the oil/water mixture flows.

Because the relative permittivity (ε_r) of water (68–80) and oil (2.5) is very different, changes in the water-cut vary the velocity of the RF, which in turn, changes the phase. Ultimately, the phenomenon of 'oscillator load pull' changes the oscillator frequency itself – depending on the water content.

* 'Oscillator load pull' is a measure of how much an oscillator changes its frequency when the load that is connected to it changes.

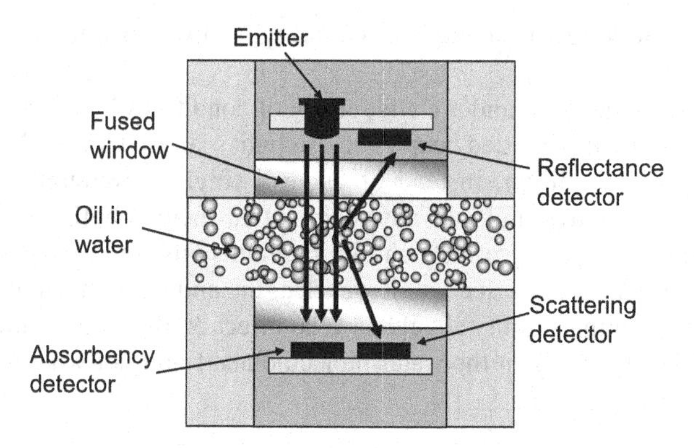

FIGURE 14.7 Red Eye water-cut meter measures the volumetric proportion of oil in a mixture of oil and water by passing a beam of IR light through the stream. (Courtesy: Weatherford.)

The analysers are available in a number of range options such as low (0–4%), full (0–100%), and high (80–100%) with repeatability in the lower ranges down to ±0.02%.

The Roxar Watercut meter again compares the permittivities of water and oil but using microwave technology in which the water molecules continuously attempt to align themselves with the changing microwave field which, in turn, slows down the propagation of the microwaves. Because hydrocarbon molecules have a much more symmetrical structure, and do not respond to the changing microwave field in the same way, they have an insignificant effect on the microwave propagation. Again, the analysers are available in a number of range options such as low (0–1%), full (0–100%), and high (15–100%) with repeatabilities in the lower ranges down to ±0.01%.

Another system, the Red Eye water-cut meter from Weatherford measures the volumetric proportion of oil in a mixture of oil and water by passing a beam of IR light through the stream (Figure 14.7).

The system relies on the preferential absorption of IR radiation in the 'near' region in which, at the operating frequency of the sensor, water is the transmitting phase while oil is the attenuating medium (Figure 14.8). Water transmits close to

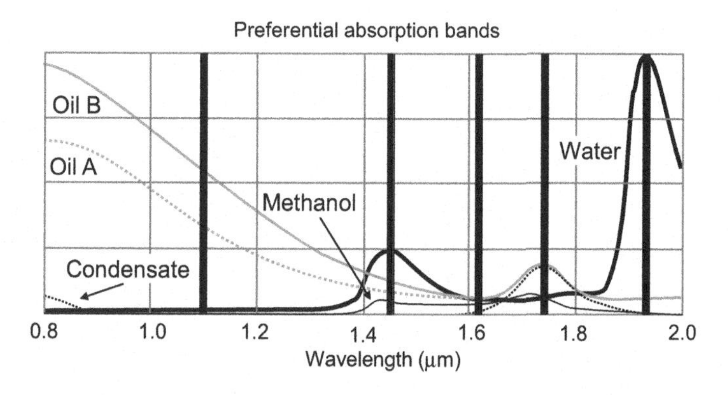

FIGURE 14.8 Spectral properties of two crude oils compared with condensate and water. (Courtesy: Weatherford.)

100% of the emitted radiation while crude oil typically transmits less than 10% of the light.

Accuracy is maintained under a wide range of conditions by taking into account not only the directly transmitted beam, but also light scattered forward and backward across the gap. In addition, instead of using a single wavelength, the analyser simultaneously measures multiple wavelengths that include both the water and oil absorbent peaks. Scattering effects caused by emulsions, sand, or gas bubbles are expected to have the same effect at all wavelengths and as such can be eliminated. Furthermore, changing salinity should have no effect on the measurement, since the water absorption is based on the water molecule itself and not what is dissolved in the water.

15 Meter Proving

15.1 INTRODUCTION

In the field of metrology the terms calibration, verification, proving, and validation are not only frequently confused but are frequently misused.

Calibration involves connecting the meter to be calibrated – usually referred to as the Meter Under Test (MUT) – to a recognised and accredited test facility (calibration rig). The results are compared and the deviations are noted in a calibration report. These corrections may, or may not, be used to adjust the instrument.

Verification, also known as *proving*, is a process carried out periodically to verify their accuracy (or original deviation) and is usually carried out in the field with the use of stationary or mobile provers (ball, small volume, or mobile). Unlike calibration, proving does not involve making any type of adjustment but rather producing a set of correction figures.

Finally, *validation* involves ensuring that all parts of the metering system are working correctly. Typically, this involves self-diagnostic programmes that perform a series of sensor and electronic tests that indicate that the unit is performing as originally calibrated at the factory.

15.2 CALIBRATION

Calibration involves the use of either volumetric or gravimetric test facilities.

15.2.1 Volumetric Calibration

In volumetric calibration the quantity of liquid flowing through the MUT is carried out by collecting a known volume of liquid in a container having a calibrated volume. As illustrated in Figure 15.1, the rig typically comprises a calibrated reservoir tank with either an integral scale to read the fill height or limit switches that stop the inflow of liquid as soon as a predetermined level is reached.

The system shown makes use of what is termed a flying-start-and-finish method in which the flow through the flow meter is first allowed to stabilise by diverting the flow return to the supply and collection tank.

At the start of the test run, the diverter valve directs the flow into the calibrated reservoir tank and, simultaneously, starts a timer. When the level reaches a predetermined volume, the flow is again diverted back to the collection tank and the timer is stopped. The accumulated values of the pulses from the MUT are then compared and the deviation or error is recorded and, if necessary, the meter is adjusted.

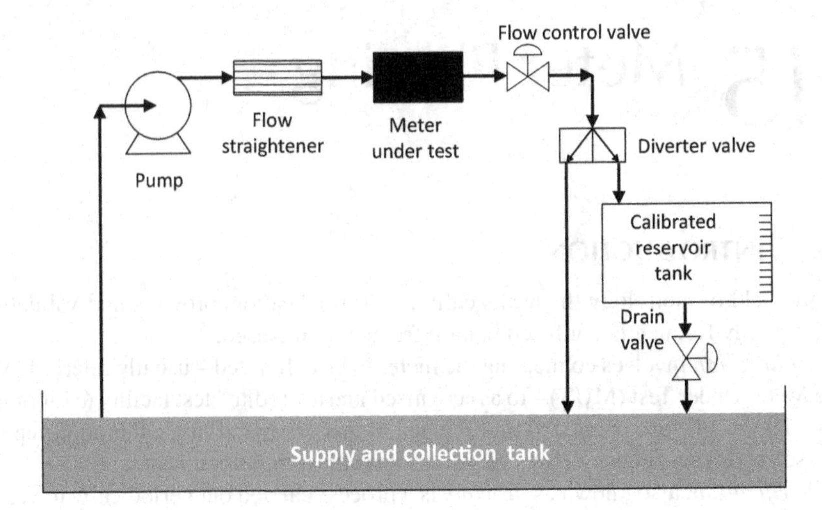

FIGURE 15.1 In volumetric calibration the quantity of liquid flowing through the MUT is carried out by collecting a known volume of liquid in a container having a calibrated volume.

15.2.2 Gravimetric Calibration

Gravimetric measurement methods are used predominantly in modern calibration facilities. A gravimetric test rig is shown in Figure 15.2 in which the quantity of liquid collected in the reservoir tank is determined by its weight rather than its volume.

A common misconception when weighing in air is that the air has an insignificant effect on the measurement result. In reality the result is influenced by the prevailing air density and the mass of the reference weight (density of 8,000 kg/m³) at 20°C should be balanced against the air (density 1.2 kg/m³).

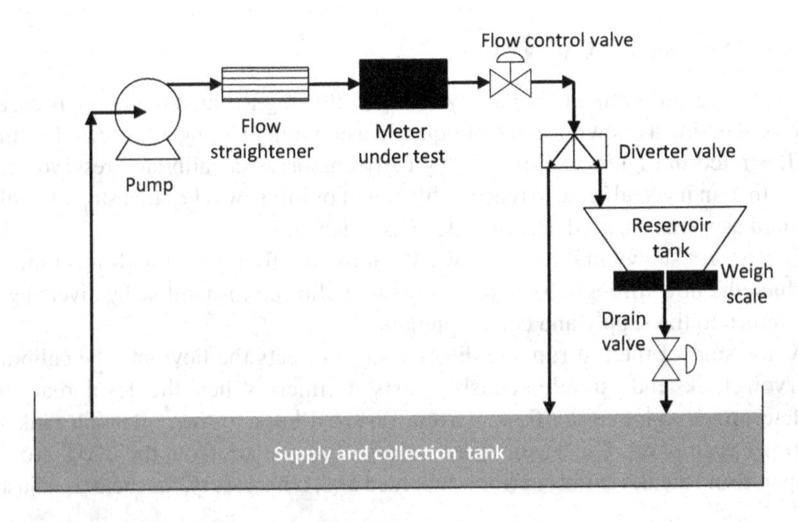

FIGURE 15.2 In a gravimetric test rig the quantity of liquid collected in the reservoir tank is determined by its weight rather than its volume.

► Consequently, the weight must be corrected for the effect of air buoyancy:

$$M = W \cdot \left[\left(1 + \rho_{air} \cdot \left(\frac{1}{\rho_f} - \frac{1}{\rho_w}\right)\right)\right] \quad (15.1)$$

where

M = the mass (kg)
W = measured weight (kg)
ρ_{air} = density of air (kg/m^3)
ρ_f = density of the fluid (kg/m^3)
ρ_w = density of calibration weights (typically 8,000 kg/m^3) ◄

The calibration procedure is similar to that used for a volumetric test. From a tank of sufficient volume, water is circulated at a flow rate equivalent to the required calibration point of the MUT. Once this nominal flow rate has been achieved, a valve diverts the water into a weigh vessel, which is mounted on load cells. Simultaneously, as the diverter valve changes over, an electronic gate is initiated and the precision frequency counters start to measure the signal derived from the weigh vessel and also the MUT.

Once a sufficient quantity of liquid has been collected in the tank, the diverter valve closes, the system goes back into re-circulation mode and the electronic gates are closed.

15.3 ON-SITE CALIBRATION OR 'PROVING'

In practice, the most desirable method would be on-site calibration in which the field device is 'proved' in its operating location.

This would indicate the installed performance of the instrument which could differ from the original manufacturer specification due to the installed effects such as asymmetrical flow profile or swirl caused by upstream bends and fittings within a flow meter.

Again, in practice, there are two on-site activities that are prevalent:

- Actual verification or proving, making use of skilled personnel, the right tools, and Standard Operating Procedures (SOP) to ensure consistency within the work; and
- Validation or 'Finger-printing' using a hand-held configuration device or PC with appropriate software that records any deterioration and allows an estimate to be made as to when the particular unit has to be removed for repair or replacement. This procedure also allows the meter characteristics to be adjusted back to the original values (within limits) and the factory calibration accuracy to be restored.

Provers are available in different forms such as:

- Tank provers
- Displacement pipe provers
- Piston provers
- Master meters

15.4 OPEN- AND CLOSED-TANK PROVERS

A tank prover is an open- or closed-volumetric measure that may have a graduated top neck or graduated bottom neck. The inside dimensions are very accurately known and the construction needs to be sturdy enough so that when it is full of liquid the steel will not lose its shape. The density and volume of the proving medium (water) are accurately known and temperature compensated.

To prove a meter with an open-tank prover as shown in Figure 15.3, the liquid flows through the meter and then into the prover. The amount registered on the meter is compared with the amount shown by a sight glass – which should be read to the nearest 0.2% of the total prover volume.

The disadvantage of using a static prover is that the entire volume of a single-proving run must be contained within the proving vessel. Consequently, the prover must be capable of containing at least 1 min of flow at the maximum-rated capacity of the meter.

Open-neck provers are commonly available up to about 2,500 L although larger sizes (up to about 4,500 L) are available. Above this size, they become too large to be practical.

Closed-tank provers may be portable or statically mounted. The major difference is that the static pressure is compensated for in a similar way to the hydrostatic pressure on a level measurement application.

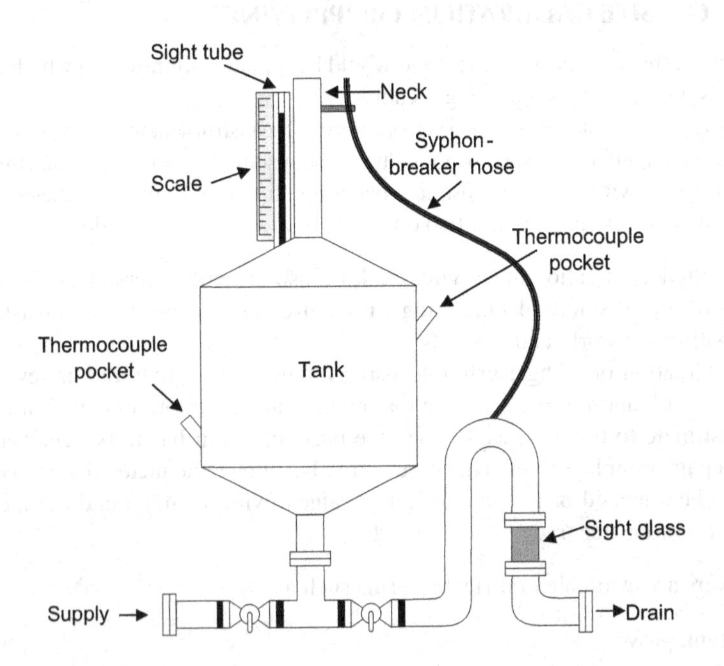

FIGURE 15.3 To prove a meter with an open-tank prover, liquid flows through the meter and then into the prover. The amount registered on the meter is compared with the amount shown by a sight glass.

15.5 DISPLACEMENT PIPE PROVERS

Essentially, a displacement pipe prover comprises a length of pipe, having an accurately determined internal volume and contains (typically) a ball displacer. During a calibration run, the meter and prover are connected in series so that the volume swept out by the ball displacer is compared with the volume registered by the meter. As the displacer enters the calibrated pipe section, it trips a detector that initiates the count of the pulses from the meter under calibration. A second detector is tripped when the displacer reaches the end of the calibrated pipe section – thus stopping the meter pulse count.

Essentially, there are two types: unidirectional and bidirectional.

The most common type is the bidirectional, U-shaped system (Figure 15.4), in which the displacer travels through the measuring section in either direction by reversal of the flow stream.

Unidirectional provers (Figure 15.5) can operate with a much higher displacer velocity and may thus use a smaller diameter, longer-calibrated section. The increased length between the detector switches improves prover repeatability by reducing the significance of the non-repeatability of the switch.

Some low frequency pulse output meters, such as helical turbines, require the prover to accumulate 10,000 meter pulses in a prover pass.

To achieve an uncertainty of ±0.01%, at least 10,000 pulses should be collected during the proving pass. It should be noted that for some low frequency pulse output meters, such as helical turbines, this resolution is not attainable – not even when using very large volume sphere provers.

To gather at least 10,000 meter pulses between the detectors, provers require a sufficient volume. Figure 15.6 illustrates the concept of detection uncertainty where:

L = run length – distance between detectors
d_e = detection uncertainty

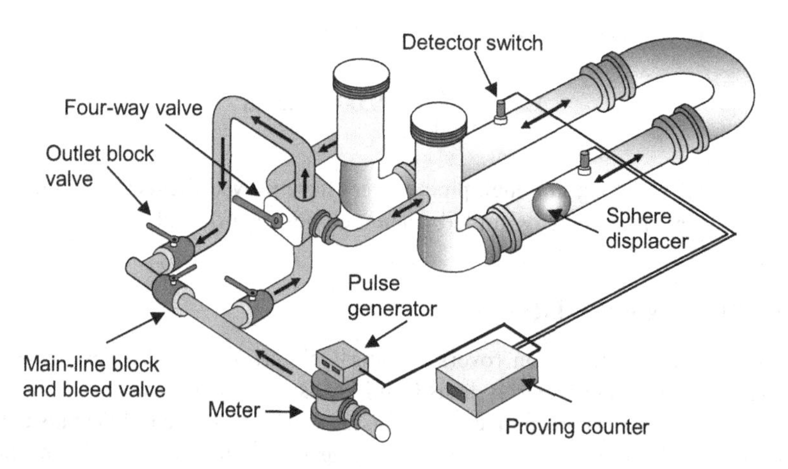

FIGURE 15.4 Most common displacement pipe prover is the bidirectional, U-shaped system.

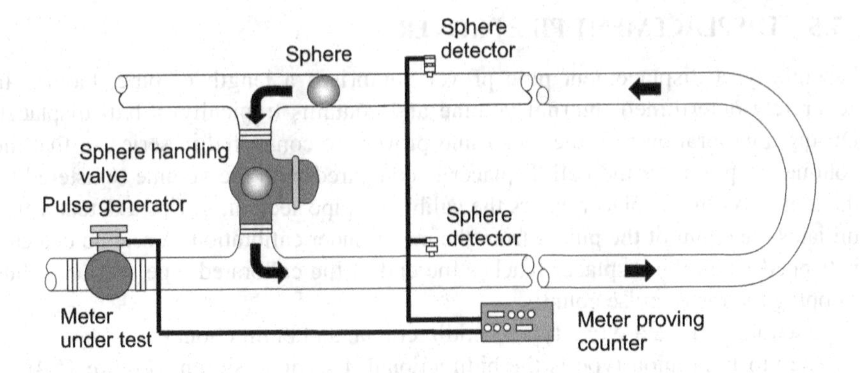

FIGURE 15.5 Unidirectional provers can operate with a much higher displacer velocity and may thus use a smaller diameter, longer-calibrated section.

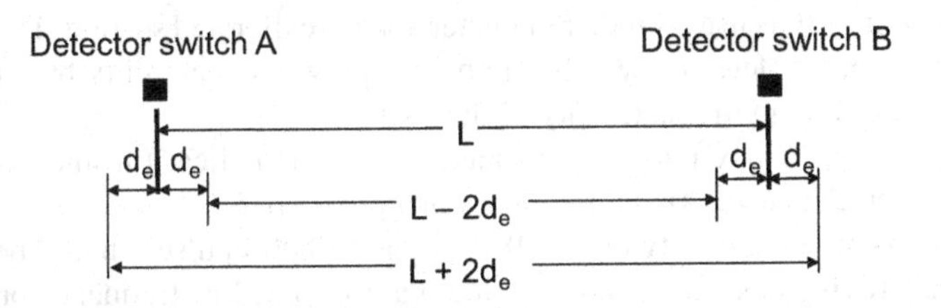

FIGURE 15.6 Concept of detection uncertainty.

What is the maximum possible difference in length (L) for a pass? Obviously, as illustrated, the maximum possible difference is $4 \cdot d_e$. And if the detection uncertainty is 0.5 mm, what is the minimum run length (L) required to obtain a 0.01% repeatability?

Total variation is:

$$4 \cdot d_e = 4 \cdot 0.5 = 2 \text{ mm} \tag{15.2}$$

Consequently, the total path length between the detectors is:

$$L = \frac{2 \text{ mm}}{0.01\%} = 20{,}000 = 20 \text{ m} \tag{15.3}$$

The consequence is that although pipe provers have proved themselves under static conditions they are extremely limited when it comes to portability.

15.6 PISTON PROVERS

The solution lies with piston rovers (formerly Small Volume Provers (SVPs) or Compact Provers) – now referred to as Captive Displacement Provers.

They comprise a piston connected to a piston rod – where the rod extends outside the barrel of the prover (Figure 15.7) and is usually fitted with precision optical switches having extremely high repeatability.

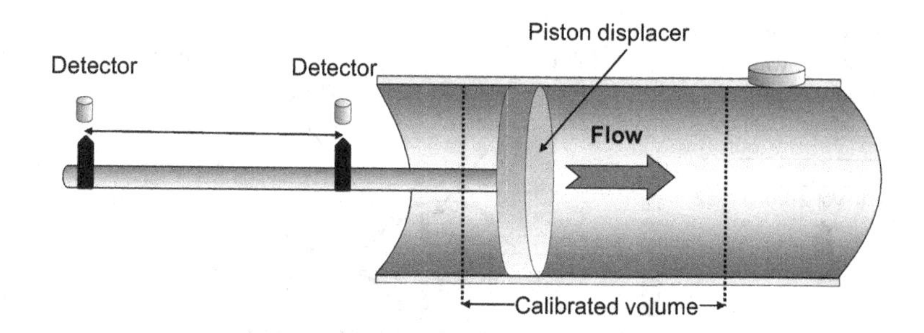

FIGURE 15.7 Piston prover comprises a piston connected to a piston rod – with the rod extending outside the barrel of the prover.

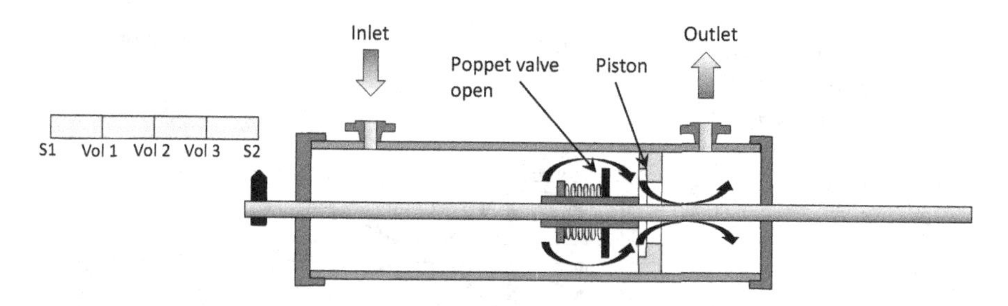

FIGURE 15.8 In the standby position the piston is in the downstream position, with the piston poppet valve held open by fluid flow through the prover.

The two detectors are mounted at a known distance (and therefore a known volume) apart. As the flow passes through the chamber, the displacer piston is moved downstream. Dividing the volume of the chamber by the time it takes for the displacer to move from one detector to the other gives the calibrated flow rate.

The operation is in several parts. In the standby position (Figure 15.8), the piston is in the downstream position, with the piston poppet valve held open by the fluid flow through the prover.

On generation of a prove command, the drive assembly starts pulling the piston assembly to the upstream position (Figure 15.9).

When the drive assembly has pulled the piston to the upstream position and comes in contact with the upstream switch S1 (Figure 15.10), a signal is sent to release the piston shaft.

Releasing the piston shaft is the start of the prove pass and, as the piston moves downstream synchronous to the fluid flow, the poppet valve closes (Figure 15.11).

The flag on the shuttle assembly actuates volume switches 1, 2, and 3 in that order (Figure 15.12). The piston then continues downstream to the standby state until another run command from the host flow computer.

15.6.1 Pulse Interpolation

As distinct from a 'conventional' prover that counts 10,000, or more, pulses from the meter being calibrated during one pass of the displacer, a 'small volume' prover

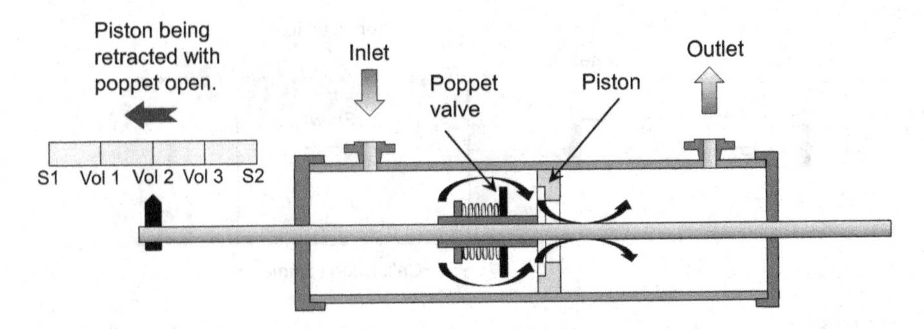

FIGURE 15.9 On generation of a prove command, the drive assembly starts pulling the piston assembly to the upstream position.

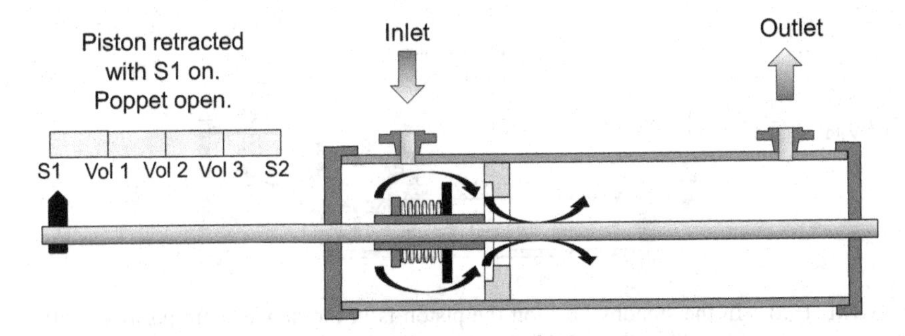

FIGURE 15.10 When the drive assembly has pulled the piston to the upstream position and comes in contact with the upstream switch S1, a signal is sent to release the piston shaft.

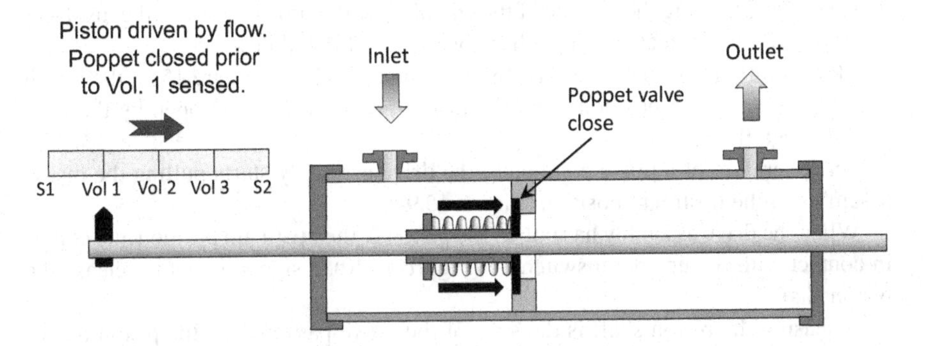

FIGURE 15.11 Releasing the piston shaft is the start of the prove pass and, as the piston moves downstream, the poppet valve closes.

is defined as one where less than 10,000 pulses are counted. Generally such provers have a volume of about one-tenth of a conventional design for the same duty and make use of pulse interpolation to provide the required 1 part in 10,000 resolution.

Pulse interpolation makes use of a timing method called double chronometry that effectively increases the resolution of a pulsed output by estimating the fraction of a pulse missed at the beginning and the end of a test.

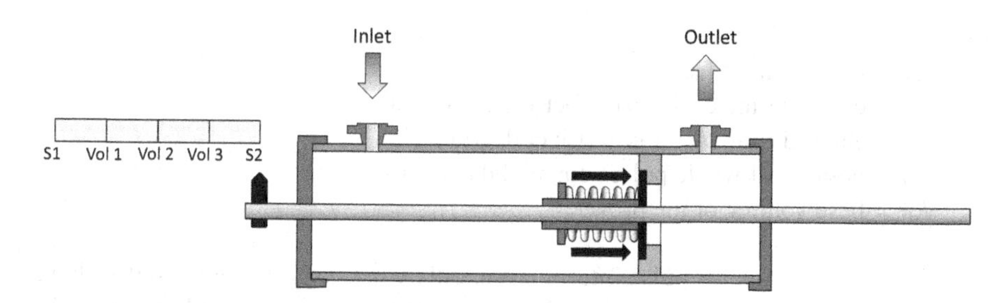

FIGURE 15.12 Flag on the shuttle assembly actuates volume switches 1, 2, and 3 in that order. The piston then continues downstream to the standby state until another run command from the host flow computer.

This is accomplished by first counting the whole number of pulses (Figure 15.13). The total whole number of pulses is then multiplied by the interpolation factor – the ratio of the time between the switches and the time between the first pulse after the start switch and the first pulse after the stop switch. The 'interpolated' pulse count is, therefore, no longer an integer but is expressed as a decimal number of pulses.

The formula to determine the number of full and partial pulses is:

$$\text{Actual pulses} = [n+1] \cdot \frac{T_1}{T_2} \tag{15.4}$$

$$K = \frac{T_1 \cdot (n+1)}{Vol \cdot T_2} \tag{15.5}$$

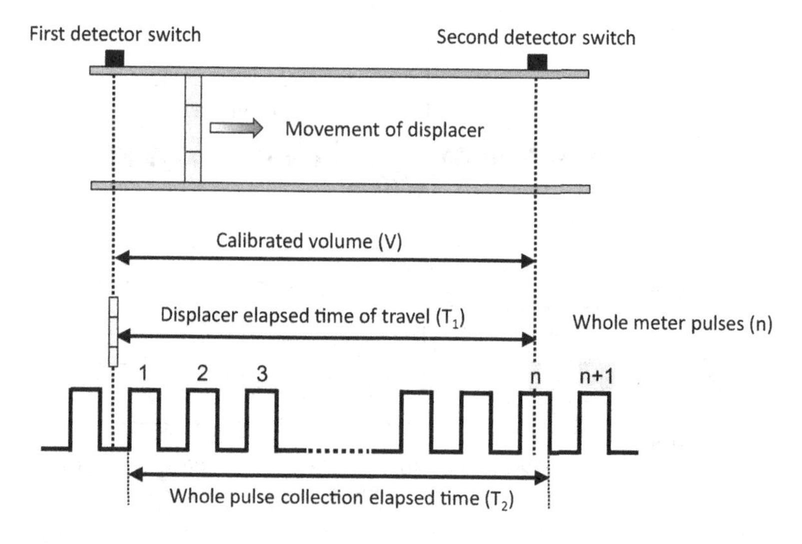

FIGURE 15.13 Pulse interpolation makes use of a timing method called double chronometry that effectively increases the resolution of a pulsed output by estimating the fraction of a pulse missed at the beginning and the end of a test.

where

K = interpolation factor
T_1 = elapsed time of displacer between detector switches
T_2 = elapsed time of whole pulse collection
n = number of whole pulses counted during T_2 timing
Vol = base volume of the prover

Double-chronometry pulse interpolation only works well when the pulses have a constant frequency or period. If the period of the individual pulses varies by more than 5–10%, lack of repeatability is found. Guidance is given in ISO 7278 Part 3.

15.6.2 Piston Launch

Although SVPs are fully accepted for *in-situ* calibration of positive displacement and turbine meters, their use in calibrating both ultrasonic and Coriolis meters has not produced satisfactory results. This is related to the principle of SVPs whereby when the piston is actively launched it creates a sudden pressure change that produces a sudden flow rate change called a 'flow rate spike' (Figure 15.14).

Due to their mechanical torque and delay in response time, mechanical flow meters cannot follow these flow rate variations and are consequently not influenced by these disturbances. However, ultrasonic and Coriolis flow meters respond immediately to any flow profile changes, and these disturbances will be 'seen' and measured.

Consequently, the short-term repeatability of ultrasonic and Coriolis meters appear poorer than for a turbine meter.

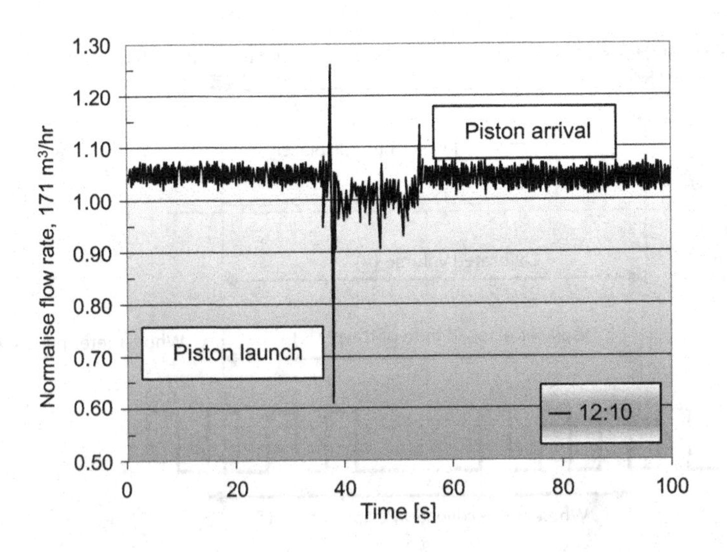

FIGURE 15.14 When the piston is actively launched in a SVPs it creates a sudden pressure change that produces a sudden flow rate change called a 'flow rate spike'. (Courtesy: Krohne.)

15.6.3 Latency

Mechanical-based meters, such as positive displacement and turbine, provide direct totalisation in terms of a direct count of the number of pulses produced per unit volume.

However, both ultrasonic and Coriolis-based meters calculate a quantity from the base technology measurement – with pulses being generated independently of the measurement and secondarily after the quantity determination.

This computational process, called *latency*, causes the meter's pulse output to be delayed beyond the actual time (real time) of the quantity being measured. Consequently, the meter pulses are out of sync with the prover indication. Commonly observed latencies vary from 250 ms to several seconds.

Any delay in meter pulses can produce a bias in the meter factor that varies depending on a number of factors – none more so than flow stability. With no flow instability during the proving process, there would be no bias, because the pulses collected would be the same before and after the proving. However, flow always varies to some degree and all provers vary the flow during their operation.

In the past, this has basically ruled out the use of SVPs for use with ultrasonic and Coriolis-based meters. However, the advent of Coriolis-based systems having latencies of the order of 10 ms has now rendered the use of SVPs a reality.

15.7 MASTER METER

Firstly we need to define our terms. The 'pay' (or 'duty') meter is the flow meter in normal operation that is used for custody or fiscal transfer measurement.

During a transfer, the pay meter may be directly compared with a series-connected 'check' flow meter having an identical performance (Figure 15.15). This would immediately reveal the presence of contamination because the 'check' meter will not have been exposed to the same degree of contamination since it is not continuously on-line.

As discussed previously, 'calibration' is the process of establishing the relationship between a 'pay' meter (the MUT) and the unit of measurement by comparing it with a 'master' flow meter having traceable characteristics to national/international defined measurement standards.

A rule of thumb is that the master meter needs to be three times more accurate than the MUT and may be either permanently installed or integrated on a mobile calibration unit.

15.7.1 Advantages of Using a Master Meter

- Installation is significantly simpler
- Overall cost reduction in the metering skid of up to 40%
- The proving run can be substantially longer using a larger volume and therefore more pulses from the units
- Maintenance requirements of the overall system are reduced
- The master meter can be removed periodically and sent back to a calibration facility

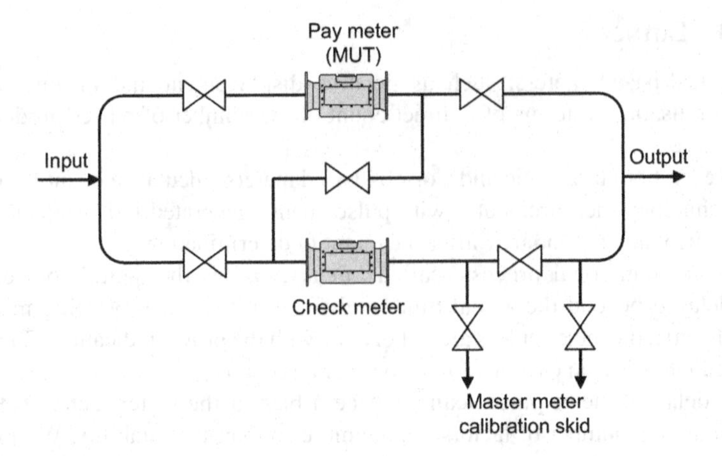

FIGURE 15.15 Pay meter directly compared with a series-connected 'check' flow meter having an identical performance.

- The overall size of the metering skid is significantly reduced
- Can be used in applications where flow cannot be interrupted
- Digital integration of information can be used to 'prove' the MUT
- Can be applied on a continuous basis so that deviation can be detected and acted upon

15.7.2 Recommendations when Using a Master Meter

An immediate concern of using a master meter in a transportable role is that it could be compromised by being transported from one calibration site to another – especially where climatic conditions can vary considerably.

The OIML R117 limits the ambient conditions to:

- Temperature: 15–35°C
- Humidity: 25–75% RH
- Barometric pressure: 84–106 kPa

15.7.2.1 Temperature

Although temperatures can exceed 40°C during the summer months in many countries, these are extremes and limited to specific areas. If calibration has to take place where temperatures in excess of 35°C are likely to be encountered, provisions can be made for such measurements to be taken in the early morning or late evening.

15.7.2.2 Humidity

Although many areas can have very high-humidity levels they rarely exceed 80% RH. The real question is what is the effect of RH on the measurement?

In fact research indicates that although high levels of humidity can lead to corrosion problems, this is unlikely to have any bearing on the main structure itself. Nonetheless, additional attention should possibly be paid to ensuring the integrity of any electrical/electronic connectors. Ideally, all printed circuit boards (PCBs) would be epoxy coated to prevent the ingress of moisture and make use of high-reliability PCB connectors. All external connectors should be rated to a minimum of IP 66.

15.7.2.3 Barometric Pressure

A barometric pressure of 84 kPa corresponds approximately to an altitude of 1,700 m at 35°C and 1,550 m at 15°C. And a barometric pressure of 106 kPa corresponds approximately to an altitude of −382 m at 15°C and −408 m at 35°C. Consequently, unless altitudes of 1,700 m or −408 m are likely to be exceeded, the effect of barometric pressure variations can be ignored.

15.7.2.4 Transportation of the Measurement Skid

Of possible concern is that careless handling during transit of the master meter skid could result in damage that is not immediately identifiable. It is recommended, therefore, that some form of impact and tilt monitoring and recording equipment is installed. This would not only alert users to any mishandling that might have occurred during the transportation process but would also raise awareness that special handling is required.

15.7.2.5 Air Eliminators/Deaerators/Air Release Traps

The amount of entrained gas is generally unknown. Since both turbine and ultrasonic meters derive the volumetric flowrate of a liquid based on the assumption that it is a single-phase fluid, any entrained gas, reducing the volume of liquid present, will give rise to a false indication of the actual liquid flow rate.

As a rule of thumb, a one-to-one relationship exists in which 0.1 vol.% of vapour bubbles give rise to an approximately 0.1% over-reading in the volume flow measurement.

Another potential problem is the reduction in strength of the received signal, by the receiving transducer, and therefore the signal-to-noise ratio.

A final potential problem is that some of the entrained gas could become trapped in the transducer ports and cause failure of the path.

With regards to turbine meters, entrained gas would again give rise to a false indication of the actual liquid flow rate. Furthermore, at high flow rates entrained gas can lead to a risk of cavitation and subsequent damage.

It is, therefore, strongly recommended to use air eliminators at each measuring point to minimise the risk of entrained gas causing problems.

16 Common Installation Practices

16.1 INTRODUCTION

In non-fiscal and non-custody transfer applications, flowmeters are rarely calibrated and are often left *in situ* for 10 or more years without any thought to their accuracy. Further, in too many instances, the initial installation is often so poorly undertaken, without any regards to basic installation practices, that it is highly unlikely that the meter in question ever met the manufacturer's stated accuracy. The data supplied by most manufacturers is based on steady flow conditions and installation in long straight pipes both upstream and downstream of the meter. In practice, most meter installations rarely meet these idealised requirements – with bends, elbows, valves, T-junctions, pumps, and other discontinuities all producing disturbances that have an adverse effect on the meter accuracy.

Both swirl and distortion of the flow profile can occur – either separately or together. Research has shown that swirl can persist for distances of up to 100 pipe diameters from a discontinuity whilst in excess of 150 pipe diameters can be required for a fully developed flow profile to form.

16.2 ENVIRONMENTAL INFLUENCES

The most important feature of a flow meter is that it should be sensitive to flow and as insensitive to environmental influences as possible. The most important environmental influences are discussed below.

16.2.1 Fluid Temperature

The temperature range of the fluid itself will vary considerably depending on the industry in which it is to be used:

- Food industry – 0–130°C to withstand CIP
- Industrial steam, water, and gases – 0–200°C
- Industrial superheated steam – up to 300°C
- Industrial outdoor usage – down to −40°C
- Cryogenics – down to −200°C

16.2.2 Pressure Pulsations

Pressure pulsations can be a problem when measuring liquids since, after they are created, they travel a long way down the pipeline without being significantly damped. In vortex meters for example such symmetrical pulsations could be detected as a vortex signal. The insensitivity to such 'common mode' pressure fluctuations should, therefore, be at least 15 Pa. Differential pressure flow measurement systems can be susceptible to common mode pressure variations if the connection systems on either side of the differential pressure cell are not identical and as short as possible.

16.2.3 Vibration

Vibration is present on any piece of pipework in industry and is of particular significance in Coriolis and vortex meters. The vortex frequencies for gas for example lie in the range of 5–500 Hz. Consequently, vibration-induced signals in this range cannot be fully filtered out. Wherever possible, therefore, the sensor itself should be insensitive to pipe vibration.

16.3 GENERAL INSTALLATION REQUIREMENTS

To ensure reliable flowmeter operation, the following check-list will help in minimising the problems:

- Install the meter in the recommended position and attitude
- Ensure that the measuring tube is completely filled at all times
- When measuring liquids, ensure that there is no air or vapour in the liquid
- When measuring gases, ensure that there are no liquid droplets in the gas
- To minimise the effects of vibration, support the pipeline on both sides of the flowmeter
- If necessary, provide filtration upstream of the meter
- Protect meters from pressure pulsations and flow surges
- Install flow control or flow limiters downstream of the meter
- Avoid strong EM fields in the vicinity of the flowmeter
- Where there is vortex or corkscrew flow, increase the inlet and outlet sections or install flow straighteners
- Install two or more meters in parallel if the flow rate is too great for one meter
- Allow for expansion of the pipework
- Make sure there is sufficient clearance for installation and maintenance work
- Wherever possible provide proving connections downstream of the meter for regular *in-situ* calibrations
- To enable meters to be removed for servicing without station shutdown, provide a by-pass line

Figures 16.1 through 16.6 illustrate a number of recommended installation practices laid down specifically for EM flowmeters. The same principles also apply to most other flow-metering devices.

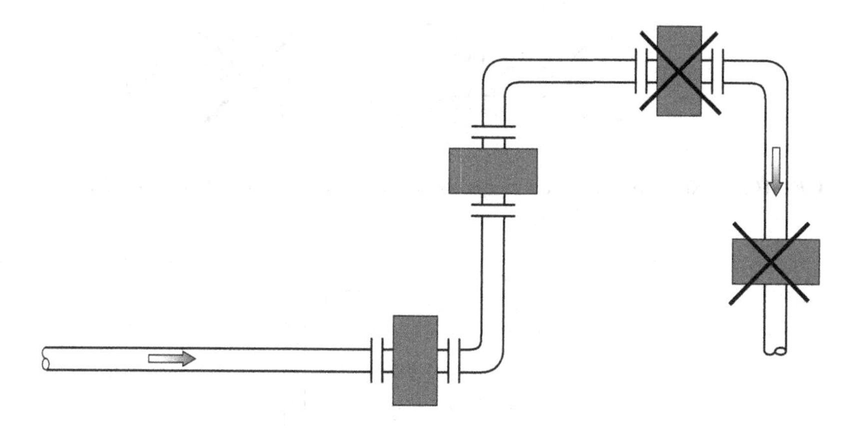

FIGURE 16.1 Preferred locations. Since air bubbles collect at the highest point on a pipe run, installation of the meter at this point could result in faulty measurements. The meter should not be installed in a downpipe where the pipe may be drained. (Courtesy: Krohne.)

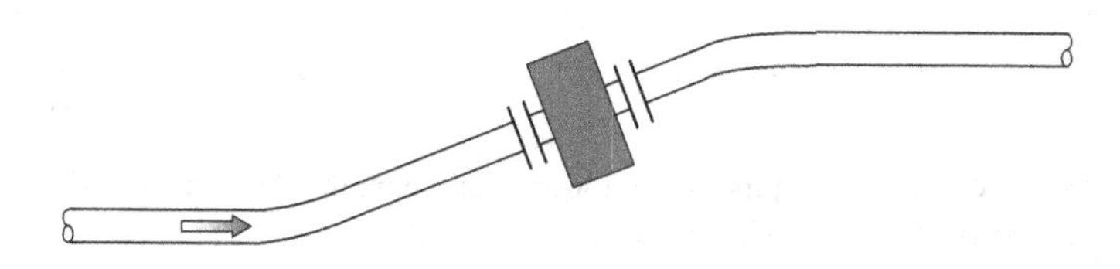

FIGURE 16.2 In a horizontal pipe run, the meter should be installed in a slightly rising pipe section. (Courtesy: Krohne.)

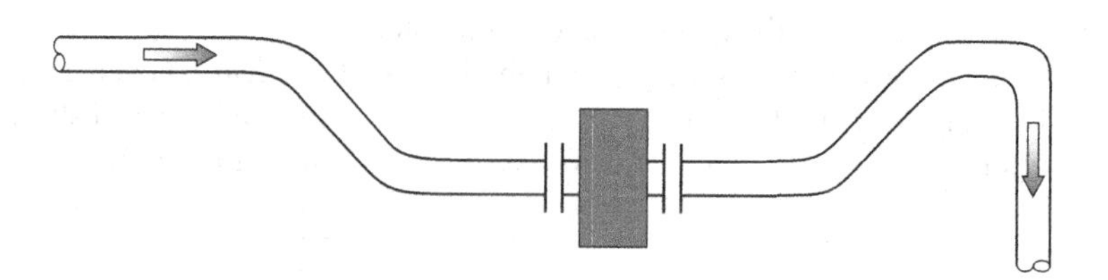

FIGURE 16.3 Where there is an open discharge, install the meter in a low section of the pipe. (Courtesy: Krohne.)

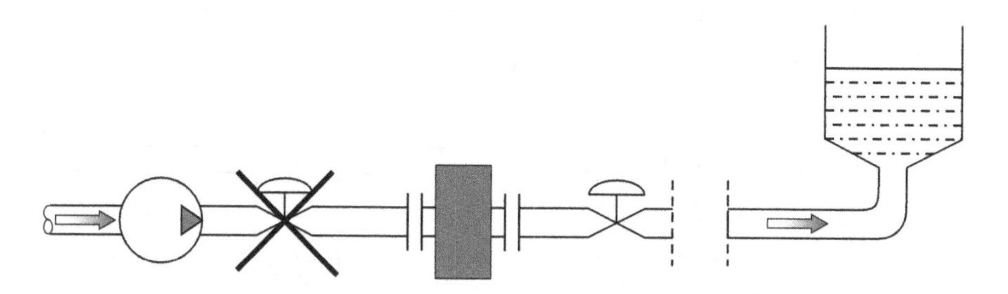

FIGURE 16.4 In long pipes, always install shutoff valves downstream of the flowmeter. (Courtesy: Krohne.)

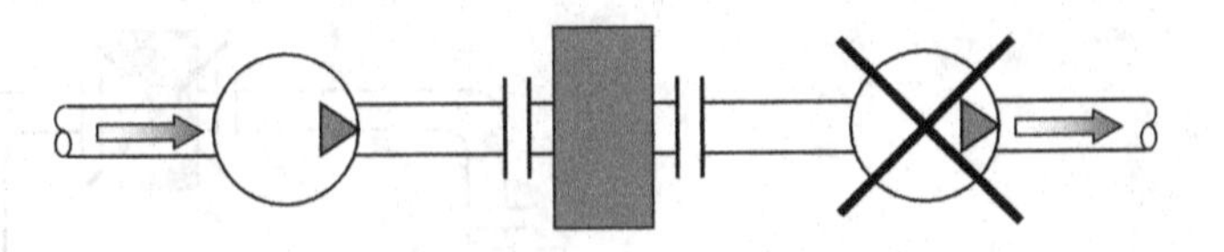

FIGURE 16.5 Never install a flowmeter on the pump suction side. (Courtesy: Krohne.)

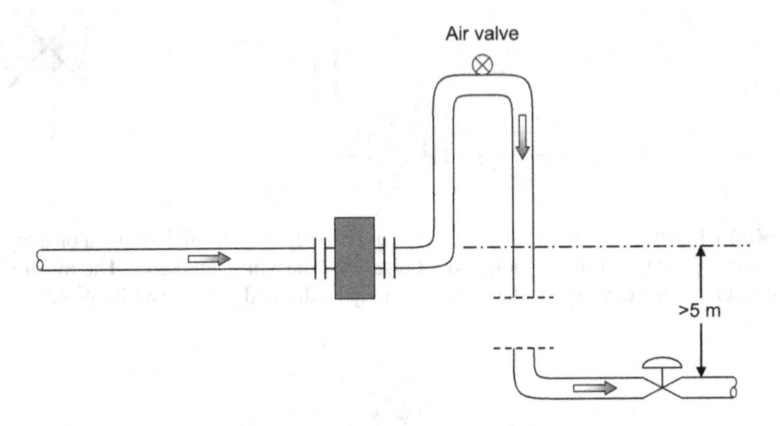

FIGURE 16.6 Where a downpipe is 5 m lower than the main inlet pipe, install an air valve at the highest point. (Courtesy: Krohne.)

16.3.1 Torquing

The role of a gasket is to form a sandwich between the flanges and ensure that the medium flowing through the meter is safely contained.

If the flange bolts are not tightened enough, the gasket will leak. If over-tightened, the gasket may become deformed – resulting in a leakage. More seriously, many gaskets (for example an O-ring) are recessed, as shown in Figure 16.7, and are

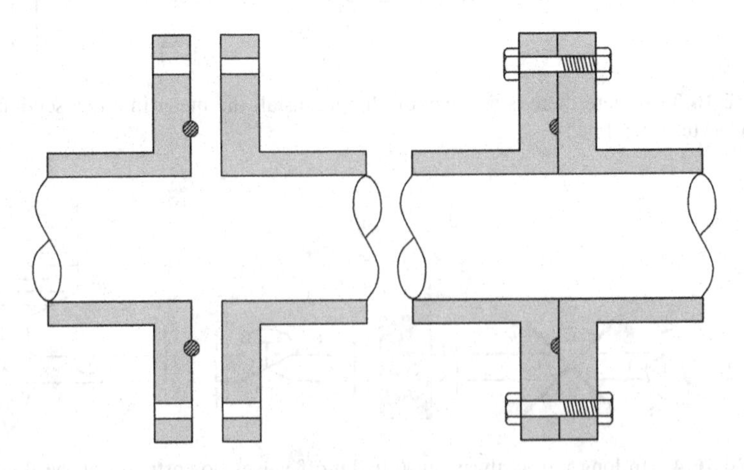

FIGURE 16.7 Recessed gaskets are normally tightened until a metal-to-metal contact occurs. (Courtesy: Endress + Hauser.)

normally tightened until a metal-to-metal contact occurs. In this case over-tightening can cause deformation of the flanges – this leading to damage to the meter itself. Ceramic liners, in particular, have been prone to damage through over-tightening as their mechanical characteristics are quite different from metals.

During commissioning or replacement of a meter, the flange bolts should be tightened only when the maximum process temperature is reached. Conversely, meters should be disconnected when the temperature is below 40°C to avoid the risk of damaging the surface of the gasket.

If a flange connection leaks, despite the fact that the bolts are tight, then they should *NOT BE TIGHTENED ANY FURTHER*. Loosen the bolts opposite the leak and tighten the bolts by the leak. If the leak persists, then the seal should be checked for foreign objects trapped in between.

The torque values given in Table 16.1 are based on greased bolts and serve as guidelines only since they depend on the material from which the bolts are manufactured.

TABLE 16.1
Torque Values Based on Greased Bolts for Various Gaskets

			Torque Values DIN in Nm		
DN	PN	Bolts	Klingerit	Soft Rubber	PTFE
15	40	4xM12			15
20		4xM12			25
25	16	4xM12	25	5	33
32		4xM16	40	8	53
40		4xM16	50	11	67
50		4xM16	64	15	84
65		4xM16	87	22	114
80		8xM16	53	14	70
100		8xM16	65	22	85
25		8xM16	80	30	103
150		8xM20	110	48	140
200		12xM20	108	53	137
250	10/16	12xM20	104/125	29/56	139/166
300		12xM20	191/170	39/78	159/227
350		16xM20	141/193	39/79	188/258
400		16xM24	191/245	59/111	255/326
450		20XM24	170/251	58/111	227/335
500		20XM24	197/347	70/152	262/463
600		20XM27	261/529	107/236	348/706
700		24xM27	312/355	122/235	
800		24xM30	417/471	173/330	
900		28XM30	399/451	183/349	
1,000		28xM33	513/644	245/470	

Source: Courtesy of Endress + Hauser.

16.4 GROUNDING AND EARTHING

First, let's clear up a source of confusion. In the UK, the term *earthing* is used almost universally for sourcing unwanted currents and for protection. In the United States, the term *grounding* is used when sourcing unwanted currents whilst *earthing* is confined only for protection.

To ensure measuring accuracy and avoid corrosion damage to the electrodes of EM flowmeters, the sensor and the process medium must be at the same electrical potential. This is achieved by grounding the primary head as well as the pipeline by any one or more of a number of methods including grounding straps, ground rings, lining protectors, and grounding electrodes.

Improper grounding is one of the most frequent causes of problems in installations. If the grounding is not symmetrical, ground-loop currents give rise to interference voltages – producing zero-point shifts.

Figures 16.8 through 16.12 show the most effective grounding configurations.

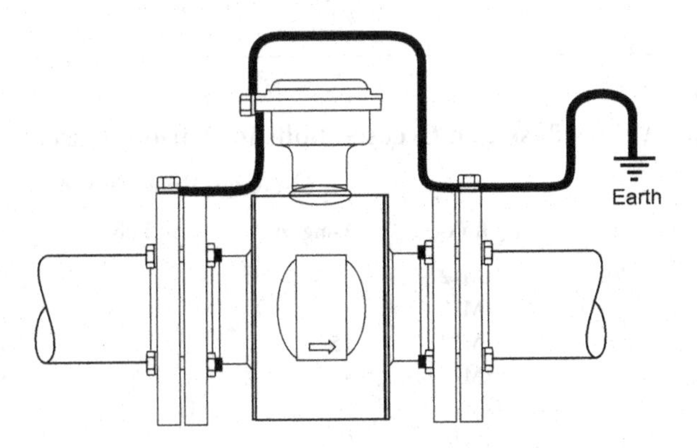

FIGURE 16.8 Grounding for conductive unlined pipe and conductive pipe with grounding electrode. (Courtesy: Emerson.)

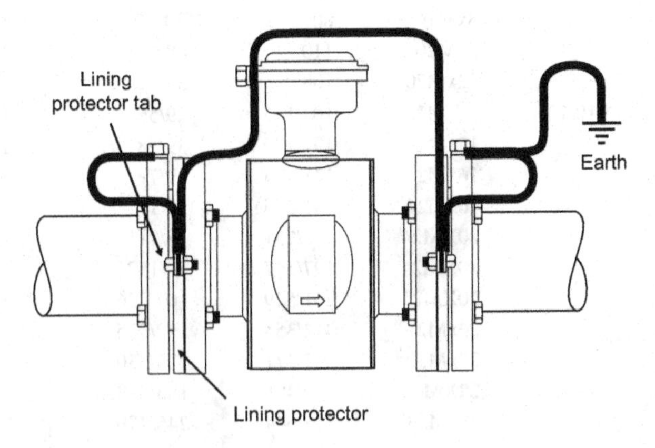

FIGURE 16.9 Grounding for conductive unlined and lined pipe with lining protectors. (Courtesy: Emerson.)

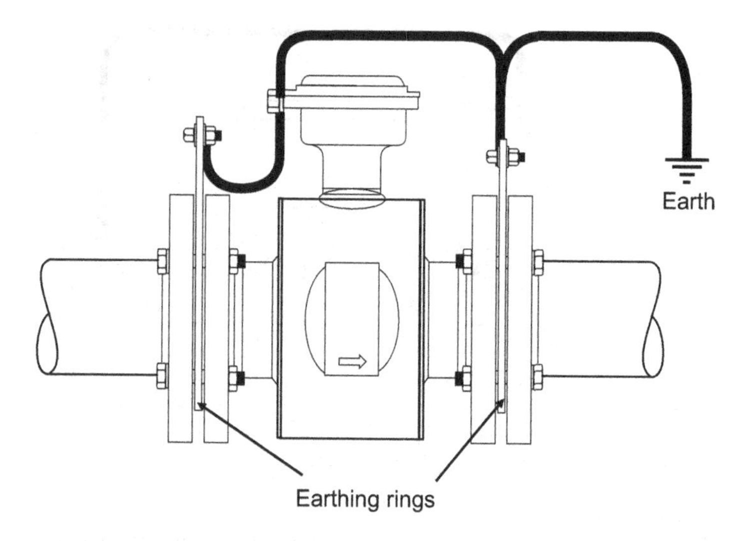

FIGURE 16.10 Grounding for non-conductive pipe with grounding rings. (Courtesy: Emerson.)

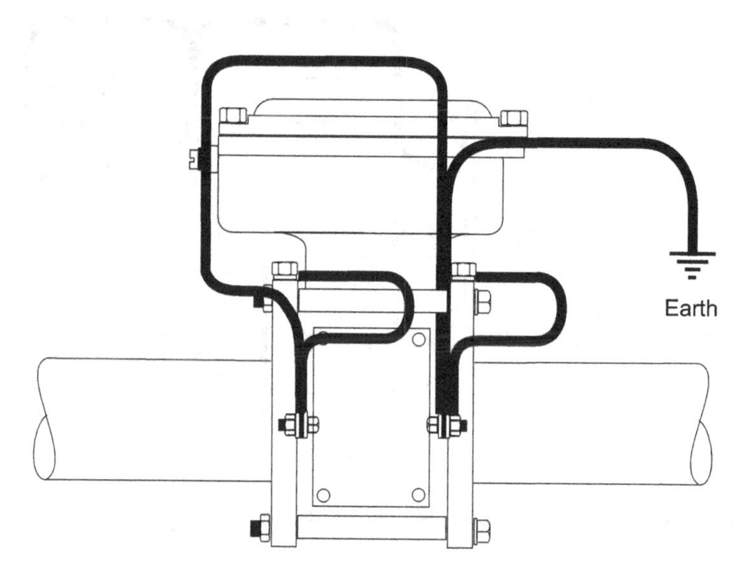

FIGURE 16.11 Grounding for conductive lined pipe with grounding rings. (Courtesy: Emerson.)

It is essential in cathodic protection installations (Figure 16.13) to ensure that there is an electrical connection between the two-piping runs using earthing rings or electrodes. It is also essential that *no* connection is made to ground.

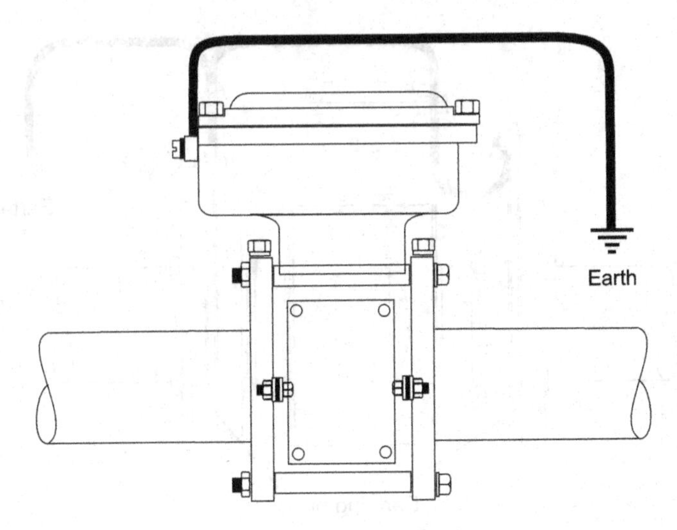

FIGURE 16.12 Grounding for non-conductive lined pipe with grounding electrodes. (Courtesy: Emerson.)

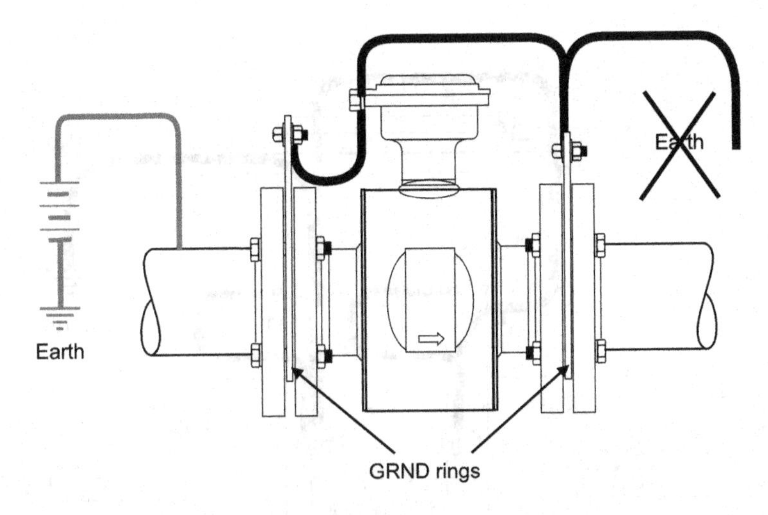

FIGURE 16.13 Cathodic protection installations. (Courtesy: Emerson.)

17 Flow Meter Selection

17.1 INTRODUCTION

As prefaced in Chapter 1, there were nearly 50 different flow-measuring technologies to choose from and making such a choice might at first appear a daunting task. However, common sense must prevail and it is unlikely that one would commit the requirement for accurate fiscal measurement on an oil pipeline to a low-cost variable area 'rotameter' any more than one would use an expensive Coriolis meter for indication purposes for a process analyser – the former a total inadequate under-kill, running to a few hundred US dollars, and the latter a total overkill running to many tens of thousands of US dollars.

But let's first look at a rough guide to application as detailed in Figure 17.1 where we've reduced the number of technologies down to 21 categories.

So what exactly are your selection criteria? Obviously, cost has got to be one of them – but it should be way down the list. And should price be brought into the decision anyway? As Aldous Huxley said in his Brave New World, 'You pays your money and you takes your choice'. In other words 'You get what you pay for'.

Probably the most important is application suitability. Can the instrument you choose perform its task to the best of its ability, within the application environment, to the required accuracy, with minimum maintenance requirements, with minimum calibration requirements, and with minimum drift? Fit and forget? Or fit and fret?

17.2 BASIC SELECTION CRITERIA

17.2.1 Choice of Supplier

Is it a good thing to keep to the same supplier to ensure complete compatibility?

Unless your supplier is making use of a unique communication strategy, where compatibility may be a problem, it shouldn't be necessary. And sticking to the same supplier could mean you are missing out on technological advances from a competitor.

A major consideration should be to take a close look at the representation and backup available from any particular supplier in your geographical area. Some European suppliers are not necessarily well represented in the United States – and vice versa.

17.2.2 Choice of Turndown Ratio

Because it is virtually impossible to know, in advance, the exact range of flows that might being encountered it is always prudent to select a flow meter that offers

	Measuring technology	Clean liquids	Dirty liquids	Corrosive liquids	Low conductivity	High temperature	LNG	Low temperature (cryogenics)	Low velocity	High viscosity	Non-Newtonian	Abrasive slurries	Fibrous slurries	Gas	Steam	Semi-filled pipe	Open channel
1	Coriolis	●	●	●	●	●	●	●	●	●	●	●	●	●	○	○	○
2	Electromagnetic	●	●	●	○	◉	○	○	●	○	○	●	●	○	○	●[1]	○
3	Flow nozzles	●	◉	◉	●	◉	○	◉	○	○	○	○	◉	●	●	○	○
4	Fluidic	●	◉	●	●	●	○	●	◉	●	◉	○	○	○	○	○	○
5	Flume	●	●	●	●	◉	○	○	●	○	○	●	●	○	○	●	●
6	Orifice plate	●	◉	●	●	●	○	○	○	○	○	○	○	●	●	○	○
7	Pitot – Averaging	●	◉	●	●	●	○	●	○	○	○	○	○	●	●	○	○
8	Positive – Lobed impeller	○	○	○	○	○	○	○	○	○	○	○	○	●	○	○	○
9	Positive – Nutating disk	●	◉	●	●	○	○	○	○	●	●	○	○	○	○	○	○
10	Positive – Oval gear	●	○	◉	●	●	○	◉	●	●	●	○	○	○	○	○	○
11	Positive – Rotary vane	●	○	◉	●	○	○	○	●	●	◉	○	○	○	○	○	○
12	Thermal mass (gas)[2]	○	○	○	○	○	○	○	○	○	○	○	○	●	○	○	○
13	Turbine	●	○	◉	●	◉	○	●	◉	○	○	○	○	●	●	○	○
14	Ultrasonic – Clamp-on	●	○	●	●	◉	○	○	●	○	○	○	○	●	○	○	○
15	Ultrasonic – Doppler	○	●	◉	●	○	○	○	◉	○	○	◉	◉	○	○	○	○
16	Ultrasonic – Transit time	●	○	●	●	○	●	●	●	○	○	○	○	●	○	○	○
17	Variable area	●	○	●	●	●	○	○	●	○	○	○	○	●	○	○	○
18	Venturi tubes	●	●	●	●	◉	○	◉	○	○	○	○	◉	●	●	○	○
19	Vortex shedding	●	●	●	●	●	○	●	○	○	○	○	○	●	●	○	○
20	Vortex precession	●	○	●	●	●	○	○	○	○	○	○	○	●	●	○	○
21	Weir	●	○	●	●	○	○	○	○	○	○	○	○	○	○	●	●

● Very suitable ◉ Applicable under certain conditions ○ Not suitable

[1] Specially designed units

[2] Liquid system also available

FIGURE 17.1 Selection guide to flow meters.

the widest possible turndown ratio to ensure that it can cover all anticipated flow variations.

17.2.3 Chemical Compatibility

Users must consider the chemical compatibility of the measured fluids with the flow meter materials:

- O-rings
- Bearings
- Gears
- Embedded ceramic magnets
- Turbines
- Rotors

17.2.4 System Parameters

Users must consider abnormal operating conditions:

- Will there be hydraulic shocks?
- Are there large variations in temperature?
- What's the highest temperature?
- What's the lowest temperature?
- Does the process liquid contain entrained gas?
- Does the process liquid contain abrasive particles?
- Do you need to consider pigging
- Do you need to consider Clean In Place (CIP)?
- Do you need to consider Steam in Place (SIP)?
- Are there contaminants in the fluid?
- Is the piping subject to vibration?
- Are there large variations in viscosity?
- Are there large variations in density?
- Is the fluid abnormally acidic/alkaline?
- Is secondary containment an issue?

17.3 DETAILED SELECTION CRITERIA

Many of the specification figures given below are typical figures and will vary from the manufacturer to manufacturer.

17.3.1 Coriolis

Main features:

Accurate mass and volumetric flow of liquids and gases
Accurate density measurement of liquids and gases
Bi-directional flow measurement

Other features:

Independent of density changes, flow profile, and flow turbulence
No routine maintenance required

Suitable for:

Clean and dirty liquids
Gases and slurries
LNG
Liquids with entrained gas ranging from 0% to 100%

Not suitable for:

Low-density gases

Application considerations:

Look to venting if secondary containment is of concern

Accuracy (mass flow):

Liquids: ±0.05%
Gases: ±0.35%

Accuracy (density):

Down to 0.0005 g/cm^3

Turndown ratio:

Up to 500:1

Pressure drop:

Negligible

Upstream/downstream requirements:

None

Relative cost:

$$$$

Effect of viscosity:

None

Moving parts:

None

Pipe size:

From 3 to 400 mm

17.3.2 Electromagnetic

Main features:

Accurate volumetric flow of conductive liquids and slurries
Bi-directional flow measurement
No obstruction to flow

Other features:

Independent of density changes, flow profile, and flow turbulence
Able to handle large particles and chemicals
No routine maintenance required
No recalibration requirements
Models available for partially filled pipes

Suitable for:

Clean and dirty liquids and slurries

Not suitable for:

Liquids having conductivities less than 5 μS/cm (low conductivity versions down to 0.1 μS/cm)
Hydrocarbons
Gases
Steam

Application considerations:

Cannot be used on pure or ultrapure water – even on meters having extended range down to 0.05 μS/cm

Accuracy:

±0.1%

Turndown ratio:

Up to 50:1

Pressure drop:

Negligible

Upstream/downstream requirements:

5D/3D

Relative cost:

$$

Effect of viscosity:

None

Moving parts:

None

Pipe size:

Typically from 10 to 1,200 mm

17.3.3 Flow Nozzles

Main features:

Used for high-velocity applications of liquids or gases
Often used when discharging into the atmosphere

Other features:

When compared with standard Venturi:
Only half the cost
Requires far less space for installation
Easier installation and maintenance
Less susceptible to wear

Suitable for:

Clean and dirty liquids and slurries
Hydrocarbons
Gases

Not suitable for:

Measurement where there is a high percentage of solids

Application considerations:

With no divergent outlet the pressure recovery is poor
Cannot be used in a low pressure head

Accuracy:

±2%

Turndown ratio:

Up to 4:1

Pressure drop:

High – between 30% and 80% of the measured differential pressure

Upstream/downstream requirements:

20D/5D

Relative cost:

$

Effect of viscosity:

Not suitable for high viscosities

Moving parts:

None

Pipe size:

Typically from 10 to 1,200 mm

17.3.4 Fluidic

Main features:

Volumetric flow of highly viscous liquids

Other features:

Rugged construction
High immunity to shock and pipe vibration
High turndown ratio (typically 30:1)
Linear output
Operation down to Re of 3,000
Operating pressure (100 mm unit) 10 bar
No routine maintenance required

Suitable for:

Highly viscous liquids
Medium temperature applications up to 176°C
Option of cryogenic service down to −196°C

Not suitable for:

Gases

Application considerations:

Requires minimum backpressure to prevent flashing

Accuracy:

Liquids: ±2%

Turndown ratio:

Up to 30:1

Pressure drop:

High – up to nearly 7 bar at a maximum flow rate of 750 L/min

Upstream/downstream requirements:

None

Relative cost:

$$

Effect of viscosity:

Determines minimum flow rate

Moving parts:

None

Pipe size:

19–38 mm

17.3.5 Flume

Main features:

Open-channel flow measurement for flows containing sediment or solids

Other features:

A higher flow rate measurement than for a comparably sized weir
A much smaller head loss than a weir
Self-cleaning
Rugged construction
Little routine maintenance required
Passes debris more readily than a weir
Wide range of styles and sizes
Off-the-shelf availability
Smaller installation footprint

Suitable for:

Water irrigation schemes
Sewage processing and effluent control
Water treatment
Mining beneficiation

Not suitable for:

Pressurised conduits
Gases

Application considerations:

Flume installation is typically more expensive than a weir

Accuracy:

±10%

Turndown ratio:

Dependent on size: 30:1 to more than 100:1

Pressure drop:

Low

Upstream/downstream requirements:

None

Relative cost:

$$

Effect of viscosity:

Not suitable for viscous materials

Moving parts:

None

Throat size:

25 mm to 4 m or more

17.3.6 Orifice Plate

Main features:

Wide range of liquids, gases, and steam

Other features:

Simple construction
Inexpensive
Robust
Large range of sizes and opening ratios
Price does not increase dramatically with size
Well understood and proven

Suitable for:

Wide range of liquids, gases, and steam

Not suitable for:

High-viscous fluids
Abrasive or fibrous slurries

Application considerations:

Accuracy is affected by density, pressure, and viscosity fluctuations
Erosion and physical damage to the restriction compromise the measurement accuracy without the operator being aware of the fact
Requires a homogeneous single-phase liquid
Output is not linearly related to the flowrate
Multiple potential leakage points
Always gives a reading irrespective of the damage to the plate
Only provides a valid reading for a fully developed flow profile

Accuracy:

Typically ±2% to ±3%

Turndown ratio:

4:1

Pressure drop:

Very high

Upstream/downstream requirements:

Typically requires from 25 to 40D upstream and 4 or 5D downstream

Relative cost:

$

Effect of viscosity:

Not suitable for viscous media

Moving parts:

None

Pipe size:

25 mm to 2.5 m

17.3.7 Pitot: Averaging

Main features:

Wide range of liquids and gases – including blast furnace gas, compressed air, and steam

Other features:

Rugged construction
Little routine maintenance required
Simple installation in a wide range of pipeline sizes
Low pressure loss
High strength
No wear
No leakage

Suitable for:

Water and steam in large pipes

Not suitable for:

High-viscous fluids
Abrasive or fibrous slurries

Application considerations:

Accurate alignment is critical. Tilting the bar forwards or backwards can affect the reading

Accuracy:

±1.0%

Turndown ratio:

Dependent on size: 14:1

Pressure drop:

Low

Upstream/downstream requirements:

2½ D downstream of a discontinuity

Relative cost:

$

Effect of viscosity:

Not suitable for viscous materials

Moving parts:

None

Pipe size:

25 mm to 4 m or more

17.3.8 Positive: Lobed Impeller

Main features:

Clean, dry gas

Other features:

No inlet and outlet sections required
No external power supply
Low pressure drop – typically 0.7 kPa
Rugged construction
Intrinsically safe

Suitable for:

Clean, dry natural gas, town gas, propane, and inert gases
Essentially designed for high-volume gas measurement up to 1,000 m^3/h

Not suitable for:

High-viscous fluids
Abrasive or fibrous slurries

Application considerations:

Pulsations caused by alternate drive action
Temperature of process medium limited to about 60°C
Wear due to moving parts

Accuracy:

±1.0%

Turndown ratio:

10:1

Pressure drop:

Low

Upstream/downstream requirements:

None

Relative cost:

$$$

Effect of viscosity:

Not suitable for viscous materials

Moving parts:

Dual rotors

Pipe size:

40–150 mm

17.3.9 Positive: Nutating Disc

Main features:

Major application in domestic and industrial potable water metering

Other features:

No inlet and outlet sections required
No external power supply
Low pressure drop – typically 0.7 kPa
Rugged construction
Easily maintained without removing from line
Intrinsically safe

Suitable for:

Clean and moderately dirty liquids, hard and soft water, oils, fuel, solvents, etc.

Not suitable for:

Abrasive or fibrous slurries
Gas or steam

Application considerations:

Upper limit of 20 bar
Wear due to moving parts

Accuracy:

±1.5%

Turndown ratio:

From 25:1 up to 60:1

Pressure drop:

Low

Upstream/downstream requirements:

None

Relative cost:

$

Effect of viscosity:

Suitable for viscous materials

Moving parts:

Single-nutating disc

Pipe size:

20–150 mm

17.3.10 Positive: Oval Gear

Main features:

Ideally suited for measurement of viscous fluids or those with varying viscosities such as oils, syrups, and fuels

Other features:

Well suited for low flow applications
Ability to handle both low- and high-viscosity products
High-operating pressures, up to 10 MPa

High temperatures, up to 300°C
Wide range of construction materials
No inlet and outlet sections required
No external power supply
Low pressure drop – typically 0.7 kPa
Rugged construction
Intrinsically safe

Suitable for:

Clean and moderately dirty liquids, hard and soft water, oils, fuel, solvents, etc.

Not suitable for:

Water or low-viscosity fluids
Abrasive or fibrous slurries

Application considerations:

Pulsations caused by alternate drive action
Wear due to moving parts

Accuracy:

±0.25%

Turndown ratio:

From 10:1 up to 25:1 – dependent on viscosity

Pressure drop:

Low – less than 20 kPa

Upstream/downstream requirements:

None

Relative cost:

$$

Effect of viscosity:

Suitable for viscous materials

Moving parts:

Dual-rotating gears

Pipe size:

Typically 6–100 mm

17.3.11 Positive: Rotary Vane

Main features:

Accurate volumetric flow of clean liquids

Other features:

No inlet and outlet sections required
No external power supply
Rugged construction
Intrinsically safe

Suitable for:

Clean liquids

Not suitable for:

Dirty liquids
Abrasive or fibrous slurries

Application considerations:

Wear due to moving parts

Accuracy:

±0.2%

Turndown ratio:

20:1 (dependent on viscosity)

Pressure drop:

Generally low – dependent on viscosity and flow rate – typically less than 0.25 bar

Upstream/downstream requirements:

None

Relative cost:

$$$

Effect of viscosity:

Viscosity affects the turndown ratio

Moving parts:

Rotating rotor and sliding vanes

Pipe size:

50–400 mm

17.3.12 Thermal Mass

Main features:

Uses the thermal or heat-conducting properties of the fluid to measure mass flow

Uses several different technologies – heat loss, internal or external temperature rise, and capillary

Other features:

Relatively low initial purchase price

Suitable for:

Mainly used for the measurement of low-density gas flow
Can also measure very low liquid flows: for example down to 30 g/h
Largely independent of flow profile, medium viscosity, and pressure

Application considerations:

System must be calibrated for each particular gas – with each mass flow/temperature sensor pair individually calibrated over entire flow range

Accuracy:

±1% to ±2%

Turndown ratio:

Up to 1,000:1

Low speed sensitivity:

60 mm/s

Pressure drop:

Low or high – dependent on technology

Upstream/downstream requirements:

20D/5D

Relative cost:

$

Effect of viscosity:

Low or high – dependent on technology

Moving parts:

None

Pipe size:

Typically up to 200 mm
Insertion types available for larger-sized pipes (but affected by flow profile, media viscosity, and pressure)

17.3.13 Turbine

Main features:

Accurate volumetric flow of clean liquids and gases

Suitable for:

Used extensively for fiscal and custody transfer
Cryogenic applications

Not suitable for:

Corrosive fluids and liquids with solids
Excessive wear with non-lubricating fluids

Application considerations:

Entrained gas in liquid affects accuracy
Magnetic drag may affect the accuracy at low flows – lower limit response may be extended by using Hall-effect devices
Swirl will affect accuracy

Accuracy:

±0.25%

Turndown ratio:

Up to 20:1

Pressure drop:

Back pressure required to avoid flashing/cavitation

Upstream/downstream requirements:

Typically 10–15D upstream and 5D downstream

Relative cost:

$$$

Effect of viscosity:

Variation in K-factor at low flow rates due to viscosity hump requires recalibration for significant changes in viscosity

Moving parts:

Rotor

Pipe size:

Typically from 25 to 600 mm

17.3.14 Ultrasonic: Clamp-On

Main features:

Clamp-on meters employ external transducers that are attached to the walls of the pipe to provide portable non-intrusive flow-measurement systems that can be installed within a few minutes to virtually any pipe

Other features:

Low installation cost
Suitable for large range of pipe diameters

Suitable for:

Clean liquids and gases
Flexural mode systems may be used in high-accuracy applications

Not suitable for:

Dirty liquids

Application considerations:

In multiphase flows the particle velocity may bear little relationship to the media velocity
Indication is heavily dependent on the flow profile
Particle size must provide sufficiently good reflections ($>\lambda/4$)

Accuracy:

Typically ±1% to ±3%
Flexural mode systems down to ±0.15%

Turndown ratio:

10:1

Pressure drop:

Low

Upstream/downstream requirements:

Requires fully developed flow profile

Relative cost:

$ to $$$

Effect of viscosity:

Negligible

Moving parts:

None

Pipe size:

Typically from 60 to 3,000 mm

17.3.15 Ultrasonic: Doppler

Main features:

Based on the frequency shift from reflective particles in the medium – directly proportional to velocity

Other features:

Low installation cost
Insertion types suitable for large range of pipe diameters

Suitable for:

Dirty liquids

Not suitable for:

Clean liquids and gases
High-accuracy applications

Application considerations:

In multiphase flows the particle velocity may bear little relationship to the media velocity
Indication is heavily dependent on the flow profile
Particle size must provide sufficiently good reflections ($>\lambda/4$)

Accuracy:

±2% to ±10% or more

Turndown ratio:

10:1

Pressure drop:

Low

Upstream/downstream requirements:

Inlet runs of 20D. Downstream not stated

Relative cost:

$

Effect of viscosity:

Unsuitable for high-viscosity fluids

Moving parts:

None

Pipe size:

Typically from 60 to 3,000 mm

17.3.16 Ultrasonic: Transit Time

Main features:

Accurate volumetric flow of clean liquids and gases

Suitable for:

Used extensively for fiscal and custody transfer
Multi-beam systems can be used to eliminate the effects of profile disturbances
Not affected by fluid properties
No pressure drop
High turndown
Bi-directional
Large line sizes

Not suitable for:

Dirty fluids

Application considerations:

Entrained gas in liquid and partially filled pipes cause signal loss
Coatings on the transducers can deflect beam and corrupt measurement
Process temperature can affect sound velocity and ultrasonic beam angle
Multiphase flow can cause error
In single-beam meters the accuracy is dependent on flow profile

Accuracy:

±0.15%

Turndown ratio:

Up to 20:1

Pressure drop:

Back pressure required to avoid flashing/cavitation

Upstream/downstream requirements:

Typically 10–15D upstream and 5D downstream

Relative cost:

$$$$

Effect of viscosity:

Unsuitable for high-viscosity fluids

Moving parts:

None

Pipe size:

Typically from 25 to 600 mm

17.3.17 Variable Area

Main features:

Low cost, direct-reading flow rate indicator

Other features:

Linear float response to flow rate change
Easy installation and maintenance
Not affected by the upstream piping configuration
Metal tubes can be used for hot and strong alkalis, fluorine, hydrofluoric acid, hot water, steam, slurries, sour gas, additives, and molten metals

Suitable for:

Clean liquids and gases

Not suitable for:

Tube is inherently self-cleaning. However, avoid fluids that could coat the float
Liquids with fibrous materials, abrasives, and large particles

Application considerations:

Operates in vertical position only*
Accessories required for remote visualisation

Accuracy:

Typically ±1% to ±3%

Turndown ratio:

10:1

Pressure drop:

Low

Upstream/downstream requirements:

None

* Spring-loaded versions are available for other orientations.

Relative cost:

$

Effect of viscosity:

Unsuitable for high-viscosity fluids

Moving parts:

Float

Pipe size:

Typically from 15 to 150 mm

17.3.18 Venturi Tube

Main features:

Primarily used on larger flows of liquids or gases

Other features:

When compared with standard orifice plate:
Permanent pressure loss only about 10% of the differential pressure
Able to handle about 60% more flow
Cost is about 20 times higher
Less susceptible to wear

Suitable for:

Clean and dirty liquids
Hydrocarbons
Gases

Not suitable for:

Abrasive media may compromise critical dimensions

Application considerations:

Bulky – requires large section for installation

Accuracy:

±0.75%

Turndown ratio:

Up to 4:1

Pressure drop:

Low – about 10% of the measured differential pressure

Upstream/downstream requirements:

20D/5D

Relative cost:

$$$

Effect of viscosity:

Not suitable for viscous media

Moving parts:

None

Pipe size:

Typically used for bore over 1,000 mm

17.3.19 Vortex Shedding

Main features:

Suitable for this high temperature liquid and gas measurement

Other features:

Used extensively for steam measurement – with built-in mass flow calculation
Low installation cost
No field calibration required

Suitable for:

Clean and dirty liquids and gases

Not suitable for:

Reynolds numbers typically lower than about 8,000
Batching operations
High-viscosity fluids

Application considerations:

Possible problems in high-vibration environments

Accuracy:

Liquids: ±0.65%
Gases: ±1.0%
Mass: ±2.0% (temperature compensated)

Turndown ratio:

>20:1 on gases and steam
>10:1 on liquids

Pressure drop:

Low

Upstream/downstream requirements:

Typically 10–15D upstream and 5D downstream

Relative cost:

$$

Effect of viscosity:

Unsuitable for high-viscosity fluids

Moving parts:

None

Pipe size:

Typically from 12 to 300 mm

17.3.20 Vortex Precession

Main features:

Although suitable for both liquids and gases, it finds its main application as a gas flow meter

Other features:

Linear flow measurement
Used extensively for steam measurement
Low installation cost
No field calibration required

Suitable for:

Media temperature up to 280°C
Clean liquids, gases, and steam

Not suitable for:

Reynolds numbers typically lower than about 17,000 for DN 100
Batching operations
High-viscosity fluids

Application considerations:

May be used in high-vibration environments

Accuracy:

±0.5% for liquids, gases, and steam

Turndown ratio:

50:1

Pressure drop:

Medium

Upstream/downstream requirements:

3D/1D

Relative cost:

$$

Effect of viscosity:

Unsuitable for high-viscosity fluids

Moving parts:

None

Pipe size:

Typically from 14 to 400 mm

17.3.21 Weir

Main features:

Open-channel flow measurement for relatively clean media

Other features:

Low cost
Rugged construction
Little routine maintenance required
Wide range of styles and sizes
Off-the-shelf availability

Suitable for:

Water irrigation schemes
Sewage processing and effluent control
Water treatment

Not suitable for:

Liquid media containing sediments
Pressurised conduits
Gases

Application considerations:

Weirs require approximately four times more head than a flume for a similar flow rate Recommended temperature range from 3.9°C to 30°C. At lower temperatures, ice may begin to form on the crest – greatly affecting the flow readings

Accuracy:

±15%

Turndown ratio:

Dependent on size: 30:1 to more than 100:1

Pressure drop:

Low

Upstream/downstream requirements:

Approaching velocity profile should be well distributed before flow reaches the point of measurement

Relative cost:

$

Effect of viscosity:

Not suitable for viscous materials

Moving parts:

None

Throat size:

25 mm to 4 m or more

References and Bibliography

REFERENCES

Bernoulli D., *'Hydrodynamica', Sive de viribus et motibus fluidorum commentarii*, Johann Reinhold Dulsseker, Strasbourg, 1738.

Karnick U., Jungowski W.M., Botros K.K., *'Effect of Turbulence on Orifice Meter Performance'*, Novacor Research & Technology Corporation, 1994.

Morrow T.B., Park J., 'Effects of tube bundle location on orifice meter error and velocity profiles', *Proceedings of the International Offshore Mechanics and Arctic Engineering Symposium*, 1992.

Shercliff J.A., *'The Theory of Electromagnetic Flow-Measurement'*, Cambridge University Press, 1962.

Thürlemann B., 'Methode zur elektrischen Geschwindigkeitsmessung in Flüssigkeiten' (Method of Electrical Velocity Measurement in Liquids), *Helvetica Physica Acta* 14, 383–419, 1941.

BIBLIOGRAPHY

ABB, 'Vortex Flowmeter FV4000 (TRIO-WIRL V) and Swirl Flowmeter FS4000 (TRIO-WIRL S)', Data sheet from ABB, 2004.

ABB, 'AquaProbe: Insertion-Type Electromagnetic Probe Flowmeter' Installation Guide, 2005.

ABB, '267CS Multivariable Transmitters', Product Specification Sheet: Data Sheet, SS/267CS/269CS, 2007.

ABB Limited, 'Differential Pressure Flow Elements', Data sheet DS/DP-EN Rev. E, ABB Limited, 2011.

Agar Corporation, *'Not All Multiphase Flowmeters Are Created Equal'*, Revision 7, 2004.

Ahmadi A., Beck S., *'Development of the Orifice Plate with a Swirler Flow Conditioner'*, University of Leeds, Sheffield and York, 2009.

Al-Allah A., 'Field experience to optimize gas lift well operations', Academic Focus, Egypt Oil, Issue 10, October 2007.

Al-Khamis M.N., Al-Bassam A.F., Saudi Aramco, Bakhteyar Z., Aftab M.N., Schlumberger, 'Evaluation of PhaseWatcher Multiphase Flow Meter (MPFM), Performance in Sour Environments', Offshore Technology Conference paper from Honeywell, 2008.

Allen L.H., *'Fundamental Principles of Rotary Meters'*, Dresser, Inc., 2007.

Annubar® Primary Flow Element Flow Test Data Book, Rosemount, Emerson, Reference Manual 00821-0100-4809, Rev BA, July 2009.

Augenstein D., Cousins T., 'The Performance of a Two Plane Multi-Path Ultrasonic Flowmeter', Caldon Inc., 2005.

AVCO Valve, 'Flow Measurement Devices', data sheet from AVCO Valve, 2004.

Baker R.C., *'Flow Measurement Handbook: Industrial Designs, Operating Principles, Performance, and Applications'*, 2nd Edition, Cambridge University Press, 2000.

Benhadj R., Ouazzane A.K., 'Flow conditioners design and their effects in reducing flow metering errors', Sensor Review, 2002.

BM Series meters, 'Precision Positive Displacement Meters', brochure from Avery-Hardoll, 2016.

Bös P.C., *'Pipelines – A Case Study'*, Siemens AG.

Brown G.J., Griffith B.W., 'A new flow conditioner for 4-path ultrasonic flowmeters', Cameron, FLOMEKO, 2013.

Brown G.J., Griffith B., Augenstein D.R., 'The influence of flow conditioning on the proving performance of liquid ultrasonic meters', *Cameron, 29th International North Sea Flow Measurement Workshop*, October 2011.

BS EN ISO 5167-1:2003, 'Measurement of fluid flow by means of pressure differential devices inserted in circular cross-section conduits running full', 2003.

Bulletin SS01014 Issue/Rev 0.8 (8/13), 'Smith Meter® PD Meters 6″ Steel Model G6 Specifications' Bulletin SS01014, Issue/Rev 0.8 (8/13).

Cada R., *'Flare and Combustor Gas Measurement: Applications, Regulations and Challenges'*, Fox Thermal Instruments, Inc., 2014.

Campbell J.M., *'Gas conditioning and processing. Volume 1: The Basic Principles', Chapter 9, Multiphase Flow Measurement*, Eight Edition, February 2004.

Cascetta F., *'Short History of Flowmetering'*, Published by Elsevier Science Ltd., Copyright ©, 1995.

Crabtree M.A., *'Mick Crabtree's Flow Handbook* ', 2nd Edition, Crown Publications, 2000.

Crabtree M.A., 'Industrial Flow Measurement', *Master's Thesis*, University of Huddersfield, 2009.

Crabtree M.A., 'Harnessing Coriolis – From Cannon Balls to Mass Flow Measurement', PetroSkills, Tip of the Month, October 2018. https://en.wikipedia.org/wiki/Coriolis_force.

Daniel, 'Profiler™ Flow Conditioning Plate', brochure from Daniel, 2001.

Daniel, 'Development of orifice meter standards', Daniel Measurement and Control White Papers, 2010.

Daniel, 'Fundamentals of Orifice Meter Measurement', Daniel Measurement and Control White Papers, 2013.

Daniel Senior® Orifice Fittings, *'Parts list and Materials, Instructions for Installation, Operation and Maintenance'*, Emerson, 2006.

Delenne B., De France G., Pritchard M., Advantica, Lezaun F.J., Enagas, Vieth D., Eon-Ruhrgas, Huppertz M., Fluxys, Ciok K., Gastra, van den Heuvel A., Gasunie Research, Mouton G., GSO, Folkestad T., Norsk Hydro, Marini G., Snam Rete Gas 'Evaluation Of Flow Conditioners – Ultrasonic Meters Combinations', 2005.

Dieball A., *'The Advantages of Multi Variable Vortex Flowmeters'*, Sierra Instruments Inc., 1999.

Elster Instromet, 'Q. Sonic Ultrasonic Gas Flow Meters', data sheet from Elster Instromet, 2009.

Emerson, 11/09, 'Annubar® Flowmeter Series', Rosemount brochure, Emerson, 00803-0100-6113 Rev CA, 11/09.

Emerson, *'Theory of DP Flow'*, Emerson, 2015.

Emerson, 'Roxar Watercut meter', data sheet from Emerson, 2016.

Falcone G., SPE, Enterprise Oil; G.F. Hewitt, Imperial College; C. Alimonti, SPE, U. of Rome La Sapienza; and B. Harrison, SPE, Enterprise Oil, 'Multiphase Flow Metering: Current Trends and Future'.

Falcone G., Texas A&M University and Bob Harrison, Soluzioni Idrocarburi Srl, 'Forecast expects continued multiphase flowmeter growth', 2012.

Faure-Herman, 'Heliflu TZN the Dedicated Turbine Flowmeter for Custody Transfer Measurement', data sheet from Faure-Herman, 2011.

Faure-Herman, 'Advantages of Master Metering Method of Proving Custody Transfer Flows', White paper from Faure-Herman, 2012, www.petro-online.com.

Fluidic Flowmeters, 'Model 140MX Flowmeter Functional Overview', PI 14-1 Rev: 1, March 2002.

FMC Technologies, 'Fundamentals of Liquid Turbine Meters', FMC Technologies, Bulletin TP0, 2001.

FMC Technologies, 'Smith Meter® Rotary Vane PD Meters', brochure from FMC Technologies.

FMC Technologies Measurement Solutions, Inc., 'Fundamentals Metering, Proving and Accuracy', PowerPoint presentation from FMC Technologies Measurement Solutions, Inc., 2007.

FMC Technologies, 'Proving Liquid Ultrasonic Flow Meters for Custody Transfer Measurement', Bulletin TPLS002, FMC Technologies, 2013.

Gallagher J.E., *'The Role of Flow Conditioners'*, Savant Measurement Corporation, 2010.

Gas Processors Suppliers Association (GPSA), 'Engineering Data Book', FPS version, Volume I & II, Sections 1–26, 1998.

Gulaga C., *'Vortex Shedding Meters'*, CB Engineering Ltd., 2005.

Heilveil E., *'Three Dirty Little Secrets about Coriolis Flow Meters'*, Siemens Industry Inc., July 2017.

Hofmann F., *'Fundamentals of Ultrasonic flow Measurement for Industrial Applications'*, Krohne Messtechnik GmbH, 2000.

Hofmann F., Schumacher B., *'Measuring Tube Construction Affects the Long-Term Stability of Magnetic Flow Meters'*, Krohne Messtechnik GmbH, 2009.

Honeywell, 'SMV 3000 Smart Multivariable Transmitter', Technical Information, 34-SM-03-01, 2009.

Howe W.H., Dinapoli L.D., Arant J.B., 'Venturi Tubes, Flow Tubes, and Flow Nozzles', 2003.

Huber C., 'MEMS-based Micro-Coriolis Density and Flow Measurement Technology', *Endress+Hauser Flowtec AG, Proceedings AMA Conferences*, 2015.

Hummel D., Atlas Pipeline Mid-Continent, 'How Not to Measure Gas', 2010.

Jukes E., *'Optimass Product Group Presentation'*, Krohne Ltd., 2018.

Kalivoda R.J., *'Understanding the Limits of Ultrasonics for Crude Oil Measurement'*, FMC Technologies, 2011.

Kopp J.G., Lipták B.G., Eren H., '2.10 Magnetic Flowmeters', 2003.

Krause J., FuE Zentrum, FH Kiel GmbH, 'Electromagnetic Flow Metering', 2019, www.intechopen.com.

Krohne, 'Altosonic V5 – Beam ultrasonic flowmeter for custody transfer of liquid hydrocarbons', Technical Datasheet from Krohne 7.02330.23.00, 04, 2006.

Krohne, 'OPTIFLUX 2070 Electromagnetic flowmeter', Technical Datasheet from Krohne, 2008.

Krohne, 'Optimass bulk flowmeter', Datasheet: 4000228103, 03/2009.

Krohne Ltd, *'New Large Line Size Coriolis Mass Flowmeters'*, Press release, Krohne Ltd., 2017.

Laird C.B., *'The Developing Role of Helical Turbine Meters'*, FMC Smith Meter Inc., 2019.

Lansing J., Daniel Measurement and Control, Inc., *'Principles of Operation for Ultrasonic Gas Flow Meters'*, Emerson, 2003.

Laws E.M., 'Flow Conditioner', International Patent Publication Number W0 91/01452, 1991.

Lowrie R., Krohne Inc., Spitzer D., Chesapeake Flow Solutions Inc. 'Partially Filled Pipe Flow Measurement Challenges and Solutions', 2012.

McCrometer, 'Installation & Maintenance Instructions', McPropeller Flowmeters MZ500, M0300, M1700, MF100, McCrometer, 2012.

McCrometer, Inc., 'FPI Mag® Full Profile Insertion Flow Meter', brochure from McCrometer, Inc.

Miller R.W., *'Flow Measurement Engineering Handbook'*, Second Edition, McGraw-Hill Publishing Company, 1989.

Mohitpour M., Szabo J., Van Hareveld T., *'Pipeline Operation and Maintenance – A Practical Approach'*, ASME Press, USA, 2005.

Najafi N., Putty M., Smith R., 'Coriolis MEMS-sensing technology for real-time fluidic measurements', Integrated Sensing Systems (ISS), reprinted from 'Flow Control' May 2017.

NFOGM, 'Handbook of Multiphase Flow Metering', produced for The Norwegian Society for Oil and Gas Measurement and The Norwegian Society of Chartered Technical and Scientific Professionals, 2005.

NIST, 'Guidelines for the Selection and Operation of Small Volume Provers (SVP) with a Micro Motion high-capacity Elite CMF Coriolis flowmeter', Specifications and Tolerances For Dynamic Small Volume Provers, NIST, 2016.

OIML R 117-1, 'Dynamic measuring systems for liquids other than water: Part 1: Metrological and technical requirements', OIML R 117-1, 2007.

Olin J.G., Sierra Instruments 'New Developments in Thermal Dispersion Mass Flow Meters', Presented at the *American Gas Association Operations Conference*, Pittsburgh, PA, May 20–23, 2014.

Paton R., National Engineering Laboratory, 'Calibration and Standards in Flow Measurement', 2005.

Perry D., 'Multiphase Flow Measurement – What is it?' PetroSkills, Tip of the Month, March 2018.

Petrol Instruments S.r.l., 'Tank Provers', data sheet from Petrol Instruments S.r.l., 2009.

Phase Dynamics, Inc. http://www.phasedynamics.com/technic.html#faq1.

Primary Flow Signal Inc., 'The PFS elbow flowmeter', a product bulletin from Primary Flow Signal Inc., 2008.

Pruysen A., Micro Motion, 'Coriolis master metering', European Flow Measurement Workshop, Lisbon, Portugal, 2014.

Rahman M.M., Biswas R., Mahfuz W.I., 'Effects of Beta Ratio and Reynolds Number on Coefficient of Discharge of Orifice Meter', Department of Irrigation and Water Management, Bangladesh Agricultural University, 2009.

Rans R., RAN Solutions, Blaine Sawchuk, Canada Pipeline Accessories, Marvin Weiss, Coanda Research & Development Corp. 'Flow Conditioning and Effects on Accuracy for Fluid Flow Measurement', 7th South East Asia Hydrocarbon Flow Measurement Workshop, March 2008.

Reader-Harris M.J., McNaught J.M., *'Best Practice Guide, Impulse Lines for Differential Pressure Flowmeters'*, NEL, 2005.

Rheonik, 'RHM 160-12 Coriolis Mass Flowmeter', Page 5 of 5, v6, April 2006.

Rieder A. (Endress+Hauser GmbH), Ceglia P. (Endress+Hauser Flowtec) AG, 'New generation vibrating tube sensor for density measurement under process conditions', 2017.

Rosemount, 'Magnetic Flowmeter Material Selection Guide', Rosemount Technical Data Sheet TDS 3033, June 1993.

Rosemount 3095 MultiVariable Mass Flow Transmitter, Product Specification Sheet: 00815-0100-4716, Rev AA, August 2005.

Rosemount, 'Plugged Impulse Line Detection Using Statistical Processing Technology', Rosemount Technical Note 00840-1800-4801, Rev AA, 2005.

Rosemount 1595 Conditioning Orifice Plate, Product Data Sheet: 00813-0100-4828, Rev EB, 2007.

Rosemount, 'Installation and Grounding of Magmeters in Typical and Special Applications', Rosemount Technical Note 00840-2400-4727, Rev BA, September 2013.

Rosemount Product Data Sheet, '8800C Series Vortex Flowmeter', 00813-0100-4003, 2006.

Rummel Th., Ketelsen B., 'Practical Electromagnetic flow measurement using non-uniform fields' (trans.) Regelungstecknik, 1966.

Sarker N.R., Razzaque M.M., Enam M.K., 'Numerical investigation on effects of deformation on accuracy of orifice meters', *Proceedings of the International Conference on*

Mechanical Engineering, Department of Mechanical Engineering, BUET, Dhaka, Bangladesh, 2011.

Saunders M.P., 'High Performance Flow Conditioners' Savant Measurement Corporation, *Proceedings of the American School of Gas Measurement Technology*, 2002.

Sawchuk D., 'Fluid flow conditioning for meter accuracy and repeatability', Canada Pipeline Accessories, 2014.

Seifert O., *'A Measuring Principle Comes of Age: Vortex Meters for Liquids, Gas and Steam'*, Endress+Hauser Flowtec AG, 2006.

Shao Z., 'Numerical and experimental evaluation of flow through Paul freighted plates', Rand Afrikaans University, 2002.

Siemens AG, 'SITRANS FUH1010 Clamp-on Flow Training'.

Sierra Instruments Inc., '220 series Innova-Flo vortex flowmeter', data sheet from Sierra Instruments Inc., 2005.

Sparreboom W., 'Miniaturization to the Extreme: Micro-Coriolis Mass Flow Sensor', Bronkhorst High-Tech B.V., January 30, 2018.

Spink L.K., *'Principles and Practice of Flow Meter Engineering'*, Ninth Edition, The Foxboro Company, Foxboro, Massachusetts, USA, 1978.

Spirax-Sarco Limited, 'Orifice plate flowmeters for steam, liquids and gases', brochure from Spirax-Sarco Limited, 2000.

Steven R., 'Orifice plate meter diagnostics', DP Diagnostics, 2012.

Stobie G.J., ConocoPhillips, 'The Selection of Multiphase Meters', PowerPoint presentation, 2004.

Swinton Technology, 'Prognosis® Differential Pressure Diagnostics', and brochure from Swinton Technology, 2016.

Tanaka M., Hayashi K., Tomiyama A., 'Hybrid multiphase-flow simulation of bubble-driven flow in complex geometry using an immersed boundary approach', Multiphase Science and Technology, Volume 21, 2009.

Tecfluid S.A., 'Series FLOMAT Insertion electromagnetic flowmeter for conductive liquids', datasheet from Tecfluid S.A., 2014.

Technical Guide, 'Daniel™ Liquid Turbine Flow Meters', Technical Guide, August 2016.

Teniou S., Meribout M., 'Multiphase flow meters principles and applications: A review', *Canadian Journal on Scientific and Industrial Research*, 2(8), November 2011.

The Petroleum Science and Technology Institute PSTI, 'Market Prospects for Multiphase Technology', THERMIE Programme action No: HC 6.3, For the European Commission Offshore Technology Park, Directorate-General for Energy (DG XVII), 1998.

Thorn R., Johansen G.A., Hjertaker B.T., 'Three-phase flow measurement in the petroleum industry'.

Toshiba, 'Capacitance type Electromagnetic flowmeter, model LF511/LF541', datasheet from Toshiba, 2015.

United Kingdom Patent, GB1263614, Published 1972-02-16, Eastech (US) 1,263,614. 'Measuring fluid-flow. Eastech Inc. May 21, 1969 (May 27, 1968), No. 25939/69. Heading G1R.

United Kingdom Patent, GB19730015259 19730329 1387380 Measuring fluid-flow Yokogawa Electric Works Ltd, 29 March 1973 (27 April 1972) 15259/73. Heading G1R.

Upp E.L., LaNasa P.J., *'Fluid Flow Measurement: A Practical Guide to Accurate Flow Measurement'*, 2nd Edition, Gulf Professional Publishing, 2002.

Vermont Technologies, 'Proving Tanks (Tank Provers)', data sheet from Vermont Technologies, 2014.

Vortab, 'Vortab® Flow Conditioners', brochure from Vortab, 2013.

Warburton P., *'Vortex Shedding Meters'*, Yokogawa Corp. of America, 2007.

Weatherford, 'Red Eye® 2G Water-cut Meter', Brochure 5313.00, 2008.

Whitman S., Coastal Flow Liquid Measurement, Inc., 'Proving Liquid Meters with Microprocessor-Based Pulse Outputs', 2015

Williams G., Flow Management Devices, 'Fundamentals of Meter Provers and Proving Methods', 2016.

Yasua O., *'Lining technology for Magnetic Flowmeters'*, Yokogawa Technical Report English Edition, No 42, 2006.

Yoder J., Flow Research, Inc. 'When Will the Vortex Flowmeter Market Pick Up Steam?', 2003.

Yoder J., *'The Paradigm Case Method of Selecting Flowmeters'*, Flow Research, Inc., 2007.

Yokogawa, 'Digital Vortex Flowmeter Digital YEWFLO', data sheet from Yokogawa, 2007.

Yokogawa, 'Impulse Line Blocking Diagnosis in DP Transmitters', Application Note from Yokogawa 2007.

Yokogawa, 'Magnetic Flowmeter Handbook', brochure from Yokogawa, 2017.

Zhu H., Rieder A., 'An Innovative Technology for Coriolis Metering under Entrained Gas Conditions', Endress+Hauser Flowtec AG, 2016.

Index

K

L

M

N

O

P

V

W

Z

Printed in the United States
by Baker & Taylor Publisher Services